国家卫生健康委员会"十三五"规划教材配套教材

全国高等学校配套教材

供基础、临床、预防、口腔医学类专业用

# 基础化学实验
## （中英对照版）

## Experiment in Basic Chemistry

### 第4版

主　审　魏祖期

主　编　李雪华　籍雪平

副主编　林　毅　胡　新　席晓岚

人民卫生出版社

图书在版编目（CIP）数据

基础化学实验/李雪华，籍雪平主编. —4版. —北京：人民卫生出版社，2019
　ISBN 978-7-117-28678-7

Ⅰ. ①基… Ⅱ. ①李… ②籍… Ⅲ. ①化学实验 Ⅳ. ①O6-3

中国版本图书馆CIP数据核字（2019）第133862号

人卫智网　www.ipmph.com　医学教育、学术、考试、健康，
　　　　　　　　　　　　　购书智慧智能综合服务平台
人卫官网　www.pmph.com　人卫官方资讯发布平台

版权所有，侵权必究！

## 基础化学实验
### 第4版

主　　编：李雪华　籍雪平
出版发行：人民卫生出版社（中继线 010-59780011）
地　　址：北京市朝阳区潘家园南里19号
邮　　编：100021
E - mail：pmph @ pmph.com
购书热线：010-59787592　010-59787584　010-65264830
印　　刷：三河市尚艺印装有限公司
经　　销：新华书店
开　　本：787×1092　1/16　印张：18
字　　数：472千字
版　　次：2005年2月第1版　2019年8月第4版
　　　　　2024年7月第4版第6次印刷（总第25次印刷）
标准书号：ISBN 978-7-117-28678-7
定　　价：39.00元

打击盗版举报电话：010-59787491　E-mail：WQ @ pmph.com
（凡属印装质量问题请与本社市场营销中心联系退换）

# 编 委

(以姓氏笔画为序)

| | |
|---|---|
| 丁　琼（武汉大学） | 宋　慧（广西医科大学） |
| 于　昆（大连医科大学） | 陈志琼（重庆医科大学） |
| 王金玲（山西医科大学） | 武世奎（内蒙古医科大学） |
| 王美玲（内蒙古医科大学） | 林　毅（武汉大学） |
| 尹计秋（大连医科大学） | 周昊霏（内蒙古医科大学） |
| 石婷婷（蚌埠医学院） | 周春艳（广西医科大学） |
| 申小爱（中国医科大学） | 赵全芹（山东大学） |
| 母昭德（重庆医科大学） | 胡　新（北京大学） |
| 乔秀文（石河子大学） | 席晓岚（贵州医科大学） |
| 庄海旗（广东医科大学） | 黄　静（广西医科大学） |
| 刘国杰（中国医科大学） | 黄燕军（广西医科大学） |
| 李　蓉（贵州医科大学） | 傅　迎（大连医科大学） |
| 李振泉（济宁医学院） | 赖泽锋（广西医科大学） |
| 李雪华（广西医科大学） | 廖传安（广西医科大学） |
| 李献锐（河北医科大学） | 魏祖期（华中科技大学） |
| 李福森（广西医科大学） | 籍雪平（河北医科大学） |
| 别子俊（蚌埠医学院） | |

**秘　书**　黄燕军（兼）

# 前 言

《基础化学实验》(第 4 版)是人民卫生出版社出版的国家卫生健康委员会"十三五"规划教材《基础化学》(第 9 版)的配套教材,供高等医学院校五年制基础、临床、预防、口腔医学类专业使用。

根据《普通高等学校本科专业类教学质量国家标准》(2018 版)《国家中长期教育改革与发展规划纲要(2010—2020 年)的意见》以及《国务院办公厅关于深化医教协同进一步推进医学教育改革与发展的意见》(国办发〔2017〕63 号)等文件精神,遵循五年制本科临床医学专业培养目标,本配套教材的编写宗旨是以学生自主学习为导向,为学生综合实验技能、创新能力的培养提供策略与手段。注重体现素质教育、创新能力与实践能力的培养,为学生知识、能力、素质协调发展创造条件。

本教材融合了全国多所高等医学院校基础化学教学改革的成果,汲取了国内外优秀教材的先进经验,在编写内容和编写方法上进行了创新。自 2004 年第 1 版教材出版以来,受到使用院校的欢迎。经过 15 年的使用,在认真听取使用单位的意见,并结合实验教学改革所取得新的进展的基础上,修订编写成《基础化学实验》(第 4 版)。

本教材以基础化学实验教学作为培养学生综合实验技能的载体之一,从教学主体、教学内容、教学手段、教学计划等多方面考虑,通过巧妙设置融入本教材中,构建有效培养学生自主学习能力、发现问题、分析问题、解决问题能力的教学手段。在实施实验教学的过程中,学生的角色准确定位,在完成《基础化学实验》课程的学习过程中逐渐获得和完善自主学习能力、创新思维能力和批判性思维等综合素质。

本教材是一本综合性基础实验教材,包含了无机化学、分析化学及物理化学课程等实验内容。教材编排上分为"基础化学实验基本知识"和"基础化学实验课题"两个部分;基础化学实验课题部分,按基本操作训练实验、滴定分析实验、分光光度法实验、化学原理实验、无机化合物制备实验、综合及研究性实验和自行设计实验 7 个课题共 22 个实验进行编写。这 22 个实验中根据内容或方法的不同可包含多个子课题的互相渗透。我们希望这样的编排分类清楚、目的明确,便于学生掌握要领。

为了加强学生外语能力的培养、便于全英语教学和双语教学,本实验教材用中、英文编写,内容基本一一对应。

新修订的教材突显了以下新的特点:

1. 更新教学和人才培养模式的观念,注重教材的启发性,合理设置实验的编写内容,从实验目的、实验预习作业、实验步骤、数据记录及结果分析到思考题,多方面强化学生的自主学习能力和培养学生综合实验技能。

2. 增加了物理化学实验及无机化合物制备实验内容,使实验内容与主教材对应,且知识点涉及更全面。

3. 修订及强化预习作业,合理地设置实验前的预习作业及思考题,通过完成课前预习作业和实践操作,培养学生的自主学习能力、理论与实践相结合的实验技能、创新思维能力和批判性思维等综合能力及综合素质。

4. 优化各实验文字表述和思考题,要求论述严谨、语言流畅简洁、层次分明、术语规范、图表直观。

5. 结合发达国家同类教材的编写模式,引进新的教学理念和内容,进一步提升教材质量。

在编写本书的过程中,编者在叙述科学性、准确性、合理性方面做了很大努力,但难免有缺点和错误,诚恳希望对书中不妥和错误之处批评指正。

李雪华　籍雪平<br>
2019 年 3 月

# 目 录

**第一篇 基础化学实验基本知识** ... 1
  第一章 实验室规则及安全知识 ... 1
    1. 基础化学实验目的 ... 1
    2. 化学实验室一般规则 ... 1
    3. 化学实验室的安全知识 ... 2
    4. 化学实验操作过程中可能发生的事故与处理 ... 2
    5. 化学实验室的防火、防电与灭火常识 ... 3
  第二章 基础化学实验常用仪器 ... 4
    1. 实验室常用仪器介绍 ... 4
    2. 玻璃仪器的洗涤和干燥 ... 5
    3. 常用仪器使用方法 ... 6
  第三章 实验结果的表示 ... 23
    1. 实验误差与有效数字 ... 23
    2. 实验数据的处理 ... 25
    3. 实验报告 ... 26

**第二篇 基础化学实验课题** ... 29
  实验课题一 基本操作训练实验 ... 29
    实验一 常用容量分析仪器操作练习 ... 29
    实验二 电子天平称量练习 ... 32
    实验三 缓冲溶液的配制与性质、溶液 pH 测定 ... 34
    实验四 胶体溶液的制备与性质 ... 37
  实验课题二 滴定分析实验 ... 42
    实验五 酸碱滴定分析法 ... 42
      1. 标准盐酸溶液浓度的标定 ... 43
      2. 标准氢氧化钠溶液浓度的标定 ... 45
      3. 食用醋中总酸度的测定 ... 47

  4. 硼砂含量的测定 ………………………………………………………………… 48

  5. 返滴定法测定阿司匹林的含量 ………………………………………………… 49

 实验六 氧化还原滴定法 ………………………………………………………………… 51

  1. 高锰酸钾法 ……………………………………………………………………… 51

  2. 碘量法 …………………………………………………………………………… 55

 实验七 配位滴定分析 …………………………………………………………………… 58

  1. 水的硬度的测定 ………………………………………………………………… 59

  2. 明矾含量的测定 ………………………………………………………………… 62

  3. 葡萄糖酸钙含量的测定 ………………………………………………………… 65

实验课题三 分光光度法实验 ………………………………………………………………… 69

 实验八 可见分光光度法测定水样中铁含量 …………………………………………… 69

  1. 邻二氮菲法 ……………………………………………………………………… 70

  2. 硫氰酸盐法 ……………………………………………………………………… 72

  3. 磺基水杨酸法 …………………………………………………………………… 74

 实验九 分光光度法测定阿司匹林药片的含量 ………………………………………… 76

 实验十 分光光度法测定磺基水杨酸合铁的组成和稳定常数 ………………………… 78

 实验十一 荧光分析法测定维生素 $B_2$ 的含量 …………………………………………… 81

 实验十二 紫外分光光度法对维生素 $B_{12}$ 的鉴别与含量测定 ………………………… 83

实验课题四 化学原理实验 …………………………………………………………………… 86

 实验十三 稀溶液的依数性及其应用 …………………………………………………… 86

  1. 凝固点降低法测定溶质的分子量 ……………………………………………… 88

  2. 利用溶液的渗透浓度测定溶质的相对分子质量 ……………………………… 89

  3. 冰点渗透压计的应用 …………………………………………………………… 90

 实验十四 置换法测定镁的原子量 ……………………………………………………… 91

 实验十五 化学反应速率与活化能的测定 ……………………………………………… 94

 实验十六 最大气泡压力法测定乙醇溶液的表面张力 ………………………………… 98

实验课题五 化合物制备实验 ………………………………………………………………… 104

 实验十七 氯化钠的精制 ………………………………………………………………… 104

 实验十八 硫酸亚铁铵的制备 …………………………………………………………… 107

 实验十九 转化法制备硝酸钾 …………………………………………………………… 109

实验课题六 综合及研究性实验 ……………………………………………………………… 112

 实验二十 醋酸解离常数的测定与食醋中 HAc 含量的测定 ………………………… 112

 实验二十一 茶叶中钙、镁及铁含量的测定 ………………………………………… 115

实验课题七 自行设计实验 …………………………………………………………………… 119

 实验二十二 实验设计及研究 ……………………………………………………………… 119

## Part I  Experiment Essentials in Basic Chemistry ········ 123

### Chapter 1  Laboratory Rules and Safety Information ········ 123

1. The Purpose of Experiments in Basic Chemistry ········ 123
2. General Rules in Laboratory ········ 123
3. Safety Information ········ 124
4. The Treatment of Accidents in Chemical Experiments ········ 125
5. Fire Prevention, Anti-electricity and Extinguishing Knowledge ········ 125

### Chapter 2  Ordinary Instruments in Chemical Experiments ········ 127

1. Introduction of Ordinary Instruments in Chemical Laboratory ········ 127
2. Cleaning and Drying Glasswares ········ 128
3. Operation of Instrument ········ 129

### Chapter 3  Experiment Results and their Expressions ········ 151

1. Experimental Error and Significant Figures ········ 151
2. Treatments of Experimental Data ········ 154
3. Experiment Reports ········ 155

## Part II  Experiments in Basic Chemistry ········ 159

### Chapter 1  Experiments of Basic Operation Practice ········ 159

Experiment 1  Operation of Volumetric Analysis ········ 159
Experiment 2  Weighing Practice with an Electronic Balance ········ 162
Experiment 3  Preparation, Properties and pH Measurement of Buffer Solutions ········ 165
Experiment 4  Preparation and Properties of Colloidal Systems ········ 168

### Chapter 2  Experiments of Titration Analysis ········ 173

Experiment 5  Acid-Base Titration Analysis ········ 173
1. Standardization of the Concentration of HCl Solution ········ 174
2. Standardization of the Concentration of NaOH Solution ········ 176
3. Determination of Acetic Acid Content in Vinegar ········ 179
4. Determination of Borax Content ········ 180
5. Determination of Aspirin Content by Back Titration ········ 182

Experiment 6  Oxidation-Reduction Titration ········ 184
1. Potassium Permanganate Method ········ 184
2. Iodimetry ········ 188

Experiment 7  Complexometric Titration ········ 192
1. Determination of Water Hardness ········ 193
2. Determination of $KAl(SO_4)_2 \cdot 12H_2O$ in Alum ········ 197

3. Determination of Calcium Gluconate ·················· 200

## Chapter 3  Spectrophotometry ·················· 204

Experiment 8  Determination of $Fe^{3+}$ ($Fe^{2+}$) Content in Water Sample with Visible Spectrophotometry ·················· 204

1. O-phenanthroline Method ·················· 205
2. Sulfocyanate Method ·················· 207
3. Sulfosalicylic Acid Method ·················· 209

Experiment 9  Determination of Aspirin Content in Tablets by Visible Spectrophotometry ·················· 211

Experiment 10  Determination of the Formula and the Stability Constant of Sulfosalicylate Iron (Ⅲ) by Spectrophotometry ·················· 213

Experiment 11  Determination of Vitamin $B_2$ by Fluorescence Analysis ·················· 217

Experiment 12  Identification and Content Determination of Vitamin $B_{12}$ by Ultraviolet Spectrophotometry ·················· 220

## Chapter 4  Chemical Principles Experiment ·················· 223

Experiment 13  Colligative Properties of Dilute Solution and Their Application ·················· 223

1. Determination of Molar Mass of Glucose by Freezing Point Depression ·················· 225
2. Determination of the Molar Mass of Glucose by Determining Osmolarity ·················· 227
3. Application of Freezing Point Osmometer ·················· 228

Experiment 14  Determination of the Relative Atomic Mass of Magnesium by Replacement Reaction ·················· 229

Experiment 15  Determination of the Rate of Chemical Reaction and Activation Energy ·················· 232

Experiment 16  The Maximum Bubble Pressure Method to Determine the Surface Tension of Ethanol Solution ·················· 237

## Chapter 5  Preparation of Compounds ·················· 244

Experiment 17  Refining Sodium Chloride ·················· 244

Experiment 18  Preparation of Ammonium Iron (Ⅱ) Sulfate ·················· 247

Experiment 19  Preparation of Potassium Nitrate by Conversion Method ·················· 250

## Chapter 6  Comprehensive Experiments ·················· 254

Experiment 20  Determination of HAc Content and Dissociation Equilibrium Constant of Content of HAc ·················· 254

Experiment 21  Determination of the Content of Calcium, Magnesium, and Iron in Tea ·················· 257

## Chapter 7  Self-Designed Experiment ·················· 262

Experiment 22  Design and Research ·················· 262

**附录** ································································································· 267
   附录Ⅰ   国际相对原子量表 ································································ 267
   附录Ⅱ   不同温度下水的饱和蒸汽压 ··················································· 269
   附录Ⅲ   危险药品的分类、性质和管理 ··············································· 270
   附录Ⅳ   标准缓冲溶液 ······································································· 273
   附录Ⅴ   参考书目 ··············································································· 274

# 第一篇 基础化学实验基本知识

## 第一章 实验室规则及安全知识

### 1. 基础化学实验目的

基础化学实验课是基础化学课程的重要组成部分。基础化学实验的目的不仅仅是印证理论知识和一些实验现象，理解和掌握实验课程内容，学会将理论与实践相结合，将理论应用于实践中的科学实验方法，培养综合实验技能和严谨的科学态度。

学生应通过实验课的严格训练，掌握规范的化学实验操作、正确记录和处理实验数据、分析实验现象、表达实验结果，学会总结实验规律，给出实验结论。通过自己动手设计和完成实验，培养独立思考和独立解决问题的能力，同时培养严谨的科学态度，从而逐步掌握科学研究方法及实验技能的应用。

### 2. 化学实验室一般规则

（1）实验前应认真预习实验教程，阅读有关教材及参考书，明确实验目的与要求，了解实验基本原理和方法，熟悉实验步骤，做好预习报告。

（2）仔细阅读仪器使用指南，按说明进行操作。不得进行未经许可的实验和操作。

（3）学生进实验室应穿白大褂，禁止穿拖鞋。实验过程应严肃认真，正确操作、认真仔细观察，并及时记录实验现象与数据。根据原始记录，写出实验报告，按时交给教师。

（4）不允许单独一人在实验室工作。实验中的任何事故无论大小均须及时向教师报告。

（5）公用仪器与试剂只能在原处使用，不得随意挪动。

（6）从试剂瓶中取出的试剂，不得再倒回原瓶中。取试剂前应两次阅读标签，以保证药品名称和浓度正确。

（7）勿将试剂瓶盖接口处接触其他表面。为避免试剂瓶盖污染试剂，建议一次只打开一瓶试剂。如果试剂瓶盖只有硬币大小，倒试剂时，随手夹住瓶盖，就不会因弄混瓶盖造成污染。

（8）禁止将食物带进实验室，勿在实验室饮食。

（9）实验中应注意安全，易燃药品应远离火源。保持实验室和桌面干净整洁，不要将杂物随地乱扔，用过的试纸、火柴等放在烧杯中，待实验结束后倒入废物缸。腐蚀性和有毒化学品必须按规定回收，切勿将固体废物、腐蚀性液体和有毒试剂倒入水槽。

(10) 实验结束前,不得擅自离开实验室。实验完毕,立即清洗仪器,整理药品、仪器及实验台,做好实验室、天平室、与仪器室的清洁。关好门、窗及水、电、气源后,方能离开实验室。

## 3. 化学实验室的安全知识

(1) 水、电、火、有毒气体防护:首次进实验室,必须了解楼层的所有逃生通道;每次进实验室,首先打开实验室窗户,并保证通风状态良好;确认实验室水、电、气的安装情况、灭火器材存放位置及使用方法,以便应急使用;使用有毒有害气体或挥发性有毒物质时,应在通风橱中操作。

(2) 危险物质防护:谨慎处理易燃、易爆、有毒及有腐蚀性和剧毒物质。使用此类物质时,应在通风条件良好的地方并远离火源。

(3) 化学操作防护:在加热试管中的液体时,先将试管外壁擦干,勿将试管对准自己或他人。不要直接加热试管底部,应倾斜试管缓缓加热液体上端到试管底部之间的部位。

(4) 化学试剂防护:打开盐酸、硝酸、氨水或过氧化氢等试剂瓶塞时小心气体冲出。嗅闻气味时,不要将鼻直接接近瓶口,而应用手扇闻。使用浓酸、浓碱和洗液时,应避免接触皮肤或溅在衣服上,特别要注意保护眼睛(佩戴防护眼镜)。

(5) 个人安全防护:使用热的或腐蚀性液体时,应穿防护外套以保护皮肤和衣物。穿皮鞋比穿帆布鞋或凉鞋更安全。

(6) 化学药品处理:保持台面清洁,及时清除溅落的酸碱。若化学药品溅到皮肤上,立即用大量清水冲洗患处,然后抹上肥皂,并用水清洗。

(7) 电气设备使用:要确保电压-电流功率匹配,切勿用湿手接触电源插头。

(8) 实验室备用急救包:学生实验室要备有急救包,常见的急救用品包括:医用纱布块、医用弹性绷带、创可贴、医用透气胶带、碘伏消毒液、医用敷贴、医用酒精棉片、清洁湿巾、烫伤药物、急救手册等,以备急救时使用。

## 4. 化学实验操作过程中可能发生的事故与处理

(1) 割伤:在伤口上涂抹碘酒后,敷贴创可贴。

(2) 烫伤:在伤口上涂抹烫伤药膏或用浓 $KMnO_4$ 溶液擦至灼伤处皮肤变为棕色,再涂上凡士林或烫伤药膏。

(3) 酸碱腐蚀:立即用大量水冲洗。酸灼伤时,用水冲洗后,用饱和碳酸氢钠、稀氨溶液或肥皂水处理;碱灼伤时,用水冲洗后,用2%~5%醋酸或3%硼酸溶液处理。若为浓硫酸烫伤,须先用脱脂棉揩净再用水冲洗。若酸溅入眼睛,首先用水冲洗,然后用1%~3%碳酸氢钠溶液处理,再用水冲洗。若碱溅入眼睛,应用水冲洗,然后用3%硼酸溶液处理,再用水冲洗。经上述处理后,立即送医院治疗。

(4) 有毒气体吸入:如果吸入溴、氯或氯化氢等有毒气体,立即离开有毒气体环境,并到室外呼吸新鲜空气,可吸入少量酒精与乙醚混合的蒸气以解毒。如果吸入硫化氢或一氧化碳气体,则立即到室外呼吸新鲜空气。

(5) 毒物入口:可内服一杯稀硫酸铜溶液,用手指触摸喉咙,促使呕吐,然后立即送医院。

(6) 触电事故:电器与水隔绝,实验桌面保持干燥、整洁。使用电器设备时,不要用湿手接触电器和插销。遇到电器着火,应立即切断电源,防止触电。

## 5. 化学实验室的防火、防电与灭火常识

（1）引起化学实验室火灾的主要原因

1）易燃物质离火源太近。实验室安全要求：不能使用明火加热。
2）电线老化、插头接触不良或电器故障等。
3）化学反应不当导燃，如化学品物性不明、操作不当等。
4）下列物质彼此混合或接触后易着火，甚至酿成火灾：
- 活性炭与硝酸铵；
- 沾染了强氧化剂（如氯酸钾）的衣物；
- 抹布与浓硫酸；
- 可燃性物质（木材或纤维等）与浓硝酸；
- 有机物与液氧；
- 铝与有机氯化物；
- 磷化氢、硅烷、烷基金属及白磷等与空气接触。

（2）灭火方法：化学实验室一旦着火或发生火灾，切勿惊慌，应冷静果断地按表1所示方法采取扑灭措施并及时报警。

**表 1　可燃物的灭火方法**

| 燃烧物 | 灭火方法 | 说明 |
| --- | --- | --- |
| 纸张、纺织品或木材 | 沙、水、灭火器 | 需降温和隔绝空气 |
| 油、苯等有机溶剂 | $CO_2$、干粉灭火器、石棉布、干沙等 | 适用于贵重仪器上的灭火，油、气等燃烧切勿用泡沫灭火器灭火 |
| 醇、醚等 | 水 | 需冲淡、降温和隔绝空气 |
| 电线、电表及仪器燃烧 | $CCl_4$、$CO_2$ 等灭火器 | 灭火材料不能导电，切勿用水和泡沫灭火器灭火 |
| 可燃性气体 | 关闭气源，使用灭火器 | 尽一切可能切断可燃气源 |
| 活泼金属（如钾、钠等）及磷化物与水接触 | 干砂土、干粉灭火器 | 绝不能使用水或泡沫、$CO_2$ 灭火器 |
| 身上的衣物 | 就地滚动，压灭火焰或脱掉衣服、用专用防火布覆盖着火处 | 切勿跑动，否则将加剧燃烧 |

（3）防触电：首先切断电源，尽快用绝缘物如干燥的木棍或竹竿等，使触电者脱离电源。必要时进行人工呼吸，并立即送医院抢救。

（籍雪平）

# 第二章
# 基础化学实验常用仪器

## 1. 实验室常用仪器介绍

实验室常用仪器主要以玻璃仪器为主,按其用途可分为容器类仪器、量器类仪器和其他类仪器。

(1)容器类:常温或加热条件下物质的反应容器、贮存容器,包括试管、烧杯、烧瓶、漏斗等,如图1所示。使用时根据用途和用量选择不同种类和不同规格的容器。

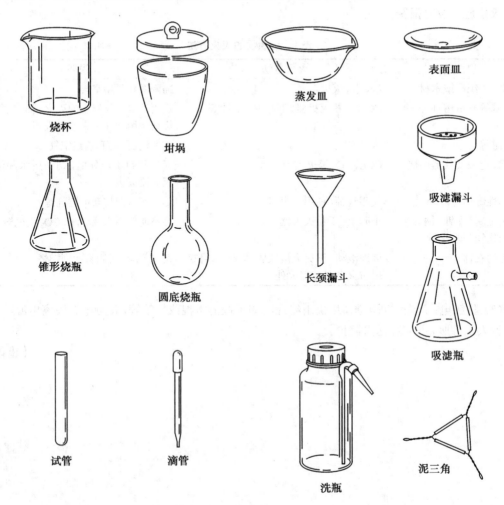

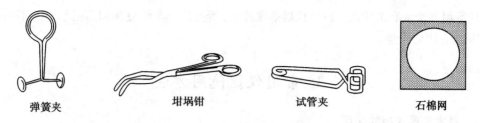

图 1 实验室常用玻璃容器

（2）量器类：用于度量或移取一定溶液体积的玻璃仪器，主要有量筒、滴定管、容量瓶、移液管和移液枪等，根据精度不同，分为精量仪器及粗量仪器。量器不可以作为实验容器，例如不可用于溶解、稀释操作、不可量取热溶液、不可加热及长期存放溶液。量器上常注明两种符号：一种为"E"，表示"量入"容器；另一种为"A"，表示"量出"容器。量器的管颈处刻有一条标线，是所移取的准确体积的标志。

其他类仪器包括玻璃仪器和非玻璃仪器。

## 2. 玻璃仪器的洗涤和干燥

### 2.1 普通玻璃仪器的洗涤

洗涤玻璃仪器，先用自来水荡洗，根据仪器形状使用相应毛刷蘸取洗涤剂等，沿器壁内外均匀刷洗，然后用自来水冲洗至无泡沫，最后用少量蒸馏水润洗三次。

若仪器沾有污垢，首先用热水或热碱液浸泡，然后用毛刷沾上洗涤剂刷洗，再用自来水冲洗至无泡沫，最后用蒸馏水润洗三次。

若仪器口小、管细、体长，使用毛刷受限时，可用烧杯或烧瓶加热铬酸洗液[*]，将洗液转移到仪器中浸泡一段时间，再用自来水连续冲洗，最后用蒸馏水润洗三次。

对于碱式滴定管的洗涤，要注意铬酸洗液不能直接接触橡胶管，洗涤前取下橡胶管。

洗涤玻璃仪器洁净的标准是，水沿器壁自然流下后均匀湿润，不挂水珠。

### 2.2 普通玻璃仪器的干燥

（1）自然晾干：敞开仪器开口并朝向下方，让水分自然流出，挥发。

需要尽快使用的玻璃仪器可用烤干、吹干、烘干、有机溶剂挥干等方法干燥。

（2）烤干：尽可能倾尽已洁净仪器内壁水分，然后用小火均匀烤干仪器。此方法适合于数量少、体积小的玻璃仪器，如：试管的干燥。

（3）吹干：尽可能倾尽已洁净仪器内壁水分，然后用电吹风或专用的气流烘干机吹干仪器。如：烧杯的干燥。

（4）烘干：尽可能倾尽已洁净仪器内壁水分，然后将仪器放入烘箱中烘干。此方法特别适合数量较多，口径较小的仪器。

（5）有机溶剂挥干：尽可能倾尽已洁净仪器内壁水分，然后用丙酮或酒精等易挥发的有机溶剂润湿仪器内壁，倒出并回收用过的有机溶剂，最后晾干或吹干仪器。

由于度量仪器具有标定的刻度，一般不用毛刷刷洗，可用铬酸洗液浸泡一段时间，再用自来水

---

[*] 铬酸洗液的配制：取 10g 工业用 $K_2Cr_2O_7$，置于烧杯中。先用少量水溶解，在不断搅拌下缓慢加入 200mL 工业用浓硫酸，待溶解并冷却后，即可保存于试剂瓶中待用。

冲洗,最后用蒸馏水润洗数次。玻璃仪器热胀冷缩,受热后不易恢复体积,因此,容量分析仪器应在标注温度范围内干燥。

# 3. 常用仪器使用方法

## 3.1 基本度量仪器的使用

### 3.1.1 滴定管

滴定管主要用于精确放取一定体积的溶液。滴定管按其构造可分为酸式、碱式或酸碱两用滴定管。酸式滴定管下端有玻璃旋塞,见图 2(a),通过自由转动旋塞以控制溶液流速,用以盛装或量取酸性及氧化性溶液;碱式滴定管下端连接一软橡胶管,内装有一玻璃珠,通过挤捏玻璃珠可控制溶液的流速,见图 2(b),用以盛装碱性溶液;酸碱两用滴定管的结构与酸式滴定管相同,但其旋塞材质是聚四氟乙烯,具有耐酸、碱的作用,因此可盛装酸性及碱性两种溶液。

常用的滴定管容积有 25mL 和 50mL,最小刻度为 0.1mL,读数可读到 0.01mL。此外,还有 10、5、2 和 1mL 的半微量或微量滴定管,最小刻度为 0.05、0.01 或 0.005mL。

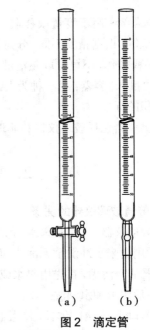

图 2 滴定管
(a)酸式(酸碱两式)滴定管;(b)碱式滴定管

(1) 检漏及装配

1) 使用酸式滴定管前,应检查旋塞转动是否灵活或漏液,否则取下旋塞,洗净后用吸水纸吸干或吹干旋塞和旋塞槽。如图 3(a),在旋塞的两端涂一层很薄的凡士林,如图 3(b)所示的 A、B 处,旋塞中部塞孔两旁只涂极薄一层的凡士林以防堵塞塞孔。将旋塞插入塞槽内并向同一方向旋转旋塞,使旋塞与塞槽接触处呈透明状且转动灵活,若油脂分布不均匀或堵塞小孔,需重新清洗、涂布,最后装水检查是否漏液。

2) 碱式滴定管应选择大小合适的玻璃珠和橡胶管,检查滴定管是否漏液,是否能灵活控制液滴流出,如不符合要求,需重新选球径合适的玻璃珠装配。

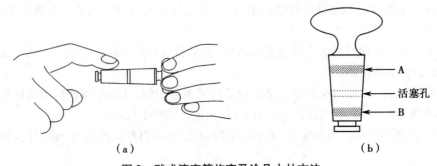

图 3 酸式滴定管旋塞及涂凡士林方法
(a)旋塞涂抹凡士林;(b)旋塞各部位

(2) 润洗：使用滴定管前，用待装液润洗 2～3 次以除去滴定管内残留水分（保证溶液浓度不变）。向滴定管注入润洗液约 10mL，两手平端滴定管，慢慢转动，使溶液流遍全管。打开酸式滴定管的旋塞或挤压碱式滴定管的玻璃珠，使润洗液一半从滴定管下端流出，余下润洗液从管口完全倒掉。

(3) 装液、赶气泡

1) 加入过量润洗液至酸式滴定管零刻度标线上方，快速旋转旋塞放液，使液体完全充满管尖，利用液流将气泡带出；或将酸式滴定管倾斜至与水平成 15°～30°，缓慢开启旋塞，浸流入管尖的液体可将气泡排出。碱式滴定管需将玻璃珠上端及下端的气泡赶尽，玻璃珠上端气泡可通过挤捏橡皮管排出，玻璃珠下端的气泡则需将橡皮管向上弯曲，挤捏橡皮管玻璃珠，如图 4 所示，使溶液从尖嘴处喷出以排出气泡。

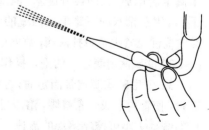

图 4　碱式滴定管排除气泡的方法

2) 气泡排出后，调整液面至零刻度或稍低于零刻度以保证具有足够的滴定剂用于滴定分析。

(4) 读数

1) 将滴定管从滴定管夹上取下，用大拇指与示指或中指轻拿滴定管上端，使其自然垂直，待溶液稳定 1～2min 后（使附着在内壁上的溶液完全流下后再读数），使视线与凹液面水平，分别读出初始读数或滴定终点时读数。读数时，管壁不应挂液滴，管尖不应有气泡或悬滴液，读数精确至 0.01mL。

2) 无色溶液或浅色溶液的凹液面清晰可见，应以凹液面下缘实线的最低点为准，如图 5(a)；有色溶液凹液面的清晰程度较差，如 $KMnO_4$、$I_2$ 溶液等，读数时应以液面两侧最高点为准。

3) 采用"蓝带"滴定管读数更准确，在其背景上有一"蓝带"，如图 5(b)。若为无色溶液，两个凹液面相交于滴定管蓝线的某一点，读数时即以此点为准；若为有色溶液，以液面两侧的最高点为准。采用读数卡可协助读数，如图 5(c)。读数卡可用涂有黑长方形的白纸制成，滴定管后面衬一张读数卡，使黑色部分在凹液面下约 1mm 处，此时即可看到凹液面的反射层成为黑色，读数时以此凹液面下缘最低点为准。

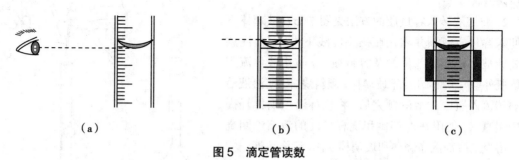

图 5　滴定管读数
(a)普通读数法；(b)蓝带滴定管读数法；(c)黑背景辅助读数法

(5) 滴定

1) 滴定管垂直固定在滴定管夹上，确定管尖是否挂液滴，若有，需用容器外壁碰掉液滴。滴定管尖距锥形瓶口 1cm 左右，样品溶液盛于锥形瓶中，锥形瓶置于一白色衬底之上。滴定时若使用酸式滴定管则采用反握法，用左手开启并控制旋塞，旋塞柄指向右方，用拇指、示指、中指轻轻向内扣住旋塞,手心空握,对旋塞施加指向手心的内扣力,以防旋塞松动或被顶出而漏液。右手握持锥形瓶,需上提锥形瓶,使滴定管下端伸入锥形瓶内约1~2cm,边滴边向同一方向作圆周旋转摇动,如图6所示(不能前后振动,否则会溅出溶液)。滴定速度一般为10mL·min$^{-1}$,即3~4滴·s$^{-1}$。滴定接近终点前,一滴一滴地加入,用少量蒸馏水冲洗黏附于锥形瓶内壁的液滴;当色晕消失缓慢,此时即将到达滴定终点,可半滴半滴地加入(即微开旋塞,液滴悬在管尖,用锥形瓶内壁将液滴碰下,再用蒸馏水冲下锥形瓶内壁的液滴),直至到达滴定终点。终点读数时需注意管尖是否挂液滴,数据记录时需减去半滴液滴体积量。

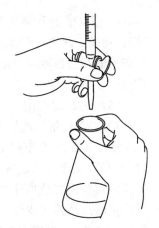

图6 滴定操作

2) 酸碱两用滴定管的滴定操作方法与酸式滴定管相似。

3) 使用碱式滴定管滴定时,左手拇指和示指挤压玻璃珠稍上缘的胶管,使管内形成一条缝隙,溶液可从玻璃管的尖嘴中流出。根据所挤压的缝隙大小可控制溶液的流速。若手放在玻璃球的下部,松手后,空气会从玻璃管尖端进入而出现气泡。

### 3.1.2 容量瓶

容量瓶是一种细梨形的平底玻璃瓶,用于配制标准溶液或试样溶液,带有玻璃磨口塞或塑料塞,颈上有一标线,一般表示指定温度下(标记于瓶上)溶液充满至标线时的容积,如图7(a)所示。通常有25、100、250、500和1 000mL等各种规格。容量瓶不能用作贮液瓶,配制或稀释后的溶液应转移到贮液瓶中,并洗净容量瓶。

(1) 检漏:使用容量瓶前,应检查是否漏液,可注入自来水至容积的2/3以上,盖好瓶塞并用右手示指顶住瓶塞,另一只手托住瓶底,将其倒立,观察瓶塞周围是否漏液。若不漏液,将瓶直立后将瓶塞转动180°,再次倒立并检查,确定容量瓶不漏液后,方可使用。

(2) 转移溶液和定容

1) 容量瓶只盛放已溶解的溶液。用固体物质配制成标准溶液时,应将准确称量的固体置于小烧杯中,用总体积10%~20%的溶剂溶解后,将溶液定量转移入容量瓶中。若溶质难溶,可加热助溶,但必须待溶液完全冷却至室温再转移,否则将造成体积误差。

2) 将溶液转入容量瓶的方法见图7(b),玻璃棒下端伸入容量瓶瓶颈并靠在瓶颈刻度线下内壁上(注意不要让玻璃棒其他部位触及容量瓶口,防止液体流到容量瓶外壁),烧杯口紧靠玻璃棒,倾斜烧杯,使溶液沿玻璃棒流入瓶中。溶液倾完后,将烧杯沿玻璃棒轻提并慢慢直立,使附在玻璃棒和烧杯口间的液滴流回烧杯。用少量溶剂洗涤烧杯和玻璃棒3次,洗涤溶液一并转移入容量瓶中。加溶剂至容量瓶总容量的一半,旋摇容量瓶,使瓶内溶液大体混匀。继续加溶剂至标线附近,改用滴管小心滴加溶剂,视线至溶液凹液面与刻度标线水平相切。盖紧瓶塞,反复倒转和摇动容量瓶,使瓶内溶液混合均匀。

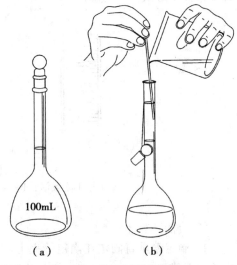

图7 容量瓶及其使用
(a)容量瓶;(b)容量瓶使用

3) 若发现静置容量瓶后液面低于刻度线,这是由于容量瓶内极少量溶液在瓶颈处润湿所致,并不影响所配制溶液的浓度,故不要在瓶内添加溶剂,否则,所配制的溶液浓度将降低。若加溶剂超过标线,需重新配制。

### 3.1.3 移液管及移液器

(1) 移液管:移液管是指在给定温度下(以 TD 标明,通常为 25℃),用于准确移取一定体积溶液的量器,精度为 0.01mL,常用的移液管有 5、10、25 和 50mL 等规格。习惯上中间为球形的称为大肚移液管,简称移液管,如图 8(a)所示,具有分刻度可移取规格内任意体积溶液的称为吸量管,如图 8(b)所示。

**移液管的操作方法**

主要有下列五个步骤:读、淋、吸、拿、放。

1) **读**:取移液管时,要明确容量的规格、类型、吸量管刻度的大小标示方向、每一分度标示的容量大小,及确认管口上是否刻有"吹"字或"快"字。

2) **淋**:移取溶液前,彻底洗净移液管,并用少量蒸馏水润洗 2～3 次,完全流干内壁水分。挤出洗耳球内空气,再把球的尖端紧接移液管口,移液管垂直插入液面 1.5cm 以下,缓慢松开洗耳球囊,吸取约占容积 1/3 体积量的溶液,双手平端移液管中后部,并不断转动,使溶液润洗全管内壁,随后使溶液从移液管管尖流出。重复吸取待移取的溶液润洗管内壁 2～3 次,以确保所移取溶液浓度不变。

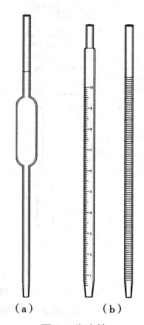

**图 8 移液管**
(a)大肚移液管;(b)吸量管

3) **吸与拿**:移取溶液时,一般用右手拇指和中指拿住管颈标线上方,将管尖适当插入溶液面下 1.5～2cm 中(勿接触底部),左手拿洗耳球先把球内空气挤出,再把球的尖端紧接移液管口,缓缓松开左手吸取溶液,随溶液在移液管内上升管尖下移,保持插入液面下 1.5～2cm 中,当溶液在管内上升至标线以上时,迅速用示指代替洗耳球按住管口,将移液管离开液面,如图 9(a)所示。

4) **放**:保持移液管垂直状态,管尖靠在倾斜的器皿液面以上的内壁上,略微放松示指,轻轻转动移液管,使过量液体流出管尖,直到溶液凹液面下缘与标线相平时立即用示指压紧管口;取出移液管,转移入承接溶液的器皿中。在放液过程中,移液管保持垂直,承接器皿稍倾斜而使移液管最大化的接触器皿内壁,使流速均匀形成均匀的内膜,松开示指,让管内全部溶液自然平稳地沿器皿壁流下。流完液体后等待 10～15s 或管尖靠壁轻轻旋转几圈取出移液管,如图 9(b)所示。若管壁上刻有"吹"或"快"字,用洗耳球将管内及管尖残留的溶液全部吹至承接器皿中;否则不可用力使其流出,因校准移液管时已考虑了管尖所保留溶液的体积。

(2) 移液器:移液器又称移液枪,是用于定量转移液体的量器,化学实验室常用的是单道微量可调手动移液器,包括 0.1μL～10mL 体积的不同规格。移液器的结构如图 10 所示,推动活塞排出空气,而后松开活塞利用大气压吸入液体;再次推动活塞排出液体,通过控制推动力度以调控被移液体的流速。

**移液器的使用方法**

1) **安装枪头**:利用旋转安装法安装枪头(也称吸头),轻轻用力下压,并左右微微转动将移液器垂直插入枪头,使其紧密结合。

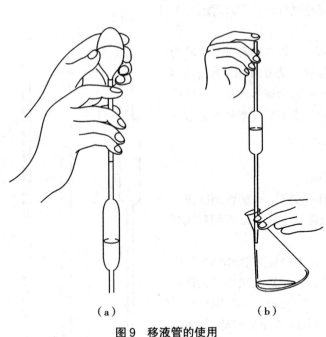
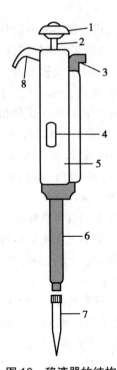

图9　移液管的使用
(a)移液管的吸液；(b)移液管的放液

图10　移液器的结构
1. 控制按钮；2. 推动杆；3. 枪头卸除按钮；4. 体积显示窗口；5. 套筒；6. 吸液杆；7. 枪头(吸头)；8. 握盖

2) 设定容量：取移液枪时，要明确其容量规格，根据需要设定其体积。将移液器从大体积调为小体积时，逆时针旋转旋钮即可；从小体积调为大体积时，可先顺时针旋转旋钮至刻度超出设定容量值的1/4圈，再回调至设定值，这样可排除机械间隙，保证设定的容量值准确。不要将按钮旋转超出量程，否则会卡住内部机械装置而损坏移液器。

3) 吸液操作：一般利用前进移液法吸液。首先连接适当的枪头，用大拇指按下控制钮至第一档，将移液器枪头垂直插入液面下1~6mm：其中0.1~10μL容量的移液器枪头插入液面下1~2mm，2~200μL容量的移液器枪头插入液面下2~3mm，1~5mL容量的移液器枪头插入液面下3~6mm。缓慢松开控制钮，等待1~3s，将移液器移出容器前，用试剂瓶口内壁将枪头外壁液体碰掉，确保枪头外壁无液体。当吸取有机溶剂或高挥发液体时，挥发性气体会在套筒室内形成负压，从而产生漏液情况，需要用待吸液预洗4~6次，让套筒室内气体达到饱和从而消除负压。

4) 放液操作：将移液枪头垂直插入承液器，缓慢将控制按钮压至第一档并等待约1~3s，继续按压至第二档，可重复几次，以确保枪头内无剩余液体。

5) 移液器使用完毕后，按压弹射键以弹除枪头。把移液器量程调至最大值，使弹簧处于松弛状态以保护弹簧，将移液器垂直放置在移液器架上。当移液器枪头里有液体时，切勿将移液器水平放置或倒置，以免液体倒流腐蚀活塞弹簧。

此外，当转移高黏液体、生物活性液体、易起泡液体或极微量的液体时，一般利用反向移液法，其原理是直接按下按钮至第二档，先吸入多于设定量值的液体，慢慢松开按钮至原位，然后将按钮按至第一档排出设定量值的液体，继续保持按钮于第一档(不再按至第二档)，取下有残留液体的枪头，弃之。

### 3.2 电子天平

电子天平为单盘称量仪器,用于称量物体质量。不同规格的电子天平,精度不同,绝对精度分度值达到0.1mg(即0.0001g)的称为万分之一天平,是目前实验室常用的电子天平,电子天平的荷载为100~200g,如图11所示。万分之一天平采用电磁平衡传感器,特点是称量准确可靠、显示快速清晰并且具有自动检测系统、简便的自动校准装置以及超载保护等装置。天平的秤盘置于电磁铁上,当样品放在秤盘上,由于样品质量和重力加速度的作用使得秤盘向下运动,天平检测到这个运动并通过电磁铁产生与此重力相抗衡的作用力,这个力与物品的质量成比例。

**称量方法**

(1) 直接称量法:直接称量法用于称量在空气中性质稳定、不吸潮的样品如金属、矿石等。称量时应注意:①不能称量热的物品;②为防止药品腐蚀托盘,所称量的物品不能直接放在托盘上,依情况置于称量纸上(称量纸四角沿对角线外延1/4处对折2次成小纸盒,纸盒大小依托盘大小或称量物品的多少而定)、表面皿上或容器中再称量。

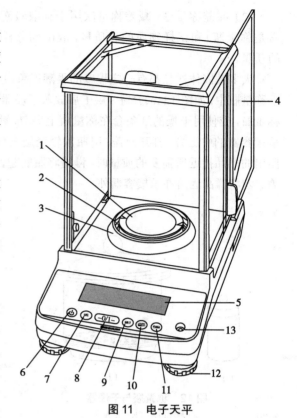

图11 电子天平

1. 托盘;2. 托盘支撑;3. 托盘圈;4. 防风罩;5. 显示器部;6. 启动键;7. "CAL"灵敏度校正键;8. "0/T"测定(零点/扣皮重)键;9. 单位切换键;10. 功能切换键;11. 数据输出菜单键;12. 水平调节支脚;13. 水准器

1) 调节天平水平:查看水准器中的气泡是否居中,否则调节垫脚使气泡居中,天平达水平。顺时针旋转水平调节垫脚时,垫脚伸长,天平上升,相反操作使天平下降。用小毛刷清扫干净托盘周围。

2) 预热:接通电源,至少预热半小时。

3) 调节灵敏度:按"CAL"灵敏度校正键,使用内置砝码的灵敏度校正自动开始,显示[END],返回。

4) 调零或去皮:调节为质量测定方式(以"g"等质量单位显示的状态)。打开防风罩的玻璃门,将容器放在托盘上,再次关闭玻璃门。显示稳定后,按"0/T"键,显示为零。

5) 测定质量:打开玻璃门,将称量器皿或折好的称量纸盒置于称台上,待显示值稳定后按"去皮"(TARE)键,天平显示"0.0"。将试剂瓶盖取下倒置于桌面,左手拿着试剂瓶,并使瓶口横向对着称量纸盒或称量器皿中心上方约5~10cm处(以防取试剂时撒落到桌面),小心用专用药勺取出试剂,轻轻抖入称量纸盒中,直至数字显示为所需要的质量,此时关闭玻璃门,显示稳定后,读取显示数值。(注:在读取天平示数时,请确认玻璃门已完全关闭。)

6) 包药:若用称量纸盛样品,样品需按药房包药的方式将药品包好,在称量纸外封写上样品的名称、质量、称量的时间及称量者的姓名。

7) 关闭天平:调零后,按住启动键不动,直到显示"OFF",关闭天平,拔掉电源。使用天平后,使用小毛刷进行清洁。

（2）减差称量法：减差称量法用于称取吸潮或挥发性的试样。称量步骤是先称量容器（通常是称量瓶）和试样物质的总质量，取出部分样品后再称量剩余质量，两者之差即为取出样品的质量。

大多数固体样品都有一定程度的吸潮现象，因此称量前需将被称物品，预先烘干 1~2h，并盛于称量瓶中，如图 12（a）所示；烘后放入干燥器内，如图 12（b）所示。不能用手直接拿取洁净的称量瓶，应使用干燥的纸条套在称量瓶上夹取，或戴上洁净的细纱手套拿取。取下称量瓶斜放在承接容器的中上方，打开瓶盖，用瓶盖轻轻敲击瓶口，使样品落入承接容器中，如图 13 所示。当倾出的样品接近所需要的质量时，慢慢将瓶竖起，轻敲瓶口，使附在瓶口的试样落在瓶内，盖好瓶盖。勿将样品洒落在承接容器外。

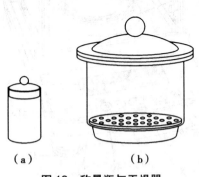

图 12　称量瓶与干燥器
（a）称量瓶；（b）干燥器

图 13　敲取试样

### 3.3　pH 计

pH 计（或称酸度计）是用来测量溶液酸碱度的仪器，它通过电位测定法，即测量浸入溶液中的指示电极和参比电极的电位差来确定溶液的 pH。

（1）原理：pH 计由参比电极与指示电极、或复合电极和精密电位计组成。

1）参比电极：参比电极有恒定的电极电位，不受溶液的影响，当有微弱电流通过时，电极电位改变微弱。常用的参比电极有饱和甘汞电极和银-氯化银电极。

**饱和甘汞电极**

饱和甘汞电极（SCE）见图 14（a）。在内管中，铂丝浸入糊状氯化亚汞（甘汞）和汞的混合物中。外管是盛有饱和氯化钾溶液的盐桥，可直接浸入待测溶液中。

甘汞电极的半电池表达式是

$$\text{Pt(s)}\,|\,\text{Hg(s)}\,|\,\text{Hg}_2\text{Cl}_2\text{(s)}\,|\,\text{KCl(饱和)}\,\|$$

在电极反应中，氯化亚汞和汞可逆地发生氧化还原反应

$$\text{Hg}_2\text{Cl}_2\text{(s)} + 2e^- \rightleftharpoons 2\text{Hg(l)} + 2\text{Cl}^-\text{(aq)}$$

由于液态 Hg 和固态 $\text{Hg}_2\text{Cl}_2$ 活度一致，电极电位的 Nernst 方程是

$$\varphi = \varphi^{\ominus}_{\text{Hg}_2\text{Cl}_2/\text{Hg}} - \frac{0.05916}{2}\lg c(\text{Cl}^-)^2 \tag{1}$$

式（1）中 $\text{Cl}^-$ 的浓度是固定的（饱和 KCl 的浓度约 4.2mol·L$^{-1}$），因此，电极电位是常数，在 25℃ 时为 0.2415V。

**银-氯化银电极**

银-氯化银电极见图14(b)，敷有氯化银的银丝浸入经氯化银饱和的氯化钾溶液中。半电池表示为

$$Ag(s)|AgCl(s)|KCl(c)\|$$

其半电池反应是

$$AgCl(s)+e^- \rightleftharpoons Ag(s)+Cl^-(aq)$$

根据Nernst方程式，电极电位仅与$Cl^-$浓度相关

$$\varphi = \varphi^{\ominus}_{AgCl/Ag^+} - \frac{0.05916}{1}\lg c(Cl^-) \tag{2}$$

如同甘汞电极，氯化钾溶液是饱和的。通常该电极体积很小，可在较高温度下使用。

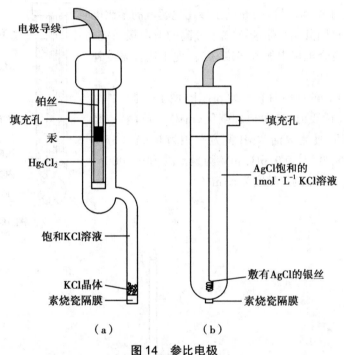

图14 参比电极
(a)饱和甘汞电极；(b)银-氯化银电极示意图

2) 指示电极：指示电极的电极电位对溶液中待测物质有响应。如玻璃pH电极是对溶液中氢离子浓度有响应的一种膜电极，当特制的玻璃膜两边氢离子浓度不等时，电极会产生一定电位。

**玻璃pH电极**

玻璃pH电极中玻璃膜结构如图15所示，膜内为银-氯化银电极，浸在经氯化银饱和的pH 7缓冲溶液中。极薄的离子敏感玻璃膜的主要成分是$SiO_2$(约70%)，其余30%由CaO、BaO、$Li_2O$和$Na_2O$混合而成。玻璃膜焊接于不敏感的玻璃管下端，测量时应浸入溶液中。当外部溶液含有氢离子时，电极反应表示为

$$Ag(s),AgCl|Cl^-(内),H_3O^+(内)|玻璃膜|H^+(外)$$

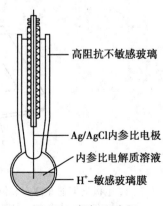

图15 玻璃pH电极

25℃时其电极电位是

$$\varphi_{glass} = K_{glass} + 0.05916\lg a(H_3O^+) = K_{glass} - 0.05916\text{pH} \tag{3}$$

式(3)中 $K_{glass}$ 是未知的玻璃电极常数，包括由 $Cl^-$ 浓度决定的内参比电极的电位、由膜内 $H_3O^+$ 浓度决定的膜电位和玻璃膜两边的不对称电位。

25℃时，用饱和甘汞电极、玻璃pH电极和待测溶液组成原电池，电池的电动势为

$$E = \varphi_{SCE} - \varphi_{glass} = 0.2415 - (K_{glass} - 0.05916\text{pH}) \tag{4}$$

因此，玻璃电极常数 $K_{glass}$ 必须用已知pH的标准缓冲溶液校准。

$$\text{pH} = \frac{E + K_{glass} - 0.2415}{0.05916} \tag{5}$$

3）复合电极：现在多数pH电极将玻璃电极和银-氯化银外参比电极组装在一个管中形成复合电极(图16)。复合电极中pH敏感玻璃电极位于中间，外围被充满了参比电解液的参比电极所包围。复合电极因其结构紧凑而优于分离的双电极系统。更便捷的电极是在复合电极中加入温度探头，便于进行温度补偿，也被称为三合一电极。

（2）FE28型pH计的使用：FE28型pH计(图17)是一种高输入阻抗的伏特计，读数为pH单位和毫伏(mV)单位，特点是操作直观、节省空间，使用灵活。pH测定范围为0~14，分辨率是0.01pH，测量精度为±0.01pH；mV测量范围为-2000~2000mV，分辨率是1mV，测量精度为±1mV。

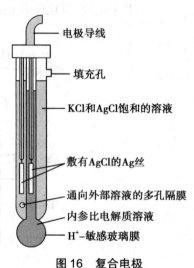

图16 复合电极

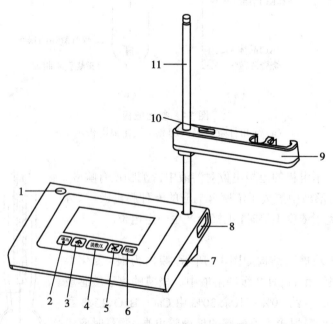

图17 FE28型pH计示意图

1. 电极支架的左侧安装位置；2. 开关按钮；3. 存储/回显；4. 读取/终点方式；5. 模式/设置；
6. 校准；7. 外壳；8. 支架杆存储空间；9. 支架臂；10. 紧固按钮；11. 电极支架杆

1）安装及开机：安装电极支架，pH 计的电源线与 100～240V 电源连接，安装电极。短按开关键打开仪器，预热仪器半小时。

2）校准

①仪器自动识别温度电极，则显示出 ATC 和样品温度，若仪器未检测到温度电极，则将自动切换到手动温度模式。长按"模式/设置"键，温度值闪烁，按上下箭头键选择温度值（默认设置为 25℃），按下"读数$\sqrt{A}$"键确认 MTC 温度补偿设置；当前标准缓冲溶液组闪烁，使用上下箭头键选择标准缓冲溶液组并确认；选择校准方式（线性，Lin. 或线段模式，Seg.）并确认；选择分辨率（0.1 或 0.01）并确认；选择温度单位（℃或℉）并确认，按下开关键，返回测量界面。

②根据需要，进行 1 点校准或多点校准（采用 1 点校准，仅调节偏移；采用 2 点校准，斜率和偏移均得到更新；采用 3 点或以上校准，斜率或零点均得以更新，并显示在显示屏的相应位置），标准缓冲溶液的 pH 与温度有关（表 2）。

表 2　不同温度下标准缓冲溶液的 pH

| 温度<br>（℃） | 邻苯二甲酸氢钾<br>（0.05mol·L$^{-1}$） | $KH_2PO_4 + Na_2HPO_4$<br>（0.025mol·L$^{-1}$ + 0.025mol·L$^{-1}$） | 硼砂<br>（0.01mol·L$^{-1}$） |
|---|---|---|---|
| 0 | 4.00 | 6.98 | 9.46 |
| 10 | 4.00 | 6.92 | 9.33 |
| 15 | 4.00 | 6.90 | 9.27 |
| 20 | 4.01 | 6.88 | 9.22 |
| 25 | 4.01 | 6.86 | 9.18 |
| 30 | 4.02 | 6.85 | 9.14 |
| 35 | 4.03 | 6.84 | 9.10 |
| 40 | 4.04 | 6.84 | 9.07 |
| 50 | 4.06 | 6.83 | 9.01 |
| 60 | 4.10 | 6.84 | 8.96 |

③清洗并擦净 pH 电极头，将电极插入标准缓冲溶液液面以下 3～5cm，短按"校准"键，开始校准仪器。通过长按"读数$\sqrt{A}$"键进行自动和手动两种终点方式的切换。显示屏上显示出样品的 pH，小数点开始闪烁，若选择了自动终点模式，当信号稳定后，显示器将自动锁定读数，出现$\sqrt{A}$，且小数点停止闪烁。若选择了手动终点模式，当信号稳定后，按下"读数"键记录缓冲液在当前温度下的 pH。

④若进行 1 点校准，短按"读数"键完成 1 点校准。

⑤若进行 2 点校准，用蒸馏水冲洗且擦干电极，将电极放入下一校准缓冲液中，短按"校准"键，根据不同终点方式完成校准。

⑥若进行 3 点校准，在 2 点校准基础上，重复步骤⑤的操作。

⑦以此类推，可对设备进行 4 或 5 点校准。（注：Seg 线段校准仅对 3 点或更多点校准有意义），按下"退出"键，返回测量界面。

3）测量 pH

①短按"模式设置"键，将测量模式调为 pH 测量模式。

②清洗并擦净 pH 电极头，将其浸入待测样液液面以下 3～5cm，短按"读数$\sqrt{A}$"键，开始测量。当信号稳定（自动终点方式，出现$\sqrt{A}$）或按下"读数$\sqrt{A}$"键（手动终点方式，出现$\sqrt{M}$）时仪器停止测量。记录溶液在当前温度下的 pH。

③将电极移上支架臂，移去被测溶液，用蒸馏水洗净擦干后放入原来的电极浸泡液中。长按

"退出"键关闭仪器。

4)测量 mV：短按"模式设置"键，将测量模式调为 mV 测量模式。测量方法与步骤"3)"测量 pH 中②和③操作相同。

### 3.4　722 型分光光度计

（1）工作原理：分光光度计的基本原理是，在光的照射下，被测物质受到激发，产生对光的吸收效应。物质对光的吸收具有选择性，各种不同的物质都具有其各自的吸收光谱。所以当某单色光通过有色溶液时（图 18）一部分光被吸光物质吸收，其余则透过和反射。有色溶液对光的吸收程度与物质的浓度、液层厚度及入射光的强度有一定的比例关系，即符合 Lambert-Beer 定律

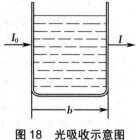

图 18　光吸收示意图

$$T = \frac{I}{I_0} \tag{6}$$

$$A = \lg \frac{1}{T} = \lg \frac{I_0}{I} \tag{7}$$

$$A = abc \tag{8}$$

式（6）中 $T$ 为透光率，$I$ 为透射光强度，$I_0$ 为入射光强度，式（7）中 $A$ 为吸光度，式（8）中 $a$ 为吸光系数，$b$ 为液层厚度，$c$ 为溶液的浓度。由上式可知，当入射光、吸光系数和溶液的厚度不变时，吸光度随溶液的浓度而变化。

（2）光学系统：722 型分光光度计是利用近紫外光和可见光（360～800nm）进行光度分析的仪器。光度分析的主要优点是具有良好的稳定性，重现性，功能完善和方法简便。其光学系统如图 19 所示。由光源灯（钨灯或卤钨灯）发出的连续辐射光线射到聚光透镜上，会聚后再经过反光镜转角 90°，反射至入射狭缝与准直镜。入射光线经过准直镜反射后，以一束平行光射向光栅（C-T 单色光器）后进行散射，并依原路稍偏转一个角度反射回来，经过准直镜会聚在出射狭缝上。光经过出射狭缝后，通过聚光透镜进行聚焦，然后通过比色皿，由光门射到光电倍增管上，产生的光电流输入显示器，可通过光电显示屏直读数据。

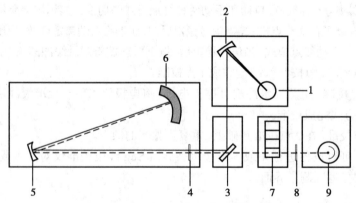

图 19　722 型分光光度计的光学系统

1．光源；2．聚光镜；3．反光镜；4．狭缝；5．准直镜；6．光栅；7．比色皿；8．光闸门；9．光电倍增管

（3）结构：图 20 是 722 型分光光度计的主要部件示意图。光源灯发射的复合光经单色光器（光栅）分解为单色光，经由色散元件和狭缝选择单一波长的光。单色光通过样品池吸收后，透过光由检测器（光电倍增管）接收及微电流放大器将信号放大，最后由读数装置显示。

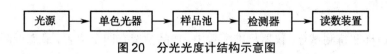

图 20　分光光度计结构示意图

（4）比色皿：比色皿，又称样品池或吸收池，它有两个平行的透光面（图 21）。常用 1cm 的比色皿，厚度更大的比色皿能装盛更多样品而增加吸光度。玻璃比色皿因对紫外光有很强吸收而只用于可见光测定，石英比色皿既可用于紫外光也可用于可见光测定。每台仪器都备有不同厚度的成套比色皿，它们有相同的光学性质。

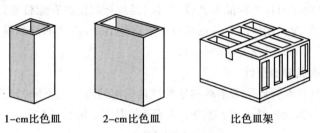

图 21　比色皿和比色皿架

取用比色皿的方法很重要，任何一点厚度变化、弯曲变形、斑点、擦毛、刮痕和裂纹都会影响测量结果。因此操作时必须遵守以下几点：

1）不要拿握透光面。

2）测量前须用待装溶液少量多次洗涤。盛装溶液不低于比色皿的二分之一。

3）用镜头纸擦拭比色皿外壁的水分或污渍，禁用抹布和毛巾擦拭。观察比色皿，外壁不应有纤维，内部不能有气泡。否则，可用硝酸浸泡，但不能超过 3 小时。

4）将比色皿置于比色皿架中时，注意不要划伤透光面。比色皿应直立，透光面正对比色皿架的窗口。

5）同时使用两只或两只以上比色皿时，需固定一只装空白溶液。

（5）722 型分光光度计的使用方法：722 型分光光度计的仪器外形如图 22 所示。

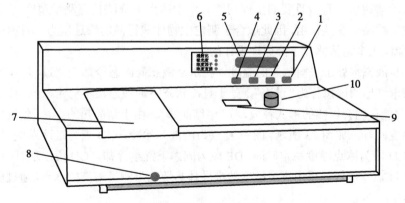

图 22　722 型分光光度计

1. 100% 键；2. 0% 键；3. 功能键；4. 模式键；5. 显示窗 4 位 LED 数字；6. 四种标尺（"透射比"、"吸光度"、"浓度因子"、"浓度直读"）；7. 样品室；8. 试样槽架拉杆；9. 波长指示窗；10. 波长调节钮

1）预热：接通仪器电源，预热半小时。

2）调整波长：使用波长调节钮调整仪器当前测试波长，具体波长由指示窗显示，读取波长时

目光垂直观察。

3）置入空白样和待测试样：仪器标准配置中，样品槽架有四个位置，最靠近测试者（最外侧）为"0"位置，往里数依次为"1""2""3"位置，一般在"0"位置放入空白样。

4）调零：打开试样盖或用不透光材料在样品室中遮断光路，按"0%"键。

5）调整100%T：通过推拉试样槽架拉杆让空白样置入样品室光路中，盖下试样盖（自动打开光门），按"100%"键（一次有误差可加按一次）。

6）测定样品吸光度：按模式键调为"吸光度"模式，通过拉试样槽架拉杆让不同样品进入光路，从"显示窗"中分别读出各个样品吸光度数值。当拉杆到位时有定位感，到位时请前后轻轻推动以确保定位正确。每次关闭样品室盖后都要重新调整100%T，再对样品进行测试。

7）仪器使用完毕后，关闭开关，切断电源。

### 3.5　FM-9X 型冰点渗透压计

（1）原理：难挥发非电解质稀薄溶液的依数性是指溶液蒸气压下降、沸点升高、凝固点降低和渗透压力，此类性质与溶质本性无关。其中，稀溶液的凝固点降低 $\Delta T_f$ 用关系式表示为

$$\Delta T_f = i K_f b_B \tag{9}$$

式（9）中 $K_f$ 为溶剂的凝固点降低常数，水的 $K_f$ 是 $1.857 \text{K·kg·mol}^{-1}$，$b_B$ 为溶质的质量摩尔浓度，单位 $\text{mol·kg}^{-1}$。校正因子 $i$ 是溶质分子解离出的粒子个数。非电解质溶质的校正因子 $i$ 为 1；在近似处理的情况下，AB 型强电解质（如 $KCl$、$CaSO_4$、$NaHCO_3$ 等）及 $AB_2$ 或 $A_2B$ 型强电解质如（$MgCl_2$、$Na_2SO_4$ 等）的校正因子 $i$ 分别为 2 和 3。

根据 van't Hoff 定律，稀溶液的渗透压力 $\Pi$ 为

$$\Pi = i c_B RT \tag{10}$$

式（10）中 $c_B$ 为溶质的物质的量浓度，单位 $\text{mol·L}^{-1}$，$R$ 为气体常数，$T$ 为绝对温度。

常温下稀的水溶液的质量摩尔浓度和物质的量浓度的数值近乎相等

$$b_B (\text{mol·kg}^{-1}) \approx c_B (\text{mol·L}^{-1}) \tag{11}$$

因此有

$$\Pi \approx i b_B RT \tag{12}$$

生物体液的渗透压力有着重要的生理功能。由于冰点下降法具有测量精度高、操作简便、样品量少和对生物样品无变性作用，因此适合于测定生物体液样品的渗透压力。目前国外生产的渗透压力测定仪器，大多是按冰点下降原理设计的。

冰点是冰与液态水处于平衡状态的温度。在水从液态向固态冷却变化过程中，温度虽达到甚至低于冰点而未结冰的现象称为"过冷"。处于过冷状态下的液态极不稳定，一经扰动便可立刻引起结晶，分子的能量就会以热的形式释放，称为"结晶热"。由于结晶热的存在，使过冷溶液在结冰形成的瞬间温度回升。如图 23 所示，图中 A 点为溶液开始冷却，B 点为过冷温度，C 点为溶液开始结冰温度，CD 段为冰点温度稳定时间，DE 段为固态下继续冷却。CD 段出现时的温度，就是冰点。FM-9X 型冰点渗透压计采用高灵敏度的半导体热敏电阻测量溶液的冰点，通过电量转化得到渗透压测定结果。

（2）FM-9X 型冰点渗透压力计的使用方法：图 24 所示为某仪器厂生产的 FM-9X 型冰点渗透压力计的外观。仪器有一套半导体制冷装置，以不冻液作为冷媒，冷却样品。为使液体在过冷后结晶，仪器还有一套过冷引晶装置，一套高精度的测温系统。通过测量溶液的冰点温度下降值来测定渗透压，选择 $300 \text{mmol·L}^{-1}$ 和 $800 \text{mmol·L}^{-1}$ 两种不同浓度标准的 NaCl 溶液对仪器进行标定。

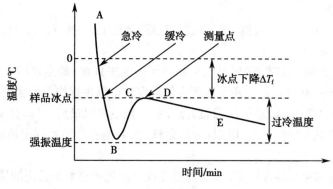

图 23　冷却曲线

1) 仪器使用的准备与调整

①在冷却槽内加入约 60mL 的不冻液，直到仪器右侧溢流杯中有不冻液排出为止。

②接通电源，仪器进入等待状态，仪器面板显示冷却槽温度(- - - - 表示温度过高)，仪器经约半小时的预热，自动平衡在温度控制点。

2) 标定

①标定要求：取一只干净且干燥的试管，用玻璃注射器从标准溶液瓶内抽取 0.5mL 的 300(或 800) mmol·L$^{-1}$ 标准溶液注入试管，对仪器进行正常的标定(早冻或不冻的结果不能作为正常标定)。注射器和试管不得污染标准溶液。

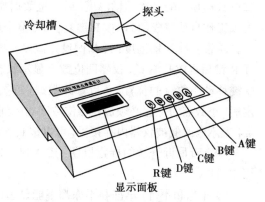

图 24　FM-9X 型冰点渗透压力计

②标定方法：将装有 0.5mL 标准溶液的试管垂直套向用软纸擦净的测量探头后置入冷却槽，按 C 键进入标定程序，仪器显示"300"。标定值须与所放仪器的标准溶液相一致，如果有错，仪器将不能进入正常测试。用 D 键可以根据所用标准溶液进行 300 和 800 变换。在确定标定液和所选标定值相同后，按 B 键执行标定功能。标定过程中显示标准溶液温度的变化，当标准溶液温度达到 -6℃时仪器自动强振，显示标定液的冰点温度，稍后显示"300E"(或"800E")，表示标定结束。按 D 键将标定值存入仪器内存，仪器显示"300P"(或"800P")。标定结束应及时将测量探头从冷却槽内取出，如长时间将探头置于冷却槽内，会使仪器测量传感器和标准溶液之间被冻结得过硬而容易损坏测量传感器。

标定中按 A 键可以退出标定，回到等待状态，显示冷却槽温度，按 B 键可以察看标定进行时间(单位：S)，按 D 键显示冷却槽温度。

3) 样品的测量

①在试管内加入 0.5mL 的被测样品溶液，将试管口套在测量探头上后置入冷却槽，按 D 键进入测量程序，测量过程中仪器显示样品温度的变化，当样品温度达到 -6℃时仪器自动强振，强振后仪器显示的样品温度迅速从 -6℃回升到样品的冰点温度，在报出测量结果后，及时将测量探头从冷却槽内取出。

②手动强振用于易产生早冻的样品，在测量过程中可进行人工干预，在刚发生早冻时，通过手动强振的方法测得样品的渗透压近似值。正常测量不需手动强振。

4) 按键辅助功能

①读出数据：在等待状态按 B 键，显示仪器最后测量的 12 个数据。按下按键，显示测量时间；放开按键，显示测量值。

②设置时间：按 R 键＋D 键，然后先放开 R 键后再放开 D 键，确定使用本功能。在此功能下，仪器四位数字显示状态为"月、日"，小数点所在的位置是可以进行修改的数据位置，用 D 键移位，按 B 键加"1"；A 键减"1"；完成设置月、日设置后，把小数点移至最后一位，按 C 键存入内存；然后仪器四位数字显示状态为"时、分"，用同样方法调整，按 D 键把小数点移至最后一位，按 C 键存入内存。

③热敏电阻温度和电压线性校正：本功能是在仪器出现内存数据丢失的情况下才使用。在测量试管内放入 0.5mL 不冻液，将试管口套在测量探头上，置入冷却槽内，按 R 键＋C 键，先放开 R 键后放开 C 键，仪器四位数字显示初始状态为"--0 0"。手推测量探头进入冷却槽到低位，仪器前二位数据显示的是冷却槽温度，后二位数据显示的是动态时间。当仪器最终显示"----"时，表示本项工作自动完成，可以按 R 键复位。（整个调整过程需要较长时间，一般在一个小时左右。）

④设置冷却槽温度：冷却槽设定温度一般为 -9.0℃。按 B 键＋R 键，确定使用本功能。先放开 R 键后再放开 B 键，仪器四位数字显示状态为"L-ＸＸ"，"-ＸＸ"是仪器原冷却槽设置温度，用 D 键移动小数点的位置，可以修改数据，每按一次 B 键可以增加 1，按 A 键减小 1，冷却槽温度设定后，小数点的位置必须放在 2 位数字的中间，按 C 键确认修改完成，在 L 位置下出现又一个小数点，表示修改值被仪器内存接受，可以按 R 键退出。

⑤查看最近一次的标定日期：按 A 键＋R 键，先放开 R 键后再放开 A 键，可以查看仪器目前标定日期。

5) 早冻和不冻：早冻和不冻都不能显示正常的测量结果。

①早冻
- 冷却槽中不冻液温度低于 -9.0℃，或被测样品的渗透压力太低；
- 试管不清洁或被测溶液中有颗粒状杂质（如样品中有结晶）；
- 仪器在有振动的环境下测试样品也容易引起早冻。

通常测量一个样品时间在 300s 以内，如果测量时间大于 300s，就应及时观察是否发生了早冻，以便及时中止。如能及时发现早冻，就可中止本次操作，如未能及时发现早冻，仪器将仍按其样品温度的下降而强振，强振后仪器显示"----"，说明本次操作失败。如果早冻未能及时观察发现，试管内的样品和热敏电阻将被冻结，这时不可急于拉下试管，应待其自然融化后取下试管，不然将会损坏热敏电阻。

②不冻
- 冷却槽中不冻液温度偏高，或被测样品的渗透压力太高；
- 强振幅度太小，振动棒在振动过程中打不到试管壁；
- 被测样品如有气泡也会产生不冻。

如果发生不冻，需稀释被测样品；或调整扳动振动棒，使振动中打到试管壁上。

### 3.6 离心机

（1）原理：离心机是通过高速旋转，对不同相(l-s、l-l)混合物进行快速分离的专用设备，见图 25(a)。如把装有固液混合物的试管放进离心机的试管套，通过离心机的运转使固体物质沉积于试管底部，可分离液体而保留固体，也可用毛细管抽提液体而遗弃固体。

在旋转离心机前，必须将一支平衡试管装在分离试管相对的位置，以避免离心机的过度振动，否则将会损坏离心机。平衡试管必须与离心试管同样大小，而且装有与待分离混合物同样体积的水。

离心分离法广泛用于医院、食品、工业、科研等单位进行化验、生化试验、分离悬浮液等工作中。

以 LDZ5-2 低速离心机为例，采用集成电路控制，控制面板见图 25(b)。具有慢启动功能，定时及转速均采用数字显示，方便、直观。该机特设超速报警及自制动系统，当旋转速度超过允许最高转速时，自动发出报警信号并停止运转。该系统对主控制电路系统故障和由于误操作引起的超速运转均具保护功能。

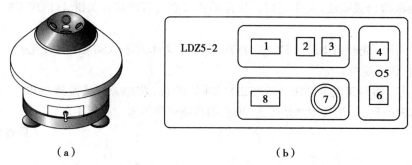

图 25　离心机和 LDZ5-2 低速离心机控制部分示意图
(a)离心机；(b)LDZ5-2 低速离心机控制部分示意图
1. 定时显示(min)；2. 定时调谐按键；3. 定时手动/自动按键；4. 启动按键；
5. 超速保护；6. 电源开关；7. 转速调节旋钮；8. 转速显示(×100r/min)

(2) 操作方法

1) 将离心机置于平稳台面上。

2) 试料配平：该机装备了自动平衡系统，试料配平较为简单，用目力观察每只试管放置溶液数量近似，然后对称放置到离心隔架内。最大不平衡量不得超过 20g，各旋转体按表 3 所示对应转速运转，不得超过 5 000r·min⁻¹(超过后机器将报警)，离心少量试液时，可在对称试管内加水平衡。

表 3　LDZ5-2 离心机的转速、离心力和容量

| 转速<br>r·min$^{-1}$ | 相对离心力<br>g | 单支试管容量<br>mL | 支数 | 总容量<br>mL | 不平衡量<br><g |
|---|---|---|---|---|---|
| 5 000 | 4 360 | 50 | 4 | 200 | 20 |
| 5 000 | 4 360 | 10 | 12 | 120 | 20 |
| 3 500 | 2 100 | 10 | 32 | 320 | 20 |

3) 检查

①调速旋钮 7(SPEED)应在 0 位。
②欲使用的旋转体应紧固，并确认安装正确。
③电源接好，插头插牢，接地使用。
④盖锁锁好。

4) 操作程序

①接通电源，将电源开关 4(POWER)接到"ON"位置，转速表显示"00"。如需定时，将定时手动自动按键 3(MAN-AUTO)，选择"AUTO"，则定时器显示出一个任意随机数并闪动，按下"定时调谐"按键 8(SET)，设定时间。

②按"启动"键6(START)，顺时针转动调整旋钮3(SPEED)，离心机开始做逆时针旋转，观察转速表，调至所需转速。

③在设定的时间内，显示表做时间减法计时，时间至"00"时，离心机停机运转，START键上指示灯灭。

④如需在此转速重复使用，可重新设置所需时间，按"启动"键6(START)后，离心机可自动在预置的转速和时间内运行。

⑤如不需使用定时，或离心时间超过59min，此时可将定时手动/自动按键设在"MAN"位，定时表显示的随机数仍做减法，但不定时。如需停止离心可将电源开关键4(POWER)搬至"OFF"位置。

⑥重复使用时，先开启电源开关键4(POWER)再按启动键6(START)，离心机可自动加速到所设置的转速（仍不定时）。

⑦离心完毕，将转速调节旋钮3(SPEED)旋回"0"位，将电源开关搬到"OFF"。按动开锁按钮，即可开盖取出样品。运转中严禁开盖，或企图在旋转惯性未完时用手制动。

（周春艳　李雪华）

# 第三章
# 实验结果的表示

## 1. 实验误差与有效数字

### 1.1 实验误差

实验误差是普遍存在的,即使严格地进行平行性测定,也并不能保证每次都能得到完全相同的结果。每次测量都有实验误差,实验误差源于以下原因。

(1) 系统误差:系统误差是由某些确定的原因造成的,有确定的值,它主要来源于方法误差、仪器误差、试剂误差和操作误差等。系统误差使得测量值总是偏高或偏低,因此可以通过校准测量仪器、改进实验方法、更换试剂或者通过对照实验来消除。

(2) 偶然误差:偶然误差是由某些难以预料的偶然因素引起的,如测定时环境温度、湿度和气压的微小波动,仪器性能的微小变化,以及分析人员对试样处理的微小差异等。即使没有系统误差,偶然误差也会使得测定结果时而比真实值高,时而又偏低。偶然误差虽然不可避免,但它对准确测定的影响可以通过多次重复的平行实验,取测定平均值的方法加以消除。

由于粗心大意或不按操作规程操作等引起的误差属于过失误差,例如错误读数或不正确操作等过失误差必须坚决避免。

### 1.2 准确度和精密度

(1) 准确度:准确度是指测定值($x_i$)与真实值($T$)符合的程度,准确度的高低用误差来衡量。误差越小,测定结果的准确度越高。

误差又称绝对误差($E$),可表示为

$$E = x_i - T \tag{13}$$

准确度也常用相对误差($E_r$)表示,即

$$E_r = \frac{E}{T} \times 100\% \tag{14}$$

由于相对误差反映出了误差在真实值中所占的比例,这对于衡量测定结果的准确度更为合理。因此,通常用相对误差来表示测定结果的准确度。

(2) 精密度:精密度指平行测定值之间的相互接近的程度,其大小用偏差来度量。偏差越小,测定结果的精密度越高。

某次测定值($x_i$)与多次测定值的算术平均值($\bar{x}$)之差称为绝对偏差($d$),即

$$d_i = x_i - \bar{x} \tag{15}$$

每次测定的绝对偏差的绝对值($|d_i|$)之和的平均值称为绝对平均偏差($\bar{d}$),设测定次数为$n$,则有

$$\bar{d} = \frac{|d_1|+|d_2|+|d_3|+\cdots+|d_n|}{n} \tag{16}$$

精密度的另一种衡量方法是相对平均偏差（$\bar{d}_r$），可表示为

$$\bar{d}_r = \frac{\bar{d}}{\bar{x}} \times 100\% \tag{17}$$

在讨论实验误差时，标准偏差（$s$）和相对标准偏差（RSD）也是时常用到的概念。
标准偏差的数学表达式为

$$s = \sqrt{\frac{d_1^2+d_2^2+d_3^2+\cdots+d_n^2}{n-1}} \tag{18}$$

标准偏差突出了单次测量中较大偏差对测定结果的影响，是表示精密度较理想的指标。
相对标准偏差可表示为

$$\text{RSD} = \frac{s}{\bar{x}} \times 100\% \tag{19}$$

在实际工作中多用相对标准偏差表示分析结果的精密度。

### 1.3 有效数字

有效数字用来表现测量值的精确程度，或者基于测量值的计算结果的精确程度。它由一些确定数字加末位不确定数字构成。例如用最小准确刻度为 mL 的量筒测量液体的体积，读到 34mL，再估读一位体积，如 0.2mL，于是液体的体积是 34.2mL。这个测量值有 3 位有效数字，误差是 ±0.1mL。

(1) 有效数字的位数：确定有效数字的位数有以下规则：

1) 数字"0"

"0"作为定位时不是有效数字。如 34.2mL 和 0.034 2mL 都只有 3 位有效数字。

数的右边有小数点，结尾的"0"都是有效数字。如 9.00cm、9.10cm、90.0cm，都有 3 位有效数字。

没有小数点的数，其结尾的"0"可能是也可能不是有效数字，不能确定。用科学计数法可以排除这种不确定性。

2) 科学计数法：科学计数法用 $A \times 10^n$ 的形式表示数，其中 $A$ 是小数点前只有一位并且是非零数字的数，$n$ 是整数。在科学计数法中，测量值 900cm 精确到两位有效数字是 $9.0 \times 10^2$cm，精确到三位有效数字是 $9.00 \times 10^2$cm。

3) 常用对数中的有效数字：常用对数其小数点前的数字对应于有理数幂的指数，因而其小数点后的数值才是有效数字。例如 pH 2.88 只有两位有效数字，它是氢离子浓度的负对数，此时氢离子的浓度是 $1.3 \times 10^{-3}$mol·L$^{-1}$。

(2) 准确数：迄今为止讨论的只是带有不确定性的数，然而也会遇到准确数。如数某一物品的数量或定义某一单位时，给出的是确定数。例如 9 只试剂瓶，那就是准确的 9，而不是 8.9 或 9.1。同样，1 英寸准确定义为 2.54cm，这里 2.54 不能说成三位有效数字，事实上它有无穷多位有效数字，只不过不可能全部写出来，因此有效数字的规则对准确数不适用。在计算中其结果的有效数字位数只取决于不确定数。例如一只试剂瓶的质量是 20.1g，9 只试剂瓶总质量的计算是

$$20.1\text{g} \times 9 = 180.9\text{g}$$

(3) 数的修约

1) 在实际测量中,常常是多种测量仪器联用,并进行多次测量,当要对多个测量数据进行处理时,需要根据误差的传递规律,对测量数据多余的数字进行取舍,即修约。

2) 修约通常按"四舍六入五留双"规则进行处理。即:当约去数为 4 时舍弃,为 6 时则进位,如用三位有效数字修约 1.214 3 和 1.216 2 的结果分别是 1.22 和 1.21;当约去数为 5 而后面无其他数字时,若保留数是偶数(包括 0)则舍去,是奇数则进位,使修约后的最后一位数字为偶数。例如用三位有效数字修约 1.215 的结果是 1.22。

(4) 有效数字的运算规则:测量值进行数据处理时,常用下面两条规则来确定计算结果的有效数字:

1) 有效数字的加减计算法:进行有效数字加减计算时,其结果以小数点后最少的位数为准。因此 34.2 + 0.21 的和是 34.4。又如,184.2g + 2.324g,计算结果为 184.2 + 2.324 = 186.524,但因为加数 184.2g 有最少的小数位数(1 位),而 2.324g 有 3 位小数,因此结果应该写成 186.5g。

2) 有效数字的乘除计算法:进行有效数字乘除计算时,其结果以有效数字位数最少的那个数为准。因此 34.2 × 0.21 的积是 7.2,两位有效数字。

做两步或两步以上的计算时,保留中间结果的非有效数字,这样能使计算中因修约而存在的小的误差不致在最后结果中出现。如果使用计算器,简单地输入数字并进行计算,最后修约。

## 2. 实验数据的处理

为了表示实验结果和分析其中的规律,需要将实验中获得的大量实验数据进行归纳和处理,常用的方法有列表法、作图法和数学方程法。

### 2.1 列表法

实验中常见的数据处理方法是列表法。将实验数据按自变量 $x$ 和因变量 $y$ 一一对应排列起来制成三线表格,以清楚地表示两者之间的关系。

每个表均应有表头,表由三线组成,不是特别需要,不能有竖线,表中需注明所示变量的名称与单位,表中数据不能再出现单位,并注意所示数据的有效数字位数。在数据处理过程中,通常选择较简单的变量(如温度、时间、浓度等)作为自变量。

### 2.2 作图法

虽然列表法有时能给出有意义的结论,但实验数据用作图来处理,常常可以更直观地显现数据之间的关联,表现数据变化的规律。根据作图还可求得斜率、截距、外推值等。下列为作图的要求:

(1) 选择合理的坐标比例,充分利用坐标纸。两根坐标轴可以有不同比例,数据的零值也可不必在原点上。坐标纸的主要分格应以 1 或 2 或 2.5 或 5 为间隔量或者为 10 的幂的倍数单位,不用 3、7 或其他非倍数值为间隔量,以免造成实验数据点在图上难以定位。

(2) 坐标轴标明所用物理量单位。习惯上以自变量为横坐标,因变量为纵坐标。

(3) 在坐标上定位实验数据,用点标出。同一实验数据组的点用同一图形符号"○""△""×"或"□"圈住,图形的面积近似等于测量误差的范围,每个点用"⊥"的长度大小表示出此值的测定标准差。

(4) 如果实验数据点看似是落在一条直线上,就用一直线表示,注意不在直线上的数据点要均匀地分布在直线两侧,这样能最佳地拟合实验数据。不能随手勾画直线,也不能用线段连接数据点而画成折线。如果数据点看似在一条光滑的曲线上，就用一条曲线表示，并使曲线尽可能地穿过各实验数据点。或用"Excel"软件拟合。

（5）给出图标题：在作图法中，直线图是处理大量数据的极为有效的方法，也容易得出可信的重要结论。尽管实验变量之间存在许多不同的关系，但线性关系最具吸引力，这不仅因为直线比曲线容易绘制，还因为直线方程的计算也更简单。

### 2.3 计算机处理

目前，利用计算机技术，使制表和制图更加方便，一些数据处理软件（如 Excel 等）的应用，可简化工作过程，提高效率。

如何利用 Excel 获得线性回归方程：

（1）将数据录入到 Excel 表中（2 列）。

（2）选中所有实验数据，然后在菜单中点击"插入"→"散点图"，即显示点状图。

（3）右击图中任意一数据点，在出现的选项中选择"添加趋势线"，即刻在图中显示直线。

（4）在上面的"趋势线选项中"，选择"显示公式"，即刻在图中显示线性回归方程，根据线性回归方程可求得直线的斜率和截距。

# 3. 实 验 报 告

实验报告表现了学生对所学化学知识的理解和掌握程度。通过参与实验、使用化学试剂和仪器，学生学习如何收集实验数据、观察实验现象，并对这些数据和现象给予解释；在准备和完成实验报告的过程中，学会设计、完成科学实验，报告研究结果。

学生进实验室之前必须做好预习，包括仔细阅读实验教程，查阅参考文献，模拟设计者的身份，简述实验设计思路、原理、步骤、实验条件及注意事项，完成预习作业。实验中应按照实验规程操作，认真观察实验现象，记录实验数据。实验结束后必须上交实验报告，包括数据和现象记录、数据处理和结论。

实验报告包括实验名称、学生姓名和学号、实验日期；目的要求、原理、仪器和试剂、实验步骤、结果和讨论，最后是思考题。书写实验报告，要求科学严谨、简洁明了、字迹工整、尊重事实、结论正确。根据实验的目的，本教程的实验从培养学生综合实验技能角度和无机化学、分析化学及物理化学等学科理论与实践相结合的角度，从不同教学层次，综合为以下三段式的化学实验训练内容：基本操作规范化及技能训练实验、以模拟设计者身份培养科学研究思维的综合性实验（容量分析实验、仪器分析、化学原理实验、无机化合物制备）、综合实验技能提升的自行设计性实验。下面是无机化合物制备和定量分析两种典型实验的报告格式。

**无机化合物制备实验**

实验 X　硫酸亚铁铵的制备

姓名＿＿＿＿＿＿　学号＿＿＿＿＿＿　日期＿＿＿＿＿＿

一、实验目的

……

二、预习作业

……

三、实验原理

……

四、仪器材料与试剂
……

五、实验步骤
……

六、数据记录与结果分析
……（此处提前做好实验数据记录的三线表格及数据处理项）

实验日期：_____ 室温：_____ ℃ 相对湿度：_____

1. 硫酸亚铁铵的制备

在台秤上称取铁屑重 $m(Fe)_1 =$ _____ g；

铁屑残渣重 $m(Fe)_2 =$ _____ g；

硫酸亚铁铵 $m[(NH_4)_2SO_4 \cdot FeSO_4 \cdot 6H_2O] =$ _____ g；

$$硫酸亚铁铵产率(\%) = \frac{产品质量(g)}{理论产量(g)} \times 100\% = \underline{\qquad}。$$

2. 产品纯度检验

| $Fe^{3+}$ 标准溶液 | 一级 | 二级 | 三级 |
|---|---|---|---|
| 硫酸亚铁铵产品级别 | | | |

七、思考题

**定量分析实验**

实验 X  盐酸溶液浓度的标定

姓名_____ 学号_____ 日期_____

一、实验目的
……

二、预习作业
……

三、实验原理
……

四、仪器材料与试剂
……

五、实验步骤
……

六、数据记录与结果分析
……

实验日期：_____ 室温：_____ ℃ 相对湿度：_____

1. 一级标准物 $Na_2CO_3$ 的称量

称量瓶 + $Na_2CO_3$ 质量 $(m_1)$ _____ g

称量瓶 + 剩余 $Na_2CO_3$ 质量 $(m_2)$ _____ g

$Na_2CO_3$ 质量 $(m_1 - m_2)$ _____ g

每次滴定的 $Na_2CO_3$ 质量 _____ g

2. 溶液配制
……

3. 数据记录及结果分析

| 样品 | 1 | 2 | 3 |
| --- | --- | --- | --- |
| 指示剂 | | | |
| 终点溶液颜色变化 | | | |
| $V(Na_2CO_3)$ / mL | | | |
| $V_{终}(HCl)$ / mL | | | |
| $V_{初}(HCl)$ / mL | | | |
| $\Delta V(HCl)$ / mL | | | |
| $c(HCl)$ / mol·L$^{-1}$ | | | |
| $\overline{c}(HCl)$ / mol·L$^{-1}$ | | | |
| 相对平均偏差 $\overline{d_r}$ / % | | | |

4. NaOH 标准溶液浓度的计算
……

七、思考题
……

（籍雪平）

# 第二篇 基础化学实验课题

## 实验课题一
## 基本操作训练实验

### 实验一 常用容量分析仪器操作练习

【实验目的】

(1) 掌握常用容量分析玻璃仪器的规范化操作、使用方法和注意事项。

(2) 练习容量瓶、移液管、吸量管、滴定管和锥形瓶等的使用，溶液配制及转移，容量仪器的洗涤与干燥，滴定分析操作。

(3) 掌握酸碱滴定分析的原理。

【预习作业】

(1) 指出玻璃仪器中的容器与量器的区别及使用要求。

(2) 简要写出本实验中常用定量分析玻璃仪器：容量瓶、移液管、吸量管、滴定管、锥形瓶规范化操作的注意要点。

(3) 查阅酸碱滴定分析相关文献，简要说明酸碱滴定分析的原理、种类及选择指示剂的原则。

【仪器材料与试剂】

**仪器** 烧杯(100mL, 250mL)，锥形瓶(250mL×3)，容量瓶(100mL)，玻璃棒，洗耳球，移液管(20mL×2)，吸量管(10mL)，酸式滴定管(25mL)，碱式滴定管(25mL)或聚四氟乙烯酸碱两用滴定管(25mL)，滴定台，万分之一电子分析天平

**材料与试剂** 0.1000 $mol·L^{-1}$ HCl 溶液，0.1000 $mol·L^{-1}$ NaOH 溶液，1.000 $mol·L^{-1}$ NaCl 溶液，$Na_2CO_3$，酚酞，甲基橙

【实验步骤】

**1. 容量仪器的洗涤和干燥练习**

参照第二章"基础化学实验常用仪器"，练习锥形瓶、容量瓶、移液管、滴定管等仪器的洗涤和干燥。

## 2. 移液管操作

### 2.1 移液管操作练习

用一支 20mL 的移液管，练习从 250mL 的烧杯中吸入自来水和调控凹液面与刻度相切，直至熟练为止。练习从 250mL 的烧杯中转移 20.00mL 水至 250mL 锥形瓶中。重复操作数次直至熟练。

### 2.2 吸量管操作练习

用一支 10mL 吸量管吸取 10.00mL 水，转移入 250mL 锥形瓶中。练习润洗、吸液及放液，要求每次放出 2.00mL 水，直至放完全部液体。重复上述操作数次直至熟练。

## 3. 容量瓶操作练习

### 3.1 验漏、转移溶液和定容

按第二章中容量瓶的操作方法，练习容量瓶的检漏、转移溶液和定容。

### 3.2 溶液配制

（1）配制 0.200 0mol·L$^{-1}$ NaCl 溶液：取一支 20mL 移液管，从试剂瓶中移取 20.00mL 1.000mol·L$^{-1}$ 的 NaCl 溶液，转移入 100mL 容量瓶中，稀释至刻线，配得最终浓度为 0.200 0mol·L$^{-1}$ 的 NaCl 溶液。重复上述操作数次直至熟练。

（2）配制 0.1mol·L$^{-1}$ Na$_2$CO$_3$ 标准溶液：在电子天平上准确称取经 105℃干燥至恒重的无水碳酸钠 1.0～1.1g（精确至 0.000 1g），置于洁净的 50mL 烧杯中，加入蒸馏水 30mL，用玻璃棒小心搅拌使之溶解。然后用玻璃棒引流将溶液转移入 100mL 容量瓶中，再用少量蒸馏水淋洗烧杯数次，淋洗液一并转移入容量瓶中。加蒸馏水至刻度标线，盖紧瓶塞上下颠倒、充分摇匀。

## 4. 锥形瓶操作练习

用移液管准确移取 20.00mL 自来水于 250mL 锥形瓶中，用右手握持锥形瓶颈，向同一方向作圆周旋转摇动练习。要求溶液在锥形瓶内旋转，不能让溶液飞溅造成损失。重复上述操作数次直至熟练。

## 5. 滴定管操作练习

按第二章中滴定管操作方法，进行滴定管漏及装配、润洗、装液、排气泡和读数等操作练习。

## 6. 滴定操作练习

### 6.1 滴定操作

按第二章中的滴定方法练习滴定操作。

### 6.2 酸碱滴定操作练习

（1）以酚酞为指示剂，用 0.1mol·L$^{-1}$ NaOH（精确至 ±0.000 1mol·L$^{-1}$）溶液作为滴定剂，滴定 0.1mol·L$^{-1}$ HCl（精确至 ±0.000 1mol·L$^{-1}$）溶液。

取洁净碱式或两用滴定管一支，用自来水检查是否漏液。若不漏液，则依次用蒸馏水和 0.100 0mol·L$^{-1}$ NaOH 溶液润洗 2～3 次，润洗液从管尖和管口弃去。将 0.100 0mol·L$^{-1}$ NaOH 注入滴定管，排除尖端的气泡，调节液面，记下滴定管初读数。

取 20mL 洁净移液管一支，用 0.100 0mol·L$^{-1}$ HCl 溶液润洗 2～3 次。准确移取 0.100 0mol·L$^{-1}$

HCl 溶液 20.00mL 于 250mL 锥形瓶中,滴加 2 滴酚酞指示剂,用 0.100 0mol·L$^{-1}$ NaOH 溶液滴定。直到溶液呈现微红色,且 30s 内不消失,即为滴定终点。从滴定管架上取下滴定管,读取并记录消耗的 NaOH 溶液的体积数(准确至 ±0.01mL)于表 1-1 中。重复滴定 3 次。

计算 NaOH 溶液与 HCl 溶液的体积比和 HCl 溶液的浓度,算出相对平均偏差。

(2) 以甲基橙为指示剂,以同样方法,用 0.100 0mol·L$^{-1}$ HCl 溶液滴定 0.100 0mol·L$^{-1}$ NaOH 溶液,记录相应数据于表 1-2 中,计算 HCl 溶液和 NaOH 溶液的体积比和 NaOH 溶液的浓度,算出相对平均偏差。

## 【数据记录与结果分析】

日期:　　　　　　温度:　　　　℃　　　　相对湿度:

表 1-1　用 0.100 0mol·L$^{-1}$ NaOH 溶液作为滴定剂滴定 0.100 0mol·L$^{-1}$ HCl 溶液

| 实验序号 | 1 | 2 | 3 |
| --- | --- | --- | --- |
| 指示剂 | 酚酞 | 酚酞 | 酚酞 |
| 滴定终点颜色变化 | | | |
| $V_{终}$(NaOH)/ mL | | | |
| $V_{初}$(NaOH)/ mL | | | |
| $\Delta V_{消耗}$(NaOH)/ mL | | | |
| $\Delta V_{平均}$(NaOH)/ mL | | | |
| $c$(NaOH)/ mol·L$^{-1}$ | | | |
| $V$(HCl)/ mL | | | |
| $c$(HCl)/ mol·L$^{-1}$ | | | |
| $\bar{c}$(HCl)/ mol·L$^{-1}$ | | | |
| $V$(HCl)/ $V$(NaOH) | | | |
| 相对平均偏差 $\overline{d_r}$ / % | | | |

表 1-2　用 0.100 0mol·L$^{-1}$ HCl 溶液作为滴定剂滴定 0.100 0mol·L$^{-1}$ NaOH 溶液

| 实验序号 | 1 | 2 | 3 |
| --- | --- | --- | --- |
| 指示剂 | 甲基橙 | 甲基橙 | 甲基橙 |
| 滴定终点颜色变化 | | | |
| $V_{终}$(HCl)/ mL | | | |
| $V_{初}$(HCl)/ mL | | | |
| $\Delta V_{消耗}$(HCl)/ mL | | | |
| $\Delta V_{平均}$(HCl)/ mL | | | |
| $c$(HCl)/ mol·L$^{-1}$ | | | |
| $V$(NaOH)/ mL | | | |
| $c$(NaOH)/ mol·L$^{-1}$ | | | |
| $\bar{c}$(NaOH)/ mol·L$^{-1}$ | | | |
| $V$(HCl)/ $V$(NaOH) | | | |
| 相对平均偏差 $\overline{d_r}$ / % | | | |

## 【思考题】

(1) 滴定管和移液管在使用前首先需用蒸馏水润洗,接着用待移液约 5mL 润洗,为什么?
(2) 为何每次滴定从零刻度或略低于零刻度开始?
(3) 滴定时,为何以甲基橙为指示剂和以酚酞作指示剂所得的 HCl 和 NaOH 的体积比不同?

(李献锐)

# 实验二 电子天平称量练习

## 【实验目的】

(1) 掌握电子天平的基本操作及常用的称量方法。
(2) 熟悉称量瓶与干燥器的使用。

## 【预习作业】

(1) 称量的方法有哪几种?直接称量法和差减法在操作和使用范围上有何不同?
(2) 简述电子天平的一般操作程序。
(3) 使用电子天平时应注意什么问题?
(4) 结合实验内容,请用流程图列出电子天平称量练习的操作步骤,标明具体的称量方法及实验的注意事项。

## 【实验原理】

电子天平的结构、原理和使用方法参见本书第一部分基础化学实验基本知识,第二章中电子天平使用的有关内容。

物质的称量有多种方法,常用的有直接称量法和差减称量法。直接称量法适用于称量洁净干燥的器皿(如小烧杯、表面皿等)以及在空气中性质稳定、不吸潮的试样(如金属、矿石等)。称量方法是首先将容器或称量纸置于天平的秤盘上称量,去除容器或称量纸质量(去皮),然后将试样放到容器或称量纸上,称出试样的质量。差减称量法是由两次称量之差得到试样质量的称量方法,并且只要求在一定的质量范围内称量。称量步骤是先称量容器(通常是称量瓶)和试样的总质量,取出部分样品后再称量剩余质量,两者之差即为取出样品的质量。

本实验练习电子天平的使用、试剂的取用及干燥器使用的规范化操作,利用常用的称量方法准确称取一定质量的试样。

## 【仪器材料与试剂】

**仪器** 万分之一电子分析天平,干燥器,称量瓶,小烧杯(50mL),药匙,小纸条,称量纸
**材料与试剂** 碳酸镁,硼砂

## 【实验步骤】

### 1. 检查天平

使用前观察水平仪,若显示天平处于非水平状态,需调整水平调节脚,使水泡位于水平仪中心。

## 2. 开启天平

接通电源,单击"ON"键,天平自检,显示屏很快出现"0.000 0g"。如果显示不是"0.000 0g",则要按一下"→ O/T ←"键。天平长时间断电之后再使用时,至少需预热 30min。

## 3. 直接称量法

准确称出小烧杯的质量和 0.20～0.25g(精确至 0.000 1g)碳酸镁。

(1) 称量容器:将干净的小烧杯轻轻放在称盘上,其质量即会自动显示,记录称量数据。

(2) 扣除皮重:轻按"→ O/T ←"键(清零/去皮),显示的数字消失,然后出现"0.000 0"字样,容器质量即被扣除。

(3) 称量样品:将碳酸镁试剂瓶口置于小烧杯或称量纸上方,用药勺取出试剂分别借助轻抖、或轻弹、或轻敲、轻刮药勺臂的方式放入小烧杯或称量纸中,直至符合称量的要求,停止加样,关闭天平门,记录称量数据于表 2-1 中。若用称量纸称量的,按药房包药方式包好,在称量纸外标注上相应样品信息及日期。

(4) 按上述方法称量,可得多份试样。

## 4. 差减称量法

准确称出 1.0～1.4g(精确至 0.000 1g)硼砂。

(1) 平移打开干燥器盖子,用小纸条将盛有硼砂的称量瓶从干燥器中夹住取出,盖好干燥器,将称量瓶置于称盘上,关好天平门。显示稳定后,按"O/T"键,去皮重并显示"0.000 0"字样。

(2) 从天平上取出称量瓶,倾斜称量瓶,用瓶盖小心敲击称量瓶口边沿,使一定量的样品抖落入烧杯中,再将称量瓶置于称盘上称量。此时显示的读数是负值,去掉负号即为倾出样品的质量。若所示质量达到要求范围,即可记录称量数据于表 2-1 中。

(3) 按上述方法称量,可得多份试样。

(4) 若用称量纸称取试样,称好的试样要求按药房包药的方式包好,并写上试剂的名称、质量、称量的时间及称量者的姓名等信息。

## 5. 称量完毕的操作

称量完毕,取下被称物,轻按"OFF"键,让天平处于待命状态。再次称量时轻按"ON"键就可使用。最后使用完毕,用毛刷清扫天平,关好天平门,拔下电源插头,盖上防尘罩,在"使用登记本"上登记。

## 【注意事项】

(1) 称量时,一切操作都要细心。读数时要注意关闭好天平门,以免气流影响使读数不准确。

(2) 天平的载重不能超过其限度,称量物体的温度必须与室温相同。

(3) 不能在天平秤盘上直接称取试样,应放在洁净的器皿中称量。有腐蚀性或吸湿性的物体都必须置于密闭容器内称量。

(4) 称量时,要用镊子或纸条夹取称量物品,也可套上医用手套进行操作,切勿用手直接拿取。

## 【数据记录及结果分析】

日期：　　　　　　　温度：　　　　℃　　　　相对湿度：

**表 2-1　电子天平称量练习记录**

| 实验序号 | | 1 | 2 |
|---|---|---|---|
| 直接称量法 | $m$(小烧杯)/g | | |
| | $m$(样品)/g | | |
| 差减称量法 | $m$(样品)/g | | |

## 【思考题】

（1）用差减法称量时，如何从称量瓶中倾出试样？
（2）影响电子天平读数稳定性的因素有哪些？

（赵全芹）

# 实验三　缓冲溶液的配制与性质、溶液 pH 测定

## 【实验目的】

（1）学习缓冲溶液及常用等渗磷酸盐缓冲溶液的配制方法。
（2）加深对缓冲溶液性质的理解。
（3）巩固吸量管的规范操作。
（4）学习移液枪的使用方法。
（5）学习使用 pH 试纸和酸度计测量溶液 pH 的方法。

## 【预习作业】

（1）普通溶液与缓冲溶液有什么不同？为什么缓冲溶液具有缓冲能力？
（2）缓冲溶液的性质有哪些？缓冲溶液的 pH 由哪些因素决定？
（3）如何衡量缓冲溶液的缓冲容量大小？缓冲溶液的缓冲容量与什么因素有关？
（4）如何检测缓冲溶液的 pH 是否发生改变？是否均需要用 pH 计？
（5）根据本实验的检测手段，说明本实验属定量分析还是定性分析或半定量分析？
（6）请用流程图阐释本实验是如何设计实验步骤，以验证缓冲溶液所具有的性质及缓冲容量的影响因素？请在流程图中标出每个实验步骤要有哪些注意事项？

## 【实验原理】

缓冲溶液由足够浓度的弱酸（HB，亦称抗碱成分）及其共轭碱（$B^-$，亦称抗酸成分）、或弱碱及其共轭酸或多元酸的酸式盐及其次级盐组成的溶液。在溶液中存在酸碱质子平衡

$$HB + H_2O \rightleftharpoons H_3O^+ + B^-$$

缓冲溶液的 pH 可用下式计算

$$pH = pK_a + \lg\frac{[B^-]}{[HB]} \tag{3-1}$$

式(3-1)中 $K_a$ 为共轭酸的酸解离常数，$\frac{[B^-]}{[HB]}$ 称为缓冲溶液的缓冲比。

当加入少量的强酸（或强碱）到缓冲溶液中时，缓冲溶液中存在的大量抗酸成分（或抗碱成分）将消耗这些外加的少量强酸（或强碱），使缓冲溶液的 pH 基本保持不变。适当稀释时，共轭酸及共轭碱受到同等程度的稀释，使 $\frac{[B^-]}{[HB]}$ 的比值基本保持不变，缓冲溶液的 pH 也基本不变。

缓冲溶液的缓冲能力大小可用缓冲容量 $\beta$（式3-2）表示

$$\beta = \frac{d\,n_{a(b)}}{V|dpH|} \tag{3-2}$$

缓冲容量的大小与缓冲溶液总浓度、缓冲比有关。当缓冲比一定时，缓冲溶液总浓度 $\{[HB]+[B^-]\}$ 越大，缓冲容量 $\beta$ 越大；当总浓度一定时，缓冲比 $\frac{[B^-]}{[HB]}$ 越趋向于1，缓冲容量 $\beta$ 越大；当缓冲比 $\frac{[B^-]}{[HB]}=1$ 时，缓冲容量 $\beta$ 达该总浓度 $\{[HB]+[B^-]\}$ 下的极大值。

缓冲溶液主要应用于需要保持稳定 pH 的溶液系统，如：生物体内的生化反应需要保持并稳定在一定的 pH 范围。等渗磷酸盐缓冲盐水（PBS）是医学上体外细胞培养常用的缓冲溶液，其渗透压和 pH 处于临床等渗和人体血浆正常 pH 范围。

本实验将学习配制缓冲溶液和医学上常用 PBS 溶液，并以普通溶液作为对照，验证缓冲溶液的抗酸、抗碱及抗适当稀释的性质。

## 【仪器材料与试剂】

**仪器**　量筒（250mL×1），烧杯（50mL×6，200mL×2，500mL×1），试管（5mL×12），比色管（10mL×3），容量瓶（50mL×1），吸量管（5mL×5）或可调移液枪（5mL×5，枪头若干），比色管架1个，试管架1个，滴管1支，玻棒1支，洗瓶1个，pH 计1台

**材料与试剂**　0.5mol·L$^{-1}$ HAc，1.0mol·L$^{-1}$ HAc，1.0mol·L$^{-1}$ NaAc，0.1mol·L$^{-1}$ HAc，0.1mol·L$^{-1}$ NaAc，1.0mol·L$^{-1}$ NaOH，1.0mol·L$^{-1}$ HCl，0.15mol·L$^{-1}$ NaCl，0.15mol·L$^{-1}$ KCl，0.15mol·L$^{-1}$ KH$_2$PO$_4$，0.10mol·L$^{-1}$ Na$_2$HPO$_4$，pH 4.00、pH 6.86、pH 9.18 的标准缓冲溶液，蒸馏水，广泛 pH 试纸，自带试样两份（体积不小于50mL）

## 【实验步骤】

### 1. 缓冲溶液的配制

**1.1　pH 7.4 的 PBS 缓冲溶液的配制**

按照表 3-1 中试剂的用量，用吸量管或移液枪分别移取相应的溶液到 50mL 容量瓶中，配制 pH 7.4 的 PBS 缓冲溶液，备用。

**1.2　HAc-NaAc 缓冲溶液的配制**

按照表 3-2 中试剂的用量，用吸量管或移液枪分别移取相应的溶液到 3 支 10mL 比色管中，分别配制 A～C 缓冲溶液，备用。

## 2. 缓冲溶液的性质及缓冲容量的影响因素测定

按照表 3-3 中试剂的用量，分别进行普通溶液（本实验为 0.5mol·L⁻¹ HAc）及缓冲溶液的抗酸、抗碱、抗稀释能力测试，记录实验数据。

通过比较分析表 3-3 各实验结果，总结以下结论：

（1）普通溶液与缓冲溶液之间的缓冲性质差异。
（2）缓冲溶液的缓冲性质。
（3）总浓度相同，缓冲比不同的缓冲溶液之间的缓冲容量差异。
（4）缓冲比相同，总浓度不同的缓冲溶液之间的缓冲容量差异。

## 3. PBS 溶液及自带试样溶液 pH 的测定

### 3.1　pH 计的校正

按"第二章基础化学实验常用仪器"中 pH 计的使用方法对 pH 计进行校正。

### 3.2　PBS 溶液及自带试样溶液 pH 的测定

将表 3-1 中配制好的 PBS 溶液和自带试样溶液分别装入 50mL 烧杯中，先用广泛 pH 试纸测量 pH，再用 pH 计测量 pH，将实验数据记录于表 3-4 中。

比较一下不同测量方法所测的相同溶液的 pH 是否与你预想的一致。根据测定结果，说明不同 pH 测定手段的精密度。

## 【数据记录与结果分析】

日期：　　　　　　温度：　　　　　　相对湿度：

表 3-1　pH 7.4 的 PBS 缓冲溶液的配制

| 试剂 | KCl | $KH_2PO_4$ | $Na_2HPO_4$ | NaCl |
|---|---|---|---|---|
| 浓度 / mol·L⁻¹ | 0.15 | 0.15 | 0.10 | 0.15 |
| 用量 / mL | 0.86 | 0.56 | 4.90 | 加至 50mL |

表 3-2　缓冲溶液配制

| 试管编号 | A | B | C |
|---|---|---|---|
| 1.0mol·L⁻¹ NaAc / mL | 2.50 | 4.00 | / |
| 1.0mol·L⁻¹ HAc / mL | 2.50 | 1.00 | / |
| 0.1mol·L⁻¹ NaAc / mL | / | / | 2.50 |
| 0.1mol·L⁻¹ HAc / mL | / | / | 2.50 |
| $V_总$/mL | | | |
| $c(\text{NaAc})$/mol·L⁻¹ | | | |
| $c(\text{HAc})$/mol·L⁻¹ | | | |
| 缓冲比 = $[\dfrac{c(\text{NaAc})}{c(\text{HAc})}]$ | | | |
| $c_总 = c(\text{NaAc}) + c(\text{HAc})$/mol·L⁻¹ | | | |

表 3-3  普通溶液及缓冲溶液的缓冲性质及缓冲容量影响因素的测定

| 测试项目<br>试剂 | 抗酸测试 | | | | 抗碱测试 | | | | 抗稀释测试 | | | |
|---|---|---|---|---|---|---|---|---|---|---|---|---|
| | 0.5mol·L$^{-1}$ HAc | A | B | C | 0.5mol·L$^{-1}$ HAc | A | B | C | 0.5mol·L$^{-1}$ HAc | A | B | C |
| 试剂用量 / mL | 1.00 | 1.00 | 1.00 | 1.00 | 1.00 | 1.00 | 1.00 | 1.00 | 1.00 | 1.00 | 1.00 | 1.00 |
| pH$_1$ | | | | | | | | | | | | |
| 1.0mol·L$^{-1}$ HCl/ 滴 | 3 | 3 | 3 | 3 | | | | | | | | |
| 1.0mol·L$^{-1}$ NaOH/ 滴 | | | | | 3 | 3 | 3 | 3 | | | | |
| H$_2$O / mL | | | | | | | | | 4.0 | 4.0 | 4.0 | 4.0 |
| pH$_2$ | | | | | | | | | | | | |
| ΔpH = \|pH$_2$ − pH$_1$\| | | | | | | | | | | | | |

表 3-4  PBS 缓冲溶液的配制及自带试样溶液 pH 的测定

| 试样 | PBS 缓冲溶液 | 自带试样溶液 1 | 自带试样溶液 2 |
|---|---|---|---|
| pH 试纸测 pH | | | |
| pH 计测 pH | | | |

## 【思考题】

(1) pH 试纸与 pH 计测量溶液 pH 的准确度如何？

(2) 为何每测量完一种溶液，复合电极需用蒸馏水洗净并吸干后才能测量另一种溶液？

(3) 如果只有 HAc 溶液和 NaOH 溶液，HCl 溶液和 NH$_3$·H$_2$O 溶液，KH$_2$PO$_4$ 溶液和 NaOH 溶液，能够进行上述实验吗？将怎样进行实验设计？

（黄燕军）

# 实验四  胶体溶液的制备与性质

## 【实验目的】

(1) 熟悉溶胶的制备方法。

(2) 验证溶胶的性质。

(3) 了解高分子溶液对溶胶的保护作用。

## 【预习作业】

(1) 如何按照分散介质粒子的大小对分散系进行分类？它们是否都能透过滤纸和半透膜？

(2) 胶体分散系是如何分类的？溶胶的基本特征有哪些？

(3) 分别写出 Fe(OH)$_3$、AgI 胶团结构式，并判断电泳时胶粒的运动方向？

(4) 为什么溶胶是热力学不稳定系统，同时溶胶又具有动力学稳定性？

(5) 试说明加热和加入电解质对溶胶的聚沉作用？

(6) 参照本实验步骤，以流程图方式给出本实验的设计方案，并在流程图中标出每个实验步骤的注意事项与依据。

## 【实验原理】

### 1. 胶体的制备

胶体是物质的一种分散体系,当物质以 1~100nm 大小的粒子分散于另一种介质中时,就成为胶体分散系。胶体分散系主要包括溶胶和高分子化合物溶液两大类。胶体通常采用化学反应或物理凝聚方法制备。

(1) 化学反应制备法

利用 $FeCl_3$ 的水解反应制备 $Fe(OH)_3$ 溶胶:将 $FeCl_3$ 溶液逐滴加入沸腾的蒸馏水中并不断搅拌,加完后根据情况可适当延长煮沸时间,得到暗红色的 $Fe(OH)_3$ 溶胶。

$$FeCl_3 + 3H_2O \longrightarrow Fe(OH)_3 + 3HCl$$

$$Fe(OH)_3(部分) + HCl \longrightarrow FeOCl + 2H_2O$$

$$FeOCl \longrightarrow FeO^+ + Cl^-$$

在本制备方式下,$Fe(OH)_3$ 胶核优先吸附与其组成类似的 $FeO^+$ 离子而形成正溶胶。

利用 $AgNO_3$ 和 KI 的复分解反应制备 AgI 溶胶:向过量的 KI 溶液中逐滴加入 $AgNO_3$ 溶液并不断搅拌,即可得乳黄色的 AgI 溶胶。

$$AgNO_3 + KI \longrightarrow AgI$$

$$KI \longrightarrow K^+ + I^-$$

由于 KI 过量,AgI 溶胶吸附 $I^-$ 离子,形成 AgI 负溶胶。

(2) 物理凝聚制备法

改换溶剂法制备硫溶胶:根据物质在不同溶剂中溶解度不同的性质,向水中滴入硫的乙醇饱和溶液,由于硫难溶于水,在水中过饱和的硫原子相互聚集,从而形成硫黄水溶胶。

### 2. 溶胶的光学及电学性质

溶胶具有较强的光散射现象,当一束光线照射溶胶时,在其垂直方向可观察到明显的光径,此现象称为 Tyndall 效应。Tyndall 效应是溶胶区别于真溶液的一个基本特征。

溶胶的另一个重要性质是其胶粒表面带有电荷。胶粒带电的原因,一是胶核选择性地吸附与其组成类似的离子,二是胶核表面分子的离解。胶粒的电性可通过将溶胶置于含有两个电极的 U 形管,即电泳装置而测定,当电流通过 U 形管时,带负电荷的胶粒向正极迁移,带正电荷的胶粒向负极迁移。在外加电场作用下,胶体分散相粒子定向迁移,产生电泳现象。如氢氧化铁胶粒将向负极移动,由此可见其带正电荷;而大部分金属硫化物将向正极移动,表明其带负电荷。

### 3. 胶体的净化

制备的溶胶中常有电解质离子存在,影响溶胶的稳定性。若除去这些杂质离子,溶胶可较长时间保持稳定。根据胶粒不能透过半透膜,而电解质离子、小分子和水等易通过之特性,分离和纯化胶体,这种半透膜称为渗析膜,这种纯化方式称为渗析法。

### 4. 溶胶的聚沉

溶胶的稳定因素有胶核带电、胶粒表面水合膜的保护作用和 Brown 运动,其中最主要的因素是胶粒带有电荷。有几种方法可引起溶胶的聚沉,最有效的方法是加入电解质,若在溶胶中加入一定量电解质,可使其电荷部分或全部被中和,溶胶易聚集变大而沉降。使溶胶聚沉的电解质离子主要是与胶粒所带电荷相反的离子(反离子);并且反离子价数越高,电解质的聚沉能力越强。如氢氧化铁溶胶是正溶胶,可加入含高价阴离子(如 $PO_4^{3-}$)的溶液使其聚沉。而上述制备的 AgI 溶胶是负溶胶,带正电荷的高价电解质离子是其聚沉的有效离子,因此,在此 AgI 溶胶中加入 $AlCl_3$ 比加入等量的 NaCl 聚沉的效果更好。

除了电解质的聚沉作用以外,两种带相反电荷的溶胶混合,也可以发生聚沉现象;另外,加热也是使溶胶聚沉的常用方法。

### 5. 高分子溶液对溶胶的保护作用

若在溶胶中加入足够量的高分子溶液时,高分子物质在胶粒周围形成高分子保护层,使胶粒不易发生聚沉,增加了溶胶的稳定性,对溶胶具有保护作用。

## 【仪器材料与试剂】

**仪器**  电泳装置,Tyndall 效应装置或激光笔,电磁加热搅拌器,酒精灯,烧杯(100mL×5),量筒(10mL×4,20mL×3),小试管,大试管,表面皿,滴管,三脚架,玻棒

**材料与试剂**  3% $FeCl_3$ 溶液,0.05mol·$L^{-1}$ $AgNO_3$ 溶液,0.05mol·$L^{-1}$ KI 溶液,硫的乙醇饱和溶液,饱和 NaCl 溶液,0.01mol·$L^{-1}$ NaCl 溶液,0.01mol·$L^{-1}$ $CaCl_2$ 溶液,0.01mol·$L^{-1}$ $AlCl_3$ 溶液,0.1mol·$L^{-1}$ $NH_3·H_2O$ 溶液,0.05mol·$L^{-1}$ 碘液,0.1mol·$L^{-1}$ KSCN 溶液,新配制的 3% 动物胶,2% $CuSO_4$ 溶液,渗析袋,淀粉溶液,pH 试纸

## 【实验步骤】

### 1. 胶体的制备

#### 1.1 水解法制备 Fe(OH)₃ 溶胶

在 100mL 烧杯中加入 30mL 蒸馏水,加热至沸,边搅拌边逐滴加入 3% $FeCl_3$ 溶液 3mL,加完后继续煮沸 2~3min,直到用激光笔照射出现明显的 Tyndall 效应,即一束光柱通过溶液,即可得 Fe(OH)₃ 溶胶,观察溶液变化,记录实验现象,写出胶团结构式,溶液冷却备用。(注意:若 Fe(OH)₃ 溶胶的 Tyndall 现象不明显,可延长制备过程,即沸腾时间,使 HCl 充分挥发,并用激光笔检测至有光柱,即可证明制备溶胶成功。)

#### 1.2 复分解法制备 AgI 溶胶

在 100mL 烧杯中加入 20mL 0.05mol·$L^{-1}$ KI 溶液,边搅拌边逐滴加入 0.05mol·$L^{-1}$ $AgNO_3$ 溶液,约 10mL,直到用激光笔照射出现明显的 Tyndall 效应,即可得乳黄色的 AgI 溶胶,观察溶液变化,记录实验现象,写出胶团结构式,溶液冷却备用。

#### 1.3 转换溶剂法制备硫溶胶

在 100mL 烧杯中加入 20mL 蒸馏水,边搅拌边逐滴加入硫的乙醇饱和溶液 2~3mL,直到用激光笔照射出现明显的 Tyndall 效应,即可得硫溶胶,观察溶液变化,记录实验现象,写出胶团结构式,溶液冷却备用。

### 2. 溶胶的性质

#### 2.1 溶胶的光学性质——Tyndall 现象

将上述实验制得的溶胶,分别置于 Tyndall 效应装置中,对准光束,从垂直于光束方向观察溶

胶的 Tyndall 现象，再观察蒸馏水和硫酸铜溶液是否具有 Tyndall 现象。

**2.2 溶胶的电学性质——电泳**

在 U 形管中注入 $Fe(OH)_3$ 溶胶。沿 U 形管两侧的管壁慢慢加入电解质溶液，使电解质溶液与溶胶之间有明显的界面，电解质层厚约 2cm，然后将金属电极分别插入电解质层，接通直流电源，调节电压在 160~180V，观察溶胶界面的移动情况，判断胶粒所带电荷的电性。

**2.3 胶体的净化——渗析**

（1）高分子溶液的净化：在渗析袋内加适量淀粉溶液，滴加 2 滴饱和 NaCl 溶液，将渗析袋浸入装有蒸馏水的小烧杯中。10min 后，取袋外液适量用 $0.05mol·L^{-1}$ $AgNO_3$ 溶液检查 $Cl^-$，再分别取少量袋内液和袋外液，各滴加 $0.05mol·L^{-1}$ 碘液，检查有无淀粉，观察实验现象，解释实验结果。

（2）$Fe(OH)_3$ 溶胶的净化：将制备的 $Fe(OH)_3$ 溶胶注入渗析袋（注意不要让溶液污染渗析袋外面，若外部附有溶液，用蒸馏水冲洗干净），扎紧袋口，置于盛有蒸馏水的烧杯中（袋口不要浸入水中）。每隔 10min，换蒸馏水一次，并取适量袋外溶液置于两小试管中，分别用 $0.05mol·L^{-1}$ $AgNO_3$ 溶液和 $0.1mol·L^{-1}$ KSCN 溶液检测 $Cl^-$ 和 $Fe^{3+}$，直至检查不出 $Cl^-$ 和 $Fe^{3+}$ 为止。将净化好的 $Fe(OH)_3$ 溶胶移入清洁干燥的容器中，老化，备用。

**2.4 溶胶的聚沉**

（1）电解质的聚沉作用：取三支干燥试管，各加入 2mL AgI 溶胶，振荡下分别滴加 $0.01mol·L^{-1}$ NaCl、$0.01mol·L^{-1}$ $CaCl_2$ 和 $0.01mol·L^{-1}$ $AlCl_3$ 溶液，至溶液刚呈现浑浊为止。记录各试管所加溶液的滴数，比较三种电解质聚沉能力大小，并说明原因。

（2）溶胶的相互聚沉：将 2mL $Fe(OH)_3$ 溶胶与 2mL AgI 溶胶充分混合，观察现象并说明原因。

（3）温度对溶胶稳定性的影响：将盛有 2mL AgI 溶胶的试管加热至沸，观察现象并说明原因。

**2.5 高分子溶液对溶胶的保护作用**

取两支大试管，一支试管中加入 2mL 蒸馏水，另一支试管中加入 2mL 新配制的 3% 动物胶溶液，然后分别在两支试管中加入 4mL AgI 溶胶，小心振摇，放置约 3min 后，分别向两支试管中滴加饱和 NaCl 溶液，观察两试管中聚沉现象的差别并说明原因。

## 【数据记录与结果分析】

日期：　　　　　温度：　　　　℃　　　　相对湿度：

表 4-1　溶胶的制备与性质

| 实验内容 | | 现象记录 | 结果 | 解释 |
| --- | --- | --- | --- | --- |
| $Fe(OH)_3$ 溶胶制备 | | | | |
| AgI 溶胶制备 | | | | |
| 溶胶的光学性质 -Tyndall 现象 | | | | |
| 溶胶的电学性质 - 电泳 | | | | |
| 胶体渗析 | 高分子溶液的净化 | | | |
| | $Fe(OH)_3$ 溶胶的净化 | | | |
| 溶胶聚沉 | 电解质的聚沉作用 | | | |
| | 溶胶的相互聚沉 | | | |
| | 温度对溶胶稳定性的影响 | | | |
| 高分子溶液对溶胶的保护作用 | | | | |

## 【思考题】

(1) 制备 $Fe(OH)_3$ 溶胶时,为什么将 $FeCl_3$ 溶液滴入沸水中?
(2) 动物胶为何能使溶胶稳定?
(3) 试解释溶胶产生 Tyndall 效应的原因。

(刘国杰  申小爱)

# 实验课题二
# 滴定分析实验

## 实验五　酸碱滴定分析法

### 【实验目的】

(1) 掌握酸碱滴定分析的基本原理及实验操作步骤。
(2) 掌握滴定规范化操作、滴定终点的判断方法。
(3) 学习酸碱标准溶液的配制及酸性物质或碱性物质的含量测定。
(4) 熟悉返滴定法。
(5) 学习阿司匹林纯度的测定方法。

### 【预习作业】

(1) 滴定测定与标定有何不同？
(2) 标准溶液的配制方法有几种？标定标准溶液的方法有几种？
(3) 酸碱滴定中常用盐酸、氢氧化钠作为标准溶液，为什么？
(4) 标定盐酸溶液可以用无水碳酸钠（$Na_2CO_3$），也可以用硼砂（$Na_2B_4O_7 \cdot 10H_2O$），用甲基橙做指示剂时，哪个更好？
(5) 酸碱滴定中指示剂选择依据是什么？
(6) 酸碱指示剂的变色范围是根据 $pK_{HIn}$ 计算出来的吗？
(7) 指示剂用量的多或少对滴定结果有影响吗？
(8) 请用流程图阐释本实验的设计步骤，并标出每步有哪些注意事项和原因。

### 【实验原理】

酸碱滴定法又称中和滴定法，是以质子转移反应为基础的滴定分析法。利用酸碱滴定法，可以测定酸性物质或碱性物质的含量，也可以测定能与酸或碱反应的其他种类物质的含量。测定酸性物质时，用已知浓度的强碱溶液与其作用，然后根据所消耗的强碱量，求出被测物质的含量。测定碱性物质，则用已知浓度的强酸溶液与之作用，然后根据所消耗掉的强酸的量来求出被测物质的含量。

对具有酸性（或碱性）且难溶于水的物质，可以通过加入过量的碱（或酸）标准溶液与其充分反应后，再用另一种酸（或碱）溶液返滴定过量的标准溶液。

滴定分析法通常分为三个步骤：标准溶液的配制，标准溶液浓度的标定，试样含量的测定。

在酸碱滴定分析中，常用盐酸和氢氧化钠溶液作为标准溶液。但浓盐酸容易挥发，氢氧化钠易吸收空气中的水分和二氧化碳,都不能直接配制成标准溶液。只能先配制成近似浓度的溶液,然后用一级标准物质或标准溶液标定其准确浓度。根据滴定所得到的 $V(\text{NaOH})/V(\text{HCl})$ 比值和一级标准物质的质量或标准溶液的浓度,可算出待标定标准溶液的浓度。

标定 HCl 溶液常用的一级标准物质有:无水碳酸钠($Na_2CO_3$)或硼砂($Na_2B_4O_7 \cdot 10H_2O$)。其中碳酸钠易制得纯品,物美价廉,但有吸湿性,会吸收空气中 $CO_2$,所以使用前必须在 270~300℃加热约 1h,稍冷后置于干燥器中冷至室温备用。硼砂的摩尔质量较大($381.4 \text{g} \cdot \text{mol}^{-1}$),但因含结晶水,须保持在相对湿度为 60% 的恒湿器中。

标定 NaOH 常用的一级标准物质有草酸($H_2C_2O_4 \cdot 2H_2O$)、邻苯二甲酸氢钾($KHC_8H_4O_4$)等。最常用的是邻苯二甲酸氢钾,它易得纯品,稳定且利于保存,且摩尔质量较大($M$ 204.2$\text{g} \cdot \text{mol}^{-1}$)。

## 1. 标准盐酸溶液浓度的标定

### 【实验原理】

用无水碳酸钠标定 HCl 溶液时,反应如下

$$Na_2CO_3 + 2HCl \rlap{=}{=} 2NaCl + H_2O + CO_2\uparrow$$

用 $0.100\,0\text{mol} \cdot \text{L}^{-1}$ HCl 溶液滴定 $0.100\,0\text{mol} \cdot \text{L}^{-1}$ $Na_2CO_3$ 溶液,计量点时 pH 3.9,滴定突跃为 pH 5.0~3.5,可选用甲基橙(变色范围 pH 3.2~4.4)或甲基红(变色范围 pH 4.8~6.0)作指示剂,滴定接近终点时,应将溶液煮沸,减少 $CO_2$ 对终点的影响。

HCl 溶液的准确浓度可通过式(5-2)计算得到

$$\frac{1}{2}n(\text{HCl}) = n(Na_2CO_3) \tag{5-1}$$

$$c(\text{HCl}) = \frac{2 \times m(Na_2CO_3)}{M(Na_2CO_3) \times V(\text{HCl})} (\text{mol} \cdot \text{L}^{-1}) \tag{5-2}$$

式(5-2)中 $m(Na_2CO_3)$ 为 $Na_2CO_3$ 质量,$V(\text{HCl})$ 为所消耗 HCl 溶液的体积。

### 【仪器材料与试剂】

**仪器** 万分之一分析天平,酸式滴定管或聚四氟乙烯酸碱两用滴定管(25mL),容量瓶(100mL, 1 000mL),移液管(20mL),锥形瓶(250mL×3),烧杯(100mL),量筒(10mL),称量瓶,滴定管架,洗瓶,玻棒

**材料与试剂** 无水碳酸钠(AR.),浓盐酸($12\text{mol} \cdot \text{L}^{-1}$),0.05% 甲基橙指示剂

### 【实验步骤】

**1.1 配制 $0.1\text{mol} \cdot \text{L}^{-1}$ 的 HCl 溶液 1 000mL**

计算配制 1 000mL $0.1\text{mol} \cdot \text{L}^{-1}$ HCl 溶液所需浓盐酸的体积。在通风橱,用 10mL 量筒量取所需浓盐酸,倒入盛有 300mL 蒸馏水的 1 000mL 的容量瓶中,用少量的蒸馏水洗涤 10mL 量筒 2~3 次,洗涤液并入 1 000mL 的容量瓶中,加蒸馏水定容至 1 000.00mL,混合均匀后转移至 1 000mL 试剂瓶中,塞好塞子,备用。

**1.2 $Na_2CO_3$ 标准溶液的配制**

在分析天平上用差减法精确称取无水 $Na_2CO_3$ 约 0.4~0.6g(准确至 0.000 1g)置于 100mL 烧杯

中,加 20~30mL 蒸馏水,用玻璃棒轻轻搅拌使其完全溶解,转移溶液至 100mL 容量瓶中,用少量蒸馏水洗涤小烧杯数次,洗涤液一并转移入容量瓶(注意控制总体积),加蒸馏水至刻度线,将溶液充分摇匀。

### 1.3 HCl 溶液的标定

取 20mL 洁净的移液管,用少量配制好的 $Na_2CO_3$ 标准溶液润洗三次,然后准确移取 $Na_2CO_3$ 溶液 20.00mL 于 250mL 锥形瓶中,用洗瓶吹入少量蒸馏水将黏附在锥形瓶内壁上的溶液冲下,加甲基橙指示剂 2 滴,混合均匀,溶液呈黄色。

用少量待标定的 HCl 溶液润洗滴定管三次,装满 HCl 溶液至"0"刻度以上 2~3cm,检查有无气泡并排除之。将凹液面调到 0.00~0.50mL 刻度之间,记下初始读数(准确至 0.01mL)。将含有 $Na_2CO_3$ 溶液的锥形瓶移至滴定管下端,从滴定管中缓慢地滴加 HCl 溶液。在滴定过程中,边滴边匀速摇动锥形瓶,待溶液由黄色刚刚变为橙色时,暂停滴定,将溶液加热至微沸 2~3min,并摇动锥形瓶以去除 $CO_2$,溶液又由橙色变为黄色,冷却后继续用 HCl 溶液滴定,当溶液颜色变成浅黄色(临近终点)时,用少量的蒸馏水冲洗锥形瓶的内上壁,继续小心地滴加 HCl 溶液直到溶液颜色刚刚变成橙色或微红,颜色保持 30s 不变即为终点。记录滴定消耗的 HCl 溶液体积于表 5-1 中。

重复上述滴定操作步骤(1.3)2~3 次。测定结果的相对平均偏差应小于 0.2%。根据滴定所消耗的 HCl 溶液的体积和实际参加反应的 $Na_2CO_3$ 质量,计算出 HCl 溶液的准确浓度。

## 【数据记录及结果分析】

日期:　　　　　　温度:　　　　℃　　　　相对湿度:

(1) 配制 $0.1mol·L^{-1}$ HCl 溶液 1 000mL 所需浓盐酸体积:_____mL。

(2) $0.1mol·L^{-1}$ HCl 溶液的标定

表 5-1　$0.1mol·L^{-1}$ HCl 溶液的标定

| 实验序号 | 1 | 2 | 3 |
|---|---|---|---|
| $m_1$(称量瓶+$Na_2CO_3$)/ g | | | |
| $m_2$(称量瓶+剩余 $Na_2CO_3$)/ g | | | |
| $m_1-m_2$($Na_2CO_3$)/ g | | | |
| $m_{测定}$($Na_2CO_3$)/ g | | | |
| 指示剂 | | | |
| 终点溶液颜色变化 | | | |
| $V$($Na_2CO_3$)/ mL | | | |
| $V_{终}$(HCl)/ mL | | | |
| $V_{初}$(HCl)/ mL | | | |
| $\Delta V_{消耗}$(HCl)/ mL | | | |
| $c$(HCl)/ $mol·L^{-1}$ | | | |
| $\bar{c}$(HCl)/ $mol·L^{-1}$ | | | |
| 相对平均偏差 $\bar{d_r}$ / % | | | |

## 【思考题】

(1) 用来配制碳酸钠溶液的容量瓶是否需要干燥,为什么?

(2) 移液管量取溶液前是否需要用待移溶液洗涤?

(3) 在用 HCl 溶液滴定碳酸钠溶液的过程中,溶液第一次变色后为什么要加热去除 $CO_2$,如果不处理,对滴定终点有什么影响?

## 2. 标准氢氧化钠溶液浓度的标定

### 【实验原理】

用邻苯二甲酸氢钾($KHC_8H_4O_4$)标定氢氧化钠(NaOH)溶液时,反应如下

$$\text{C}_6\text{H}_4(\text{COOH})(\text{COOK}) + \text{NaOH} = \text{C}_6\text{H}_4(\text{COONa})(\text{COOK}) + \text{H}_2\text{O}$$

计量点溶液的 pH 9.1,可选用酚酞(变色范围 pH 8.2～10.0)作指示剂。
NaOH 溶液的准确浓度可通过式(5-4)计算得到

$$n(\text{NaOH}) = n(\text{KHC}_8\text{H}_4\text{O}_4) \tag{5-3}$$

$$c(\text{NaOH}) = \frac{m(\text{KHC}_8\text{H}_4\text{O}_4)}{M_r(\text{KHC}_8\text{H}_4\text{O}_4) \times V(\text{NaOH})} (\text{mol} \cdot \text{L}^{-1}) \tag{5-4}$$

式(5-4)中 $m(\text{KHC}_8\text{H}_4\text{O}_4)$ 为所用的邻苯二甲酸氢钾的质量,$V(\text{NaOH})$ 为消耗的 NaOH 溶液的体积。

### 【仪器材料与试剂】

**仪器** 万分之一分析天平,百分之一电子天平,碱式滴定管(25mL)或聚四氟乙烯酸碱两用滴定管(25mL),容量瓶(100mL,250mL),移液管(20mL),锥形瓶(250mL×3),烧杯(100mL),称量瓶,试剂瓶(500mL),滴定管架,洗瓶

**材料与试剂** 固体 NaOH,邻苯二甲酸氢钾($KHC_8H_4O_4$,AR.),0.1% 酚酞指示剂

### 【实验步骤】

**2.1 配制近似浓度 0.1mol·L$^{-1}$ NaOH 溶液 250mL**

计算配制 250mL 0.1mol·L$^{-1}$ NaOH 溶液所需的固体 NaOH 的质量。用电子天平快速称出所需的 NaOH,置于 100mL 小烧杯中,加 50mL 蒸馏水溶解,放冷后移入 250mL 容量瓶中,用少量蒸馏水洗涤小烧杯数次,洗涤液一并移入容量瓶,加蒸馏水定容至 250.00mL,充分摇匀,再移至 500mL 试剂瓶中,塞入橡皮塞,备用。

**2.2 邻苯二甲酸氢钾($KHC_8H_4O_4$)标准溶液的配制**

于分析天平上用差减法称取 $KHC_8H_4O_4$ 约 1.0～1.3g(准确至 0.0001g),置于 100mL 小烧杯中,加 20～30mL 蒸馏水并用玻璃棒轻轻搅拌使其完全溶解,将溶液转移至 100mL 的容量瓶中,用少量蒸馏水洗涤小烧杯数次,洗涤液一并移入容量瓶,加蒸馏水至刻度线,塞入塞子,将溶液充分摇匀。

**2.3 NaOH 溶液的标定**

取 20mL 洁净的移液管,用少量配制好的邻苯二甲酸氢钾溶液润洗 2～3 次,然后准确移取邻苯二甲酸氢钾溶液 20.00mL 放入锥形瓶中,用洗瓶吹入少量蒸馏水将黏附在锥形瓶内壁上的溶液冲下,加酚酞指示剂 2 滴。

用少量 NaOH 标准溶液润洗两用滴定管（或碱式滴定管）3 次，装满 NaOH 溶液至"0"刻度以上 2～3cm，检查有无气泡并排之。将凹液面调到 0.00～0.50mL 刻度之间，记下初始读数（准确至 0.01mL）。将含有邻苯二甲酸氢钾溶液的锥形瓶移至滴定管下端，从滴定管中缓慢的滴加 NaOH 溶液。在滴定过程中，边滴边匀速摇动锥形瓶，接近终点时，用少量蒸馏水冲洗锥形瓶内壁，继续小心滴加 NaOH 溶液直到溶液颜色刚刚变成浅红色，颜色保持 30s 不变即为终点。记录滴定消耗的 NaOH 溶液的体积于表 5-2 中。

重复滴定操作（步骤 2.3）2～3 次，测定的结果相对平均偏差应小于 0.2%。

根据消耗的 NaOH 溶液的体积和实际参加反应的邻苯二甲酸氢钾质量，计算碱标准溶液的浓度。

### 【注意事项】

（1）NaOH 具有强腐蚀性，称量过程中，不要接触到皮肤、衣物等，尽量不要洒在操作台上，如有洒落，应及时处理。

（2）配制好的 NaOH 标准溶液须存放于塑料试剂瓶中，而不能存放于玻璃试剂瓶中。

### 【数据记录及结果分析】

日期：　　　　　温度：　　　℃　　　　相对湿度：

（1）配制 0.1mol·L$^{-1}$ NaOH 溶液 250mL 所需的固体 NaOH 的质量：_____g。

（2）0.1mol·L$^{-1}$ NaOH 溶液的标定

表 5-2　0.1mol·L$^{-1}$ NaOH 溶液的标定

| 实验序号 | 1 | 2 | 3 |
|---|---|---|---|
| $m_1$（称量瓶 + KHC$_8$H$_4$O$_4$）/ g | | | |
| $m_2$（称量瓶 + 剩余 KHC$_8$H$_4$O$_4$）/ g | | | |
| $m_1 - m_2$（KHC$_8$H$_4$O$_4$）/ g | | | |
| $m_{测定}$（KHC$_8$H$_4$O$_4$）/ g | | | |
| 指示剂 | | | |
| 终点溶液颜色变化 | | | |
| $V$（KHC$_8$H$_4$O$_4$）/ mL | | | |
| $V_{终}$（NaOH）/ mL | | | |
| $V_{初}$（NaOH）/ mL | | | |
| $\Delta V_{消耗}$（NaOH）/ mL | | | |
| $c$（NaOH）/ mol·L$^{-1}$ | | | |
| $\bar{c}$（NaOH）/ mol·L$^{-1}$ | | | |
| 相对平均偏差 $\overline{d_r}$ / % | | | |

### 【思考题】

（1）一级标准物质应满足什么条件？

（2）滴定接近终点时，为什么用蒸馏水淋洗锥形瓶内上壁？

（别子俊）

## 3. 食用醋中总酸度的测定

### 【预习作业】

（1）为什么需要将食醋稀释后再测定其酸度？
（2）滴定完成后，如果滴定管下端玻璃尖嘴外留有液滴对实验结果是否有影响？

### 【实验原理】

食醋中的主要成分是醋酸，含量约为 30～50 g·L$^{-1}$。尽管还存在一些其他有机酸，但含量是以醋酸来计算。醋酸是一元弱酸，它的解离常数 $K_a=1.75\times10^{-5}$，用 NaOH 标准溶液滴定时，发生如下反应：

$$HAc + NaOH \Longleftrightarrow NaAc + H_2O$$

计量点时 pH 8.73，滴定突跃范围 pH 7.75～9.70，可选用酚酞作指示剂。食醋中醋酸的质量浓度可根据下式进行计算

$$\rho_{(HAc)} = \frac{c_{(NaOH)} \times V_{(NaOH)} \times M_{(HAc)}}{V_{(样品)}} (g \cdot L^{-1}) \tag{5-5}$$

式中(5-5) $c$(NaOH) 为 NaOH 标准溶液的浓度（mol·L$^{-1}$），$V$(NaOH) 为标定时用去 NaOH 标准溶液的体积（mL），$M$(HAc) 为 HAc 的摩尔质量（g·mol$^{-1}$），$V$(样品) 为每次滴定所用食醋的体积。

### 【仪器材料与试剂】

**仪器**　碱式滴定管或两用滴定管（25mL），容量瓶（100mL），移液管（20mL，25mL），锥形瓶（250mL×3），滴定管架，洗瓶，玻棒

**材料与试剂**　食醋（样品），NaOH 标准溶液，0.1% 酚酞指示剂

### 【实验步骤】

取 1 支 25mL 洁净的移液管，用少量待测的食醋润洗 3 次，吸取食醋 25.00mL，移入 100mL 的容量瓶中，加蒸馏水稀释至刻度线，塞入塞子，将溶液充分混匀。

取 1 支 20mL 洁净的移液管，用稀释后的食醋润洗移液管 3 次。移取稀释后的食醋 20.00mL 置 250mL 锥形瓶中，用洗瓶吹入少量蒸馏水将锥形瓶内壁上的溶液冲下，加酚酞指示剂 2 滴。

用 NaOH 标准溶液滴定至溶液呈浅粉红色，且 30s 内不褪色即为终点，记录滴定所消耗的 NaOH 标准溶液体积于表 5-3 中。重复滴定操作 2～3 次，三次测定结果的相对平均偏差应小于 0.2%。由消耗的 NaOH 体积和浓度计算食醋中醋酸的含量。

### 【数据记录及结果分析】

日期：　　　　　　室温：　　　℃　　　　　相对湿度：

表 5-3　食醋中醋酸含量的测定

| 实验序号 | 1 | 2 | 3 |
| --- | --- | --- | --- |
| $V_{样}$（食醋）/mL | | | |
| $V_{总}$（稀释后食醋）/mL | | | |
| 指示剂 | | | |

续表

| 实验序号 | 1 | 2 | 3 |
|---|---|---|---|
| 终点溶液颜色变化 | | | |
| $V_{测定}$(稀释后食醋)/mL | | | |
| $V_{终}$(NaOH)/mL | | | |
| $V_{初}$(NaOH)/mL | | | |
| $\Delta V_{消耗}$(NaOH)/mL | | | |
| $c$(NaOH)/mol·L$^{-1}$ | | | |
| $\rho$(HAc)/g·L$^{-1}$ | | | |
| $\bar{\rho}$(HAc)/g·L$^{-1}$ | | | |
| $\bar{d_r}$/% | | | |

### 【思考题】

(1) 能用甲基红或甲基橙指示剂代替酚酞指示剂测定食醋中醋酸含量吗？为什么？

(2) 移取醋酸溶液前必须用醋酸润洗移液管3次，滴定用的锥形瓶是否需要润洗？

## 4. 硼砂含量的测定

### 【预习作业】

(1) 滴定硼砂时使用酸式还是碱式滴定管，为什么？

(2) 本实验采用何种滴定方式，直接还是间接滴定？

(3) 温度会影响本实验指示剂变色敏锐性吗？如果有影响宜选择什么温度？

### 【实验原理】

硼砂($Na_2B_4O_7·10H_2O$)溶于水，解离出$Na^+$离子和$B_4O_7^{2-}$离子。根据Brönsted酸碱质子理论，$B_4O_7^{2-}$是质子的接受体，与水发生质子转移反应，溶液呈碱性，因此可利用酸碱滴定法来测定其含量。

用盐酸标准溶液测定硼砂时，发生的反应如下：

$$Na_2B_4O_7·10H_2O + 2HCl = 2NaCl + 4H_3BO_3 + 5H_2O$$

或 $B_4O_7^{2-} + 2H_3O^+ + 3H_2O = 4H_3BO_3$

由滴定反应式可知，在计量点时，有如下关系式

$$\frac{1}{2}n(HCl) = n(Na_2B_4O_7·10H_2O) \tag{5-6}$$

$$\omega_{(HAc)} = \frac{c_{(HCl)} \times V_{(HCl)} \times M_{(Na_2B_4O_7·H_2O)}}{m_{(样品)} \times 2 \times 1000}(g·g^{-1}) \tag{5-7}$$

式(5-7)中，$c$(HCl)为滴定所消耗的标准HCl溶液的浓度(mol·L$^{-1}$)，$V$(HCl)为滴定所消耗的标准HCl溶液的体积(mL)，$m$(样品)为每次滴定中硼砂的实际质量(g)，硼砂的摩尔质量$M$为301.37g·mol$^{-1}$。

用标准盐酸溶液滴定硼砂溶液，滴定到达计量点时，溶液的pH 5.1，可选用甲基红作指示剂。

### 【仪器材料与试剂】

**仪器** 分析天平，酸式滴定管或两用滴定管(25mL)，移液管(20mL)，锥形瓶(250mL×3)，容

量瓶(100mL),烧杯(100mL),称量瓶,滴定架,洗瓶

**材料与试剂** 硼砂样品,标准 HCl 溶液,0.1% 甲基红指示剂

## 【实验步骤】

在分析天平上称取硼砂样品约 1.0~1.4g(准确至 0.000 1g)置于 100mL 小烧杯中,加 20~30mL 蒸馏水,加热并用玻璃棒轻轻搅拌使其完全溶解,将溶液转移至 100mL 的容量瓶中,用少量蒸馏水洗涤小烧杯数次,洗涤液一并转移入容量瓶,加蒸馏水至刻度线,将溶液充分摇匀。

用少量的待测硼砂溶液润洗移液管 2~3 次,移取硼砂溶液 20.00mL 置于 250mL 锥形瓶中,用洗瓶吹入少量蒸馏水将锥形瓶内壁上的溶液冲下,加甲基红指示剂 2 滴,溶液呈黄色。

用标准 HCl 溶液滴定待测硼砂溶液至颜色呈浅红色,并保持 30s 不变即为终点。记录滴定所消耗的标准 HCl 溶液的体积于表 5-4 中。重复测定 2~3 次,测定的结果相对平均偏差应小于 0.2%。计算硼砂的含量 $\omega$(质量分数)。

## 【数据记录及结果分析】

日期: 室温: ℃ 相对湿度:

**表 5-4 硼砂含量(质量分数 $\omega$)的测定**

| 实验序号 | 1 | 2 | 3 |
|---|---|---|---|
| $m_{样品}$(硼砂)/ g | | | |
| $m_{测定}$(硼砂)/ g | | | |
| 指示剂 | | | |
| 终点溶液颜色变化 | | | |
| $V$(硼砂)/ mL | | | |
| $V_{终}$(HCl)/ mL | | | |
| $V_{初}$(HCl)/ mL | | | |
| $\Delta V_{消耗}$(HCl)/ mL | | | |
| $c$(HCl)/ mol·L$^{-1}$ | | | |
| $\omega$(Na$_2$B$_4$O$_7$·10H$_2$O)/g·g$^{-1}$ | | | |
| $\bar{\omega}$(Na$_2$B$_4$O$_7$·10H$_2$O)/g·g$^{-1}$ | | | |
| $\overline{d_r}$ /% | | | |

## 【思考题】

用标准盐酸溶液滴定硼砂时,最好的指示剂是甲基红还是甲基橙,为什么?

(石婷婷)

# 5. 返滴定法测定阿司匹林的含量

## 【实验目的】

(1)学习返滴定法的原理与操作。
(2)熟悉利用酸碱滴定法测定药片中阿司匹林含量的方法。

## 【预习作业】

（1）本实验中，为什么不宜采用直接滴定法测定阿司匹林的含量？
（2）在阿司匹林与过量 NaOH 的反应中，水浴加热样品 15min 后要迅速用流水冷却，试述该操作的原因。
（3）请用流程图简述本实验的操作步骤，并标明实验过程中的注意事项及依据。

## 【实验原理】

阿司匹林曾经是广泛使用的解热镇痛药，它的主要成分是乙酰水杨酸。乙酰水杨酸是有机弱酸（$K_a = 1 \times 10^{-3}$，$M_r = 180.16$），微溶于水，易溶于乙醇，在强碱性溶液中水解为水杨酸和乙酸盐，反应式为

$$\text{邻-COOH, OAc} + OH^- \xrightarrow{快} \text{邻-COO}^-, \text{OAc} + H_2O$$

$$\text{邻-COO}^-, \text{OAc} + OH^- \xrightarrow{慢} \text{邻-COO}^-, \text{OH} + CH_3COO^-$$

由于药片中一般都添加了一定量的赋形剂，如硬脂酸镁、淀粉等不溶物，不宜直接滴定，可采用返滴定法进行测定。

将药片研磨成粉状后定量加入过量的 NaOH 标准溶液 $[n(\text{NaOH})]$，加热一段时间使乙酰基水解完全。以酚酞为指示剂，用 HCl 标准溶液 $[n(\text{HCl})]$ 返滴定过量的 NaOH，至粉红色刚刚消失即为终点。在这一反应中，1mol 乙酰水杨酸消耗 2mol NaOH。根据最初加入的碱量 $n(\text{NaOH})$ 和消耗 HCl 溶液的量 $n(\text{HCl})$，即可知道阿司匹林实际消耗碱的量，从而计算出样品中阿司匹林的物质的量

$$n(\text{阿司匹林}) = \frac{1}{2}[n(\text{NaOH}) - n(\text{HCl})] \tag{5-8}$$

## 【仪器材料与试剂】

**仪器**　万分之一电子分析天平，碱式滴定管（50mL），酸式滴定管（50mL），两性滴定管（50mL），锥形瓶（250mL×3），恒温水浴锅，量筒（25mL），洗瓶，表面皿，研钵。

**材料与试剂**　阿司匹林药片，$0.1\text{mol}\cdot\text{L}^{-1}$ NaOH 标准溶液（精确至 $0.0001\text{mol}\cdot\text{L}^{-1}$），$0.1\text{mol}\cdot\text{L}^{-1}$ HCl 标准溶液（精确至 $0.0001\text{mol}\cdot\text{L}^{-1}$），酚酞指示剂（$2\text{g}\cdot\text{L}^{-1}$ 乙醇溶液），乙醇（95%），沸石。

## 【实验步骤】

### 5.1　试样的制备

用研钵将阿司匹林药片研成粉末。分别准确称取三份阿司匹林粉末 0.27～0.33g（精确至 0.0001g）于 3 个 250mL 锥形瓶中，分别加入 20.00mL 乙醇和 3 滴酚酞指示剂，轻轻振荡锥形瓶溶解样品。

### 5.2　阿司匹林与过量 NaOH 标准溶液的反应

采用两性滴定管或碱式滴定管分别准确加入 40.00mL NaOH 标准溶液于三个盛有阿司匹林样

品溶液的锥形瓶中,准确记录各锥形瓶中加入 NaOH 溶液的体积 $V_{总}$(NaOH)。

向锥形瓶中各加入 2~3 粒沸石,水浴加热样品以促进阿司匹林水解反应。注意加盖并避免溶液沸腾,不时轻轻旋转锥形瓶。15min 后停止加热,迅速用流动自来水淋洗锥形瓶外壁,将溶液冷却到室温。

### 5.3 采用 HCl 标准溶液进行返滴定

用 HCl 标准溶液分别滴定锥形瓶中过量的 NaOH 溶液,至溶液的粉红颜色刚刚消失即为终点。准确记录所消耗的 HCl 标准溶液体积 $V$(HCl)于表 5-5 中。

## 【数据记录及结果分析】

日期:　　　　　温度:　　　℃　　　相对湿度:

表 5-5　阿司匹林含量测定

| 实验序号 | 1 | 2 | 3 |
|---|---|---|---|
| $c$(HCl)/ mol·L$^{-1}$ | | | |
| $c$(NaOH)/ mol·L$^{-1}$ | | | |
| $m$(样品)/ g | | | |
| $V_{始}$(NaOH)/ mL | | | |
| $V_{终}$(NaOH)/ mL | | | |
| $V_{总}$(NaOH)/ mL | | | |
| $V_{始}$(HCl)/ mL | | | |
| $V_{末}$(HCl)/ mL | | | |
| $V$(HCl)/ mL | | | |
| $\omega$(阿司匹林) | | | |
| $\bar{\omega}$(阿司匹林) | | | |
| 相对平均偏差 $\bar{d_r}$ / % | | | |

## 【思考题】

(1) 试分析本实验的主要误差来源。

(2) 实验中用于溶解阿司匹林的乙醇是一种弱酸,可以与 NaOH 反应而消耗一定量的 NaOH。请设计一个空白实验,以测定乙醇消耗 NaOH 的量。如何利用空白的结果消除实验误差?

(3) 为何采用乙醇溶解试样?试述溶解后溶液呈混浊状态的原因。

(林　毅)

# 实验六　氧化还原滴定法

## 1. 高锰酸钾法

## 【实验目的】

(1) 掌握高锰酸钾溶液的配制和标定方法。

(2) 学会高锰酸钾法测定草酸钠和过氧化氢含量的原理及方法。
(3) 巩固分析天平、滴定管、移液管的规范操作。

## 【预习作业】

(1) 过氧化氢（或双氧水）对人体的危害有哪些？为何还可医用？
(2) 双氧水中过氧化氢含量的测定原理是什么？检测过氧化氢含量的方法有哪些？
(3) 在配制 $KMnO_4$ 标准溶液过程中要用烧结玻璃漏斗过滤，能否用滤纸？为什么？
(4) 为何在 $KMnO_4$ 溶液的标定开始时先将约 10mL 的 $KMnO_4$ 溶液加入锥形瓶中与 $Na_2C_2O_4$ 作用？可以先不加入 $KMnO_4$ 溶液吗？
(5) 用滴定管装有色溶液时如何准确读数？
(6) 如何通过实验措施解决引起误差的原因？保证实验结果的准确性？
(7) 根据氧化还原滴定原理及相应的 $KMnO_4$ 滴定法原理，参照本实验步骤，以流程图方式给出本实验的设计思路，并在流程图中标注上滴定反应条件、操作条件及保证实验结果的准确度所采取的可行性实验措施及依据。

## 【实验原理】

高锰酸钾是强氧化剂，是氧化还原滴定法中常用的标准溶液，在强酸性条件下可被还原为 $Mn^{2+}$，其半反应及电极电势如下

$$MnO_4^- + 8H_3O^+ + 5e \Longrightarrow Mn^{2+} + 12H_2O \qquad \varphi^\ominus = 1.507V$$

市售 $KMnO_4$ 试剂不稳定，易被还原性物质还原，而且常含有少量的 $MnO_2$ 和其他杂质，因此 $KMnO_4$ 不能直接配制成标准溶液，只能配成近似浓度的溶液，然后再用一级标准物质标定，为使 $KMnO_4$ 溶液浓度达到稳定，常将配好的 $KMnO_4$ 加热至微沸，保持微沸 1h，放置 2~3 天，用砂芯漏斗过滤后，置于棕色瓶中保存。

常用来标定 $KMnO_4$ 溶液的一级标准物质是 $Na_2C_2O_4$（$M_r = 134.0$），在酸性条件下其反应为

$$2MnO_4^- + 5C_2O_4^{2-} + 16H_3O^+ \Longrightarrow 2Mn^{2+} + 10CO_2\uparrow + 24H_2O$$

使用 $KMnO_4$ 标准溶液滴定还原性物质时，其酸性条件常用 $H_2SO_4$ 溶液来控制，$H_2SO_4$ 的适宜酸度是 0.5~1.0 mol·L$^{-1}$，若酸度过高，会引起 $KMnO_4$ 的分解

$$4MnO_4^- + 12H_3O^+ \Longrightarrow 4Mn^{2+} + 5O_2\uparrow + 18H_2O$$

$H_2O_2$ 是医药上常用的消毒剂，市售的双氧水含 $H_2O_2$ 约 3% 或 30%（g·mL$^{-1}$），在酸性溶液中，$KMnO_4$ 能使 $H_2O_2$ 氧化成 $H_2O$ 和 $O_2$，而本身被还原成 $Mn^{2+}$ 离子，其反应式如下

$$2MnO_4^- + 5H_2O_2 + 6H_3O^+ \Longrightarrow 2Mn^{2+} + 14H_2O + 5O_2\uparrow$$

上述的两个反应在室温下是一个慢反应，但可通过对溶液加热和加入催化剂 $Mn^{2+}$ 来加快化学反应速率。在 $KMnO_4$ 与 $Na_2C_2O_4$ 的滴定反应中，预先加入约 10mL 的 $KMnO_4$ 溶液与 $Na_2C_2O_4$ 作用，稍加热，使之产生的产物 $Mn^{2+}$ 成为催化剂，以提高其滴定速率，温度应控制小于 75℃，不能超过 90℃，否则 $Na_2C_2O_4$ 会分解。滴定开始时，反应较慢，$KMnO_4$ 溶液必须逐滴加入，如果滴加过快，$KMnO_4$ 在热溶液中未来得及与 $Na_2C_2O_4$ 作用自身部分分解而造成误差。而与 $H_2O_2$ 的反应中，却不宜加热提高反应速度，因为 $H_2O_2$ 加热易发生分解。

$$4KMnO_4 + 6H_2SO_4 \Longrightarrow 2K_2SO_4 + 4MnSO_4 + 6H_2O + 5O_2\uparrow$$

在滴定过程中，溶液中逐渐产生的产物 $Mn^{2+}$ 对反应起催化剂作用，使反应速度逐渐加快，因此，无需另外加入催化剂，且随着反应的进行，滴定速度可稍加快些。当紫红色的 $KMnO_4$ 滴加到酸性的 $Na_2C_2O_4$ 或 $H_2O_2$ 溶液中时，其紫红色将褪掉，直到反应完全，此时加入过量的一滴 $KMnO_4$ 将使溶液显淡红色，即为滴定终点。根据滴定消耗的 $KMnO_4$ 溶液体积和相应物质的量，可计算 $KMnO_4$ 溶液的准确浓度及样品中 $H_2O_2$ 的含量。

## 【仪器材料与试剂】

**仪器** 万分之一分析天平，酸式滴定管（25mL）或聚四氟乙烯酸碱两用滴定管（25mL），容量瓶（100mL×2），锥形瓶（250mL×3），烧杯（200mL），移液管（20mL），吸量管（1mL），量筒（10mL），洗瓶，吸耳球，称量瓶，酒精灯，石棉网，玻棒，滴管

**材料与试剂** 分析纯 $Na_2C_2O_4$，分析纯 $KMnO_4$，$6mol·L^{-1}$ $H_2SO_4$，市售双氧水溶液（约3%或30%）。

## 【实验步骤】

### 1.1 $KMnO_4$ 溶液的标定

（1）$0.004mol·L^{-1}$ $KMnO_4$ 溶液的配制：用台秤称取 $KMnO_4$（$M_r$=158 约0.16g）置于烧杯中，加蒸馏水溶解，稀释至250.0mL，盖上表面皿加热至沸并保持微沸1h，放置2~3天后，然后用砂芯漏斗过滤，滤液储存于棕色瓶中，在暗处密闭保存，以待标定。

（2）$KMnO_4$ 标准溶液的标定：准确称取 105℃ 干燥至恒重的 $Na_2C_2O_4$（$M_r$=134）一级标准物质 0.13~0.14g（精确到 0.0001g），置于小烧杯中，加 20.0mL 蒸馏水使之溶解，转移至 100mL 容量瓶中，润洗烧杯 2~3 次，一并转移入容量瓶中，加水定容，摇匀。

（3）用 20mL 移液管准确移取 20.00mL $Na_2C_2O_4$ 溶液于 250mL 锥形瓶中，加 $6mol·L^{-1}$ $H_2SO_4$ 5.0mL，加热至 70℃左右，用 $KMnO_4$ 标准溶液滴定至出现粉红色且保持 30s 内不褪色即达到滴定终点，记录结果于表 6-1 中。

平行测定三次，根据消耗的 $KMnO_4$ 溶液体积和 $Na_2C_2O_4$ 的质量，按式（6-1）计算 $KMnO_4$ 标准溶液的准确浓度

$$c(KMnO_4) = \frac{2 \times m_{总}(Na_2C_2O_4) \times \frac{20.00}{100.00} \times 1000}{5 \times V(KMnO_4) \times M_r(Na_2C_2O_4)} (mol·L^{-1}) \quad (6-1)$$

$$M_r(Na_2C_2O_4) = 134.0$$

### 1.2 市售双氧水中 $H_2O_2$ 含量的测定

用 1mL 吸量管准确移取 1.00mL 市售 $H_2O_2$ 溶液于 100mL 容量瓶中，加蒸馏水至刻度，摇匀。然后用移液管吸取 20.00mL 稀释后的双氧水待测液，置于 250mL 锥形瓶中，加入 $6mol·L^{-1}$ $H_2SO_4$ 5.0mL，用已标定过的 $KMnO_4$ 标准溶液滴定，至溶液呈淡红色并在 30s 内不褪色，即达滴定终点。记录结果于表 6-2 中，平行测定三次。按式（6-2）计算 $H_2O_2$ 的含量

$$\rho(H_2O_2) = \frac{5 \times c(KMnO_4) \times V(KMnO_4) \times M_r(H_2O_2)}{2 \times 1.00 \times \frac{20.00}{100.00}} (g·L^{-1}) \quad (6-2)$$

## 【数据记录及结果分析】

日期：　　　　　温度：　　　℃　　　　相对湿度：

表 6-1　$KMnO_4$ 溶液的标定

| 实验序号 | 1 | 2 | 3 |
|---|---|---|---|
| 指示剂 | | | |
| 终点颜色变化 | | | |
| $m_{总}(Na_2C_2O_4)/g$ | | | |
| $V(Na_2C_2O_4)/mL$ | | | |
| $V_{终}(KMnO_4)/mL$ | | | |
| $V_{初}(KMnO_4)/mL$ | | | |
| $\Delta V_{消耗}(KMnO_4)/mL$ | | | |
| $c(KMnO_4)/mol·L^{-1}$ | | | |
| $\bar{c}(KMnO_4)/mol·L^{-1}$ | | | |
| 相对平均偏差 $\overline{d_r}/\%$ | | | |

表 6-2　市售双氧水中 $H_2O_2$ 含量的测定

| 实验序号 | 1 | 2 | 3 |
|---|---|---|---|
| 指示剂 | | | |
| 终点颜色变化 | | | |
| $V_{市售}(H_2O_2)/mL$ | | | |
| $V_{测定}(稀释后H_2O_2)/mL$ | | | |
| $c(KMnO_4)/mol·L^{-1}$ | | | |
| $V_{终}(KMnO_4)/mL$ | | | |
| $V_{初}(KMnO_4)/mL$ | | | |
| $\Delta V_{消耗}(KMnO_4)/mL$ | | | |
| $\rho(H_2O_2)/g·L^{-1}$ | | | |
| $\bar{\rho}(H_2O_2)/g·L^{-1}$ | | | |
| 相对平均偏差 $\overline{d_r}/\%$ | | | |

## 【思考题】

(1) 配制 $KMnO_4$ 溶液应注意些什么？

(2) 在标定高锰酸钾溶液时，为什么要将溶液加热小于75℃？能否加热到90℃以上？为什么？

(3) 用 $KMnO_4$ 溶液滴定 $H_2O_2$ 时，溶液能否加热？为什么？

（席晓岚）

## 2. 碘 量 法

### 【实验目的】

（1）了解直接碘量法测定维生素 C 的原理和方法。
（2）掌握碘标准溶液、硫代硫酸钠标准溶液的配制和标定方法。
（3）学会用淀粉指示剂指示终点的方法。

### 【预习作业】

（1）维生素 C 有哪些药理作用和临床应用？
（2）有哪些方法可用于标定 $Na_2S_2O_3$ 标准溶液？
（3）碘量法测定维生素 C 实验为什么要在弱酸性溶液中进行？
（4）请用流程图阐释本实验是如何设计实验步骤，并在流程图中标出每一实验步骤的设计原由及实验注意事项？

### 【实验原理】

碘量法是以 $I_2$ 做氧化剂或以碘化物做还原剂进行氧化-还原滴定的分析方法。其氧化还原半反应式为

$$I_2 + 2e \rightleftharpoons 2I^- \qquad \varphi^{\ominus} = 0.54\text{V}$$

标准电极电位比 $\varphi^{\ominus}(I_2/I^-)$ 低的还原性物质可用 $I_2$ 标准溶液直接滴定，这种方法称为直接碘量法。直接碘量法只能在酸性、中性及弱碱性溶液中进行。

固体碘易挥发，难溶于水，但易溶于 KI 溶液，$I_2$ 标准溶液的配制通常是将 $I_2$ 溶解于 KI 溶液并保存于棕色磨口瓶中，用一级标准物质 $As_2O_3$ 标定或用已标定的 $Na_2S_2O_3$ 标准溶液标定。

市售 $Na_2S_2O_3 \cdot 5H_2O$ 常含有杂质，在空气中易风化或潮解，故不能直接配制标准溶液。通常需加入 $Na_2CO_3$ 作稳定剂，使溶液的 pH 保持在 9~10，用新煮沸并放冷的蒸馏水配制后放置 7~10d，再用 $KIO_3$ 一级标准物质进行标定。

$KIO_3$ 在酸性溶液中与过量的 KI 反应，定量地析出 $I_2$，用待标定的 $Na_2S_2O_3$ 溶液滴定析出的 $I_2$，便可求出 $Na_2S_2O_3$ 溶液的准确浓度。其反应式为

$$IO_3^- + 5I^- + 6H^+ \rightleftharpoons 3I_2 + 3H_2O$$

$$I_2 + 2S_2O_3^{2-} \rightleftharpoons 2I^- + S_4O_6^{2-}$$

由以上反应可知，存在如下的定量关系

$$n(Na_2S_2O_3) = 6n(KIO_3) \tag{6-3}$$

$Na_2S_2O_3$ 标准溶液的准确浓度可计算为

$$c(Na_2S_2O_3) = \frac{6W(KIO_3) \times 1\,000}{M_r(KIO_3) \times V(Na_2S_2O_3)} (\text{mol} \cdot \text{L}^{-1}) \tag{6-4}$$

上述滴定反应用淀粉溶液作指示剂，注意淀粉指示剂应在近终点时（溶液呈浅黄色时）加入，滴定至蓝色变为无色即为滴定终点。

维生素 C 又称抗坏血酸，化学式为 $C_6H_8O_6$（$M_r = 176.12$），属于水溶性维生素，它广泛存在于水果和蔬菜中。

维生素C具有还原性，分子中的烯二醇结构易被氧化成二酮基，可直接用$I_2$标准溶液滴定。反应如下：

$$\text{C-C=C-C-C-CH}_2\text{OH} + I_2 \longrightarrow \text{C-C-C-C-C-CH}_2\text{OH} + 2HI$$

从式中可知1分子维生素C与1分子$I_2$完全反应，即反应摩尔比为1:1。维生素C易被空气氧化，在碱性溶液中氧化更快，所以滴定常在弱酸性条件（pH 3~4）下进行。用淀粉溶液作指示剂，终点时，过量的$I_2$与淀粉生成蓝色的配合物。维生素C的含量可用式(6-5)计算

$$\text{维生素C}(\%) = \frac{c(I_2)V(I_2)M_r(C_6H_8O_6)}{W(\text{样品}) \times 1000} \times 100\% (\text{g} \cdot \text{g}^{-1}) \tag{6-5}$$

式(6-5)中$W$(样品)为每次滴定的维生素C样品的质量。

【仪器材料与试剂】

**仪器**　万分之一分析天平，台式天平，酸式滴定管或聚四氟乙烯酸碱两用滴定管(50mL)，移液管(25mL×2)，容量瓶(250mL)，棕色试剂瓶(500mL×2)，量筒(10mL，50mL，100mL)，锥形瓶(250mL×3)，烧杯(100mL，250mL)，洗耳球

**材料与试剂**　维生素C片剂，果蔬样品（如西红柿、橙子、草莓等），$KIO_3$基准物质，$I_2$(s, AR)，$Na_2S_2O_3 \cdot 5H_2O$(s, AR)，$Na_2CO_3$(s, AR)，0.2%淀粉水溶液（新配置的），2mol·$L^{-1}$ HAc水溶液，1mol·$L^{-1}$ $H_2SO_4$水溶液，20% KI溶液

【实验步骤】

**2.1　0.1mol·$L^{-1}$ $Na_2S_2O_3$标准溶液的配制与标定**

用台式天平称取$Na_2S_2O_3 \cdot 5H_2O$($M_r$=248.17)12.5g溶于适量刚煮沸并放冷的蒸馏水中，加入$Na_2CO_3$固体0.1g，搅拌，稀释至500.0mL，转移至棕色瓶中，放置7~10d后标定。用分析天平准确称取0.9~1.0g（精确至0.0001g）$KIO_3$基准物质于烧杯中，加水溶解后，定量转入250mL容量瓶中，加水稀释至刻度线，充分摇匀。用移液管吸取25.00mL $KIO_3$标准溶液3份，分别置于250mL锥形瓶中，用合适量程的量筒依次加入20.0mL 20% KI溶液，5.0mL 1mol·$L^{-1}$ $H_2SO_4$溶液，50.0mL蒸馏水，立即用待标定的$Na_2S_2O_3$溶液滴定至浅黄色，然后再用量筒加入5.0mL 0.2%淀粉溶液，继续滴入标准溶液至蓝色恰好变为无色即为滴定终点，记录结果于表6-3中。平行三份并计算相对偏差。

**2.2　0.05mol·$L^{-1}$ $I_2$标准溶液的配制及标定**

用电子天平称取6.6g $I_2$和10.0g KI，置于研钵中，加少量蒸馏水，在通风橱中研磨。待$I_2$全部溶解后，将溶液转入500mL烧杯中，加水稀释至500mL，充分搅拌，再将溶液转入棕色试剂瓶中，放置暗处保存。用移液管移取25.00mL $Na_2S_2O_3$标准溶液于250mL锥形瓶中，依次用量筒加入50.0mL蒸馏水、5.0mL 0.2%淀粉溶液，然后用$I_2$标准溶液滴定至溶液呈浅蓝色，30s内不褪色即为终点。平行标定三份并计算$I_2$标准溶液的浓度，记录结果于表6-4中。

**2.3　0.005mol·$L^{-1}$ $I_2$标准溶液的配制**

用移液管移取25.00mL $I_2$标准溶液于250mL容量瓶中，加蒸馏水至刻度线，充分摇匀，备用。

**2.4　维生素C片剂中Vc含量的测定**

在分析天平上准确称取维生素C药片粉末2.3~2.7g（精确至0.0001g），置于250mL锥形瓶

中,用量筒依次加入 100.0mL 新煮沸并冷却的蒸馏水、10.0mL 2mol·L$^{-1}$ HAc 溶液和 5.0mL 0.2% 淀粉溶液,立即用 $I_2$ 标准溶液滴定至出现稳定的浅蓝色,且在 30s 内不褪色即为终点,记下消耗的 $I_2$ 溶液体积于表 6-5 中。平行测定三份,计算维生素 C 片剂中抗坏血酸的质量分数。

### 2.5 果蔬样品中 Vc 含量的测定

用 100mL 干燥小烧杯准确称取 50g(精确至 0.000 1g)左右的果蔬均浆样品(如草莓,用绞碎机打成糊状),将其转入 250mL 锥形瓶中,用水冲洗小烧杯 1~2 次。用量筒向锥形瓶中加入 10.0mL 2.0mol·L$^{-1}$ HAc 溶液和 5.0mL 0.2% 淀粉溶液,然后用 $5.000 \times 10^{-3}$ mol·L$^{-1}$ 的 $I_2$ 标准溶液滴定至试液由红色变为蓝紫色即为终点,记下消耗的 $I_2$ 溶液体积于表 6-6 中,平行测定三份,计算 Vc 的含量 $\omega(V_c)\%(mg \cdot g^{-1})$。

## 【注意事项】

(1)在标定 $Na_2S_2O_3$ 溶液时,淀粉指示剂一定要在临近终点(即 $I_2$ 的黄色接近褪去)时加入,否则会有较多的 $I_2$ 被淀粉包藏导致终点滞后。

(2)果蔬样品的前处理应根据样品的实际情况进行选择。如橘子、柚子等应除去果皮、膜、丝络和籽,只取果肉部分进行均浆处理,否则存在于待测液中的这些含纤维较多的部分会吸附维生素 C,影响终点的判断。同时终点的颜色变化要考虑到果肉本身所具有的颜色,不能一概而论。

## 【数据记录及结果分析】

日期:　　　　　　温度:　　　　℃　　　　相对湿度:

表 6-3　$Na_2S_2O_3$ 溶液的标定

| 实验序号 | 1 | 2 | 3 |
|---|---|---|---|
| $m(KIO_3)$ / g | | | |
| $c(KIO_3)$ / mol·L$^{-1}$ | | | |
| $V(Na_2S_2O_3)$ / mL | | | |
| $c(Na_2S_2O_3)$ / mol·L$^{-1}$ | | | |
| $\bar{c}(Na_2S_2O_3)$ / mol·L$^{-1}$ | | | |
| 相对平均偏差 $\bar{d_r}$ / % | | | |

表 6-4　$I_2$ 溶液的标定

| 实验序号 | 1 | 2 | 3 |
|---|---|---|---|
| $c(Na_2S_2O_3)$ / mol·L$^{-1}$ | | | |
| $V(Na_2S_2O_3)$ / mL | | | |
| $V(I_2)$ / mL | | | |
| $c(I_2)$ / mol·L$^{-1}$ | | | |
| $\bar{c}(I_2)$ / mol·L$^{-1}$ | | | |
| 相对平均偏差 $\bar{d_r}$ / % | | | |

表 6-5 维生素 C 片剂中 Vc 含量的测定

| 实验序号 | 1 | 2 | 3 |
|---|---|---|---|
| $m(V_C)$ / g | | | |
| $V(I_2)$ / mL | | | |
| $\omega(V_C)\%$ / mg·g$^{-1}$ | | | |
| $\bar{\omega}(V_C)\%$ / mg·g$^{-1}$ | | | |
| 相对平均偏差 $\bar{d_r}$ / % | | | |

表 6-6 果蔬试样中 Vc 含量的测定

| 实验序号 | 1 | 2 | 3 |
|---|---|---|---|
| $m$(果蔬试样) / g | | | |
| $V(I_2)$ / mL | | | |
| $\omega(V_C)\%$ / mg·g$^{-1}$ | | | |
| $\bar{\omega}(V_C)\%$ / mg·g$^{-1}$ | | | |
| 相对平均偏差 $\bar{d_r}$ / % | | | |

## 【思考题】

(1) 溶解 $I_2$ 时,加入过量 KI 的作用是什么?
(2) 维生素 C 固体试样溶解时为何要加入新煮沸并冷却的蒸馏水?
(3) 碘量法的误差来源有哪些?应采取哪些措施减小误差?

(丁 琼)

# 实验七　配位滴定分析

## 【实验目的】

(1) 掌握使用铬黑 T 或二甲酚橙指示剂的条件和方法。
(2) 学习 EDTA 标准溶液的配制和标定方法。
(3) 了解配位滴定的基本过程。
(4) 学习应用配位滴定法测定样品中金属离子含量的方法。

## 【实验原理】

EDTA 是乙二胺四乙酸钠盐的简称,可分别用 $H_4Y$ 和 $Na_2H_2Y$ 表示。$H_4Y$ 难溶于水,通常采用易溶于水的二钠盐($Na_2H_2Y·2H_2O$)来配制标准溶液。EDTA 的二钠盐可精制成一级标准物质,但通常因蒸馏水中含有杂质,会使 EDTA 标准溶液的浓度改变。因此,在要求准确度较高的实验中,常采用间接法配制 EDTA 标准溶液,再用一级标准物质标定。为防止 EDTA 与玻璃成分中的金属离子作用，配好的 EDTA 标准溶液应贮存在聚乙烯塑料瓶中。

标定 EDTA 标准溶液的一级标准物质有纯锌、铜、铋、氧化锌、碳酸钙、碳酸镁、硫酸镁和硫酸锌等。最常用的一级标准物质是纯 Zn、ZnO 和 $CaCO_3$。为了减少系统误差，标定 EDTA 标准溶液的条件应尽可能与测定条件相同。例如测定 $Bi^{3+}$、$Zn^{2+}$、$Pb^{2+}$、$Al^{3+}$ 等离子时，宜用 ZnO 或金属锌或 $ZnSO_4 \cdot 7H_2O$ 作一级标准物质，在 pH 5~6 的条件下以二甲酚橙作指示剂进行标定。若测定 $Ca^{2+}$、$Mg^{2+}$，一般可使用 $CaCO_3$ 作一级标准物质，在 pH 10 的 $NH_3$-$NH_4Cl$ 缓冲溶液中，用铬黑 T 作指示剂进行标定。但铬黑 T 指示剂与 $Ca^{2+}$ 的显色灵敏度较差，终点变色不敏锐，而 $Mg^{2+}$ 与铬黑 T 显色灵敏。因此，可直接用 $MgCO_3$ 作一级标准物质。

EDTA 能与大多数金属离子配位形成稳定的配合物，元素周期表中，绝大多数金属元素可以用配位滴定法进行直接或间接滴定。

测定水样中的 $Ca^{2+}$、$Mg^{2+}$ 含量或总硬度时，如有 $Fe^{3+}$、$Al^{3+}$、$Cu^{2+}$、$Zn^{2+}$、$Pb^{2+}$ 等离子存在时，对指示剂铬黑 T 有封闭作用，出现终点变色延长的不敏锐现象。为此，可加入三乙醇胺溶液掩蔽 $Fe^{3+}$、$Al^{3+}$ 及加入 $Na_2S$ 溶液掩蔽少量的 $Cu^{2+}$、$Zn^{2+}$、$Pb^{2+}$ 等重金属离子。

当复杂试样中的金属离子与 EDTA 配合反应较缓慢时（例如 $Al^{3+}$），可用返滴定法或置换滴定法来测定。

## 1. 水的硬度的测定

### 【预习作业】

(1) 何为水的硬度？何为水的总硬度？水的总硬度如何表示？生活用水的总硬度多大时不可饮用？

(2) 在测定水样硬度时，可用哪些一级标准物质标定 EDTA 标准溶液？

(3) 在水样硬度的测定时，为什么要滴加 1~2 滴 $3mol \cdot L^{-1}$ HCl 溶液酸化溶液？为什么要加入三乙醇胺和 $Na_2S$ 溶液？

(4) 测定水的总硬度使用何种指示剂来指示滴定的终点？需要用什么溶液控制待测溶液的酸度？

(5) 分别测定水样 $Ca^{2+}$、$Mg^{2+}$ 含量时，为何要先将待测溶液的 pH 调到 12？

(6) 试述测定水的总硬度时，指示剂铬黑 T 与 EDTA 反应的颜色变化过程。

(7) 试述分别测定水中钙、镁含量时，钙指示剂与 EDTA 反应的颜色变化过程。

(8) 请用流程图阐释本实验是如何设计实验步骤，以测定水硬度及钙、镁各自的含量？请在流程图中标出每个实验步骤的注意事项及原因。

### 【实验原理】

水硬度的测定分为水的总硬度的测定和钙、镁含量的测定两种。按国际标准，水的总硬度是指水中 $Ca^{2+}$、$Mg^{2+}$ 的总含量，用 $c_{总硬度}$ / $mmol \cdot L^{-1}$ 表示。工农业生产用水、生活饮用水等对水的硬度都有一定的要求，当水的总硬度 $c_{总硬度} > 10 mmol \cdot L^{-1}$ 时，就不可饮用。

测定水的总硬度是在 pH 10 的 $NH_3$-$NH_4Cl$ 缓冲溶液中，以铬黑 T 为指示剂，用 EDTA 标准溶液滴定至待测溶液由酒红色变为纯蓝色即为终点。

而要分别测定水中钙、镁含量时，首先将待测溶液 pH 调至 12，使 $Mg^{2+}$ 被沉淀为 $Mg(OH)_2$，然后再加入钙指示剂，钙指示剂与少量 $Ca^{2+}$ 配位呈现酒红色，当用 EDTA 标准溶液进行滴定时，EDTA 首先与游离的 $Ca^{2+}$ 配位，当溶液中游离的 $Ca^{2+}$ 耗尽时，EDTA 将从指示剂与 $Ca^{2+}$ 配位的配

合物中夺取 $Ca^{2+}$,使指示剂游离出来,恢复其原来的蓝色,即溶液由酒红色变成纯蓝色即为终点。$Mg^{2+}$ 含量则由 $Ca^{2+}$、$Mg^{2+}$ 总量与 $Ca^{2+}$ 测定值之差求得。

为了减少系统误差,本实验中所用的 EDTA 标准溶液用 $MgCO_3$ 一级标准物质来标定。

## 【仪器材料与试剂】

**仪器** 万分之一电子分析天平,台秤,酸式滴定管(25mL)或聚四氟乙烯酸碱两用滴定管(25mL),锥形瓶(250mL×3),移液管(20mL,25mL,100mL),容量瓶(25mL,250mL,1 000mL),烧杯(150mL,250mL),聚乙烯塑料瓶(1 000mL),量筒(5mL,10mL×2,50mL),玻棒

**材料与试剂** $Na_2H_2Y \cdot 2H_2O$(s,A.R),3mol·L$^{-1}$ HCl 溶液,$MgCO_3$(s,A.R),10%NaOH 溶液,pH 10 的 $NH_3$-$NH_4Cl$ 缓冲溶液,2%$Na_2S$ 溶液,0.5% 铬黑 T,20% 三乙醇胺水溶液,钙指示剂[1-(2-羟基-4-磺基-1-萘基偶氮)-2-羟基-3-萘甲酸],广泛 pH 试纸

## 【实验步骤】

### 1.1　0.01mol·L$^{-1}$ EDTA 溶液的配制

在台秤上称取约 3.8g $Na_2H_2Y \cdot 2H_2O$($M_r$=372.26)于 250mL 烧杯中,加适量蒸馏水溶解后,稀释至 1.0L,贮存于聚乙烯塑料瓶中,摇匀。

### 1.2　EDTA 标准溶液的标定

准确称取一级标准物质 $MgCO_3$($M_r$= 84.32)(110℃干燥 2h 至恒重)0.20～0.25g(精确至 0.000 1g)于 150mL 烧杯中,加 5 滴蒸馏水润湿,盖上表面皿,从烧杯嘴处慢慢滴加 3mol·L$^{-1}$ HCl 溶液 3.0mL,至 $MgCO_3$ 完全溶解后加热微沸几分钟以除去 $CO_2$,冷却后,用洗瓶以蒸馏水冲洗表面皿,冲洗液与烧杯内液合并,将烧杯内溶液转移到 250mL 容量瓶中,用少量蒸馏水润洗烧杯 2～3 次,合并转移入容量瓶中加蒸馏水稀释至刻度,摇匀。

准确移取 20.00mL $MgCO_3$ 标准溶液于 250mL 锥形瓶中,加入 10.0mL pH 10 的 $NH_3$-$NH_4Cl$ 缓冲溶液和 2～3 滴铬黑 T 溶液,用 EDTA 标准溶液滴定至溶液由酒红色变为纯蓝色,颜色保持 30s 不褪即为终点。平行测定三次,数据记录于表 7-1 中,根据所消耗的 EDTA 体积 $V$(EDTA)按下式(7-1)计算出 EDTA 标准溶液的浓度 $c$(EDTA)。

$$c(\text{EDTA}) = \frac{m(\text{MgCO}_3) \times \dfrac{20.00}{250.00} \times 1\,000}{M_r(\text{MgCO}_3) \times V(\text{EDTA})} \,(\text{mol}\cdot\text{L}^{-1}) \tag{7-1}$$

### 1.3　水样硬度的测定

(1) 总硬度的测定　用移液管准确移取水样 100.00mL 于 250mL 锥形瓶中,加入 1～2 滴 3mol·L$^{-1}$ HCl 使溶液酸化,加热微沸几分钟以除去 $CO_2$,冷却后加入 5.0mL 三乙醇胺[*]和 10.0mL pH 10 的 $NH_3$-$NH_4Cl$ 缓冲溶液及 10 滴 $Na_2S$ 溶液,再加 2～3 滴铬黑 T 指示剂,用 EDTA 标准溶液滴定至溶液由酒红色变为纯蓝色,颜色保持 30s 不褪即为终点。平行测定三次,数据记录于表 7-2 中。根据 EDTA 的用量 $V_1$(EDTA),可按式(7-2)计算出水样的总硬度 $c$(总硬度)。

$$c(\text{总硬度}) = \frac{c(\text{EDTA}) \times V_1(\text{EDTA})}{V(\text{水样})} \times 1\,000\,(\text{mmol}\cdot\text{L}^{-1}) \tag{7-2}$$

(2) 水样中 $Ca^{2+}$ 含量及 $Mg^{2+}$ 含量的测定　用移液管准确移取 100.00mL 水样于 250mL 锥形瓶中,加入 6 滴 3mol·L$^{-1}$ HCl 使溶液酸化,加热微沸几分钟以除去 $CO_2$,冷却后加入 5.0mL 三乙醇胺

---

[*] 三乙醇胺作掩蔽剂掩蔽 $Fe^{3+}$、$Al^{3+}$ 时,必须在酸性溶液中加入,然后再调节溶液 pH 至碱性,否则达不到掩蔽效果。

溶液和 10.0mL 10% 的 NaOH 溶液** 以及用药匙加入适量的钙指示剂（与干燥的 NaCl 混合的紫黑色固体，一般取绿豆大小体积量），用 EDTA 标准溶液慢慢滴定，并用力摇动，至溶液由酒红色变为纯蓝色，颜色保持 30s 不褪即为终点。平行测定三次，数据记录于表 7-3 中。根据 EDTA 的用量 $V_2$(EDTA)，可按式（7-3）、式（7-4）分别计算出 $Ca^{2+}$ 和 $Mg^{2+}$ 的浓度。

$$c(Ca^{2+}) = \frac{c(EDTA) \times V_2(EDTA)}{V(水样)} \times 1000 \, (mmol \cdot L^{-1}) \quad (7\text{-}3)$$

$$c(Mg^{2+}) = \frac{c(EDTA) \times [V_1(EDTA) - V_2(EDTA)]}{V(水样)} \times 1000 \, (mmol \cdot L^{-1}) \quad (7\text{-}4)$$

或

$$c(Mg^{2+}) = c(总硬度) - c(Ca^{2+}) \, (mmol \cdot L^{-1}) \quad (7\text{-}5)$$

## 【数据记录及结果分析】

日期：　　　　　温度：　　　℃　　　　相对湿度：

表 7-1　EDTA 标准溶液的标定

| 实验序号 | 1 | 2 | 3 |
|---|---|---|---|
| $m_1$（称量瓶 + $MgCO_3$）/ g | | | |
| $m_2$（称量瓶 + 剩余 $MgCO_3$）/ g | | | |
| $m_1 - m_2$（$MgCO_3$ 质量）/ g | | | |
| $m_{测定}$（$MgCO_3$ 质量）/ g | | | |
| 指示剂 | | | |
| 终点颜色变化 | | | |
| $V(MgCO_3)$ / mL | | | |
| $V_终(EDTA)$ / mL | | | |
| $V_初(EDTA)$ / mL | | | |
| $\Delta V_{消耗}(EDTA)$ / mL | | | |
| $c(EDTA)$ / $mmol \cdot L^{-1}$ | | | |
| $\overline{c}(EDTA)$ / $mmol \cdot L^{-1}$ | | | |
| 相对平均偏差 $\overline{d_r}$ / % | | | |

表 7-2　水样总硬度的测定

| 实验序号 | | 1 | 2 | 3 |
|---|---|---|---|---|
| 指示剂 | | | | |
| 终点前后颜色变化 | | | | |
| $V$（水样）/ mL | | 100.00 | 100.00 | 100.00 |
| 水样处理 | $3mol \cdot L^{-1}$ HCl / 滴 | | | |
| | 加热微沸除 $CO_2$ | | | |
| | 三乙醇胺 / mL | | | |
| | $NH_3$-$NH_4Cl$ 缓冲溶液 / mL | | | |
| | $Na_2S$ / 滴 | | | |

---

** 测定 $Ca^{2+}$ 时，需加 NaOH 使 $Mg^{2+}$ 生成 $Mg(OH)_2$ 沉淀，若沉淀量过多则可能吸附 $Ca^{2+}$，使 $Ca^{2+}$ 的测定结果偏低。若遇到这种情况，可按上述方法预滴一份。再取试样 2 份，先滴加 EDTA 溶液至比预滴时所消耗的 EDTA 体积少 1mL 左右时，再加入 NaOH 和指示剂，然后继续用 EDTA 溶液滴定至终点。

| 实验序号 | | 1 | 2 | 3 |
|---|---|---|---|---|
| 测定 | $V_{终}$(EDTA)/mL | | | |
| | $V_{初}$(EDTA)/mL | | | |
| | $\Delta V_{消耗}$(EDTA)/mL | | | |
| | $c$(总硬度)/mmol·L$^{-1}$ | | | |
| | $\bar{c}$(总硬度)/mmol·L$^{-1}$ | | | |
| 相对平均偏差 $\overline{d_r}$/% | | | | |

表 7-3 水样中 $Ca^{2+}$ 含量及 $Mg^{2+}$ 含量的测定

| 实验序号 | 1 | 2 | 3 |
|---|---|---|---|
| 指示剂 | | | |
| 终点颜色变化 | | | |
| $V$(水样)/mL | 100.00 | 100.00 | 100.00 |
| $V_{终}$(EDTA)/mL | | | |
| $V_{初}$(EDTA)/mL | | | |
| $\Delta V_{消耗}$(EDTA)/mL | | | |
| $c$($Ca^{2+}$)/mmol·L$^{-1}$ | | | |
| $c$($Mg^{2+}$)/mmol·L$^{-1}$ | | | |
| $\bar{c}$($Ca^{2+}$)/mmol·L$^{-1}$ | | | |
| $\bar{c}$($Mg^{2+}$)/mmol·L$^{-1}$ | | | |
| 相对平均偏差 $\overline{d_r}$/% | | | |

【思考题】

(1) 测定水样总硬度时，为什么要加 pH 10 的 $NH_3$-$NH_4Cl$ 缓冲溶液？

(2) 本实验测定钙、镁含量时，试样中存在少量 $Fe^{3+}$、$Al^{3+}$、$Cu^{2+}$、$Zn^{2+}$、$Pb^{2+}$ 等杂质离子，对测定有干扰吗？用什么方法可消除 $Fe^{3+}$、$Al^{3+}$、$Cu^{2+}$、$Zn^{2+}$、$Pb^{2+}$ 干扰？

(3) 试述配位滴定法测定石灰石中 $Ca^{2+}$、$Mg^{2+}$ 含量的原理。

(庄海旗)

## 2. 明矾含量的测定

【目的要求】

(1) 了解返滴定的基本过程。
(2) 学习 EDTA 标准溶液的配制和标定方法。
(3) 掌握使用二甲酚橙指示剂的条件和方法。
(4) 学习配位滴定法测定明矾含量的原理和方法。

## 【预习作业】

(1) 明矾含量的测定方法有哪些？
(2) 哪些情况下宜采用返滴定法？
(3) 为什么明矾含量的测定采用返滴定法？
(4) 二甲酚橙指示剂在配位滴定中的使用条件是什么？
(5) 根据配位滴定原理，参照本实验步骤，以流程图方式给出本实验的设计思路，并在流程图中标注上滴定反应条件、操作条件及保证实验结果的准确度所采取的可行性实验措施及依据。

## 【实验原理】

明矾又名白矾，是明矾石的提炼品。明矾的含量测定一般都是先测定其组成中 $Al^{3+}$ 的含量，再换算成明矾的含量。$Al^{3+}$ 的含量测定可采用配位滴定法，但 $Al^{3+}$ 与 EDTA 的配位反应不仅速度缓慢，而且对二甲酚橙指示剂还具有封闭作用，当酸度不高时，$Al^{3+}$ 易水解形成多种多核羟基配合物，因此，$Al^{3+}$ 的含量测定不能用直接滴定法而用返滴定法。方法是：在含 $Al^{3+}$ 的试液中，先加入一定量过量的 EDTA 标准溶液，煮沸以加速 $Al^{3+}$ 与 EDTA 的反应，冷却后，加入 HAc-NaAc 缓冲溶液调节 pH 至 5～6，以保证 $Al^{3+}$ 与 EDTA 的定量配合，然后以二甲酚橙作指示剂（$Al^{3+}$ 已形成 $AlY^-$ 配合物，不再封闭指示剂），用 $Zn^{2+}$ 标准溶液滴定过量的 EDTA，$Zn^{2+}$ 与过量的 EDTA 反应完全后，$Zn^{2+}$ 开始与二甲酚橙结合转化为紫红色配合物 $ZnH_3In^{2-}$，由于二甲酚橙为黄色，两者的混合色为橙色，所以溶液颜色变为橙色时，终点到达。由两种标准溶液的浓度和用量，可以求得 $Al^{3+}$ 的含量。反应过程如下

$$Al^{3+} + H_2Y^{2-}(过量) + 2H_2O \Longrightarrow AlY^- + 2H_3O^+$$

$$H_2Y^{2-}(剩余) + Zn^{2+} + 2H_2O \Longrightarrow ZnY^{2-} + 2H_3O^+$$

$$Zn^{2+} + H_3In^{4-} \Longrightarrow ZnH_3In^{2-}$$

（黄色）　　（紫红色）

滴定过程各配合物的稳定顺序：$AlY^- > ZnY^{2-} > ZnH_3In^{2-}$。

## 【仪器材料与试剂】

**仪器** 电子天平，酸式滴定管或两性滴定管（25mL），锥形瓶（250mL×3），移液管（25mL×3），容量瓶（100mL，250mL），烧杯（100mL×2），试剂瓶（250mL），量筒（10mL×3，100mL），玻棒

**材料与试剂** $0.05mol \cdot L^{-1}$ EDTA 溶液，HAc-NaAc 缓冲溶液（pH 4.5），$2mol \cdot L^{-1}$ HCl 溶液，$ZnSO_4 \cdot 7H_2O$（A.R 或 C.P），0.5% 的二甲酚橙溶液，明矾试样

## 【实验步骤】

### 2.1　$0.05mol \cdot L^{-1}$ $Zn^{2+}$ 标准溶液的配制

准确称取 3.6～3.8g（精确至 ±0.0001g）一级标准物质 $ZnSO_4 \cdot 7H_2O$（$M_r$=287.56）于小烧杯中，用量筒加 2.0mL $2mol \cdot L^{-1}$ HCl 溶液和少量蒸馏水，溶解后定量转移至 250mL 容量瓶中，稀释至刻度，摇匀。根据下式(7-6)计算 $Zn^{2+}$ 标准溶液浓度

$$c(ZnSO_4 \cdot 7H_2O) = \frac{m(ZnSO_4 \cdot 7H_2O)}{M_r(ZnSO_4 \cdot 7H_2O) \times \dfrac{250.00}{1000}} (mol \cdot L^{-1}) \tag{7-6}$$

### 2.2  0.05mol·L⁻¹ EDTA 溶液浓度的测定

用移液管准确移取 25.00mL EDTA 溶液于 250mL 锥形瓶中,加入 100mL 蒸馏水,5.0mL HAc-NaAc 缓冲溶液及 1mL 二甲酚橙指示剂。用 $Zn^{2+}$ 标准溶液滴定至溶液由黄色变为橙色即为终点。平行滴定三次,数据记录于表 7-4,按式(7-7)计算 EDTA 的准确浓度。

$$c(\text{EDTA}) = \frac{c(\text{ZnSO}_4) \times V(\text{ZnSO}_4)}{V(\text{EDTA})} (\text{mol} \cdot \text{L}^{-1}) \tag{7-7}$$

### 2.3  明矾含量的测定

准确称取 1.3~1.4g(精确至 ±0.0001g)已研细的明矾[$M_r$(KAl(SO$_4$)$_2$·12H$_2$O) = 474.4]试样于小烧杯中,加蒸馏水溶解,定量转移至 100mL 容量瓶中,用蒸馏水稀释至刻度,摇匀。用移液管准确移取 25.00mL 明矾待测液于锥形瓶中,另用移液管吸取 25.00mL 已标定的 EDTA 溶液与之混合均匀,在沸水浴中加热 10min,冷却至室温,加入 100mL 蒸馏水,5.0mL HAc-NaAc 缓冲溶液及 1.0mL 二甲酚橙指示剂,用 $Zn^{2+}$ 标准溶液滴定至溶液由黄色变为橙色即为终点。平行测定三次,数据记录于表 7-5,按式(7-8)计算明矾的含量

$$\omega[\text{KAl}(\text{SO}_4)_2 \cdot 12\text{H}_2\text{O}] = \frac{[c(\text{EDTA})V(\text{EDTA}) - c(\text{ZnSO}_4)V(\text{ZnSO}_4)] \times M_r[\text{KAl}(\text{SO}_4)_2 \cdot 12\text{H}_2\text{O}] \times 10^{-3}}{m(\text{样品}) \times \dfrac{25.00}{100.00}} (\text{g} \cdot \text{g}^{-1}) \tag{7-8}$$

## 【注意事项】

(1)在 $Al^{3+}$ 的试液中加入过量 EDTA 标准溶液后,一定要煮沸以加速 $Al^{3+}$ 与 EDTA 的反应。

(2)$Al^{3+}$ 与 EDTA 的反应液充分冷却后,调节 pH 至 5~6,以保证 $Al^{3+}$ 与 EDTA 的定量配合。

(3)$Al^{3+}$ 与 EDTA 的反应完毕后才加入二甲酚橙作指示剂,用 $Zn^{2+}$ 标准溶液滴定过量的 EDTA。

## 【数据记录及结果分析】

日期:　　　　　　　温度:　　　℃　　　　相对湿度:

表 7-4  0.05mol·L⁻¹ EDTA 溶液浓度的测定

| 实验序号 | 1 | 2 | 3 |
|---|---|---|---|
| $m_{称量}$(ZnSO$_4$·7H$_2$O)/ g | | | |
| $m_{测定}$(ZnSO$_4$·7H$_2$O)/ g | | | |
| 指示剂 | | | |
| 终点颜色变化 | | | |
| $V$(ZnSO$_4$·7H$_2$O)/ mL | | | |
| $V_{终}$(EDTA)/ mL | | | |
| $V_{初}$(EDTA)/ mL | | | |
| $\Delta V_{消耗}$(EDTA)/ mL | | | |
| $c$(EDTA)/ mol·L⁻¹ | | | |
| $\bar{c}$(EDTA)/ mol·L⁻¹ | | | |
| 相对平均偏差 $\bar{d_r}$ / % | | | |

表 7-5　明矾含量的测定

| 实验序号 | | 1 | 2 | 3 |
|---|---|---|---|---|
| $m_{称量}$（明矾）/g | | | | |
| $m_{测定}$（明矾）/g | | | | |
| $V$（明矾）/mL | | | | |
| 样品处理 | 0.5mol·L$^{-1}$ EDTA / mL | | | |
| | 沸水浴加热 / min | | | |
| | 蒸馏水 / mL | | | |
| | HAc-NaAc / mL | | | |
| 指示剂 | | | | |
| 终点颜色变化 | | | | |
| $V_{终}$（ZnSO$_4$·7H$_2$O）/ mL | | | | |
| $V_{初}$（ZnSO$_4$·7H$_2$O）/ mL | | | | |
| $\Delta V_{消耗}$（ZnSO$_4$·7H$_2$O）/ mL | | | | |
| $\omega$（明矾）/ g·g$^{-1}$ | | | | |
| $\bar{\omega}$（明矾）/ g·g$^{-1}$ | | | | |
| 相对平均偏差 $\bar{d_r}$ / % | | | | |

## 【思考题】

（1）用 EDTA 测定 Al$^{3+}$ 时，能采用直接滴定法吗？

（2）用 EDTA 测定 Al$^{3+}$ 时，能否用铬黑 T 为指示剂？为什么？

（3）Al$^{3+}$ 对二甲酚橙有封闭作用，为什么在用 EDTA 标准溶液测定铝含量的返滴定法中还能采用二甲酚橙作指示剂？

## 3．葡萄糖酸钙含量的测定

### 【目的要求】

（1）了解配位滴定的基本过程。

（2）学习 EDTA 标准溶液的直接配制方法。

（3）掌握使用铬黑 T 指示剂的条件和方法。

（4）学习应用配位滴定法测定葡萄糖酸钙含量的原理和方法。

### 【预习作业】

（1）如何用直接配制法获得 EDTA 标准溶液？

（2）葡萄糖酸钙的含量测定原理是什么？

（3）配位滴定法测定葡萄糖酸钙的含量时采用何种指示剂？

（4）铬黑 T 指示剂的使用条件是什么？本实验中为何要加入辅助指示剂？

（5）根据配位滴定原理，参照本实验步骤，以流程图方式给出本实验的设计思路，并在流程图中标注上滴定反应条件、操作条件及保证实验结果的准确度所采取的可行性实验措施及依据。

## 【实验原理】

临床上常用的葡萄糖酸钙药物有片剂和针剂两种剂型，其含量的测定常用螯合滴定法。方法是：在葡萄糖酸钙的溶液中，用 pH 10 的 $NH_3$-$NH_4Cl$ 缓冲溶液来控制溶液的酸度，以铬黑 T 作指示剂，用 EDTA 标准溶液直接滴定。但因 $Ca^{2+}$ 与铬黑 T 形成的螯合物（$CaIn^-$）不够稳定，单独使用铬黑 T 会使终点过早到达，为此可先加入少量的 $Mg^{2+}$ 作辅助指示剂，以提高终点变色的敏锐性。为使加入的 $Mg^{2+}$ 不影响测定的结果，本实验在加入葡萄糖酸钙样品之前，先将加入的 $Mg^{2+}$ 用 EDTA 溶液滴定至终点，再加入葡萄糖酸钙样品溶液，然后用 EDTA 标准溶液再次滴定至终点，则滴定葡萄糖酸钙消耗的 EDTA 标准溶液的体积为滴定管中两次滴定终点读数之差，并由此可计算出葡萄糖酸钙的含量。葡萄糖酸钙含量测定的实验步骤如下：

（1）先将 $Mg^{2+}$ 溶液和铬黑 T 加入 $NH_3$-$NH_4Cl$ 缓冲溶液中，用 EDTA 标准溶液滴定至溶液由酒红色恰好变为纯蓝色，为第一滴定终点。

（2）在所得溶液中加入葡萄糖酸钙样品溶液，用 EDTA 标准溶液滴定至溶液由酒红色恰好变为纯蓝色，为第二滴定终点。

用于滴定葡萄糖酸钙所消耗的 EDTA 标准溶液与第一滴定无关，故滴定葡萄糖酸钙消耗的 EDTA 标准溶液的体积为滴定管中两次滴定终点体积读数之差。

## 【仪器材料与试剂】

**仪器**　电子天平，碱式滴定管或两性滴定管（25mL），锥形瓶（250mL×3），移液管（10mL，20mL），容量瓶（250mL×3），烧杯（250mL×2），量筒（10mL×2，50mL），洗瓶

**材料与试剂**　$Na_2H_2Y \cdot 2H_2O$（s, A.R），1%$MgSO_4$ 溶液，pH 10 的 $NH_3$-$NH_4Cl$ 缓冲溶液，葡萄糖酸钙（片剂或针剂），0.5% 铬黑 T 溶液

## 【实验步骤】

### 3.1　0.01mol·$L^{-1}$ EDTA 标准溶液的配制

在分析天平上准确称取 1.0～1.1g（精确至 ±0.000 1g）一级标准物质 $Na_2H_2Y \cdot 2H_2O$（$M_r$ = 372.26）于 250mL 烧杯中，用 50.0mL 蒸馏水溶解，定量转移至 250mL 容量瓶中，加蒸馏水稀释至刻度，摇匀。

$$c(\text{EDTA}) = \frac{m(\text{EDTA}) \times 1\,000}{M_r(\text{EDTA}) \times 250.00} \quad (\text{mol} \cdot \text{L}^{-1}) \tag{7-9}$$

### 3.2　葡萄糖酸钙待测溶液的配制

在分析天平上准确称取葡萄糖酸钙（$M_r$ = 448.6）片剂 1.0～1.1g（精确至 ±0.000 1g）于 250mL 烧杯中，加 50.0mL 蒸馏水并微热使其溶解，冷却后定量转移至 250mL 容量瓶中，加蒸馏水稀释至刻度，摇匀。

或用移液管吸取 10.00mL 针剂葡萄糖酸钙溶液于 250mL 容量瓶中，加蒸馏水稀释至刻度，摇匀。

### 3.3　葡萄糖酸钙含量的测定

取蒸馏水 10.0mL 于 250mL 锥形瓶中，加入 10.0mL pH 10 的 $NH_3$-$NH_4Cl$ 缓冲溶液，1～2 滴 1% $MgSO_4$ 和 2～3 滴铬黑 T，摇匀。用 EDTA 标准溶液滴定至溶液由酒红色恰好变为纯蓝色，记

录滴定管中 EDTA 溶液的读数 $V_1$。用移液管吸取 20.00mL 葡萄糖酸钙待测液于上述锥形瓶中,继续用 EDTA 标准溶液滴定至溶液由酒红色恰好变为纯蓝色,记录滴定管中 EDTA 溶液的读数 $V_2$ 于表 7-6 中。平行测定三次。由式(7-10)计算葡萄糖酸钙的质量分数 $\omega$(片剂)和式(7-11)计算质量浓度 $\rho$(针剂)。

$$片剂:\omega(葡萄糖酸钙)=\frac{c(\text{EDTA})\times[V_2(\text{EDTA})-V_1(\text{EDTA})]\times M_r(葡萄糖酸钙)}{m(葡萄糖酸钙)\times\frac{20.00}{250.00}}(\text{mg}\cdot\text{g}^{-1}) \quad (7\text{-}10)$$

式(7-10)中 $m$(葡萄糖酸钙)为片剂的质量。

$$针剂:\rho(葡萄糖酸钙)=\frac{c(\text{EDTA})\times[V_2(\text{EDTA})-V_1(\text{EDTA})]\times M_r(葡萄糖酸钙)}{10.00\times\frac{20.00}{250.00}}(\text{g}\cdot\text{L}^{-1}) \quad (7\text{-}11)$$

### 【注意事项】

(1) 在 pH 10 的 $NH_3$-$NH_4Cl$ 缓冲溶液中使用铬黑 T。
(2) 加入少量 $MgSO_4$ 作为辅助指示剂以免因 $Ca^{2+}$ 与铬黑 T 形成的螯合物不够稳定而使滴定终点提前。
(3) 实验中有两个滴定终点,$Ca^{2+}$ 消耗的 EDTA 是第一滴定终点至第二滴定终点之体积。

### 【数据记录及结果分析】

日期:　　　　　　温度:　　　　℃　　　　相对湿度:

表 7-6　针剂及片剂中葡萄糖酸钙含量的测定

| 实验序号 | 1 | 2 | 3 |
| --- | --- | --- | --- |
| $m(\text{Na}_2\text{H}_2\text{Y}\cdot2\text{H}_2\text{O})/\text{g}$ | | | |
| $c(\text{Na}_2\text{H}_2\text{Y}\cdot2\text{H}_2\text{O})/\text{mol}\cdot\text{L}^{-1}$ | | | |
| 第一终点指示剂 | | | |
| 第二终点指示剂 | | | |
| 第一终点颜色变化 | | | |
| 第二终点颜色变化 | | | |
| $V$(葡萄糖酸钙)/ mL | | | |
| $V_{终点1}(\text{EDTA})/\text{mL}$ | | | |
| $V_{终点2}(\text{EDTA})/\text{mL}$ | | | |
| $[V_{终点2}(\text{EDTA})-V_{终点1}(\text{EDTA})]/\text{mL}$ | | | |
| $\omega$(葡萄糖酸钙)/ $\text{mg}\cdot\text{g}^{-1}$ | | | |
| $\bar{\omega}$(葡萄糖酸钙)/ $\text{mg}\cdot\text{g}^{-1}$ | | | |
| $\rho$(葡萄糖酸钙)/ $\text{g}\cdot\text{L}^{-1}$ | | | |
| $\bar{\rho}$(葡萄糖酸钙)/ $\text{g}\cdot\text{L}^{-1}$ | | | |
| 相对平均偏差 $\overline{d_r}$ / % | | | |

**【思考题】**

（1）在测定葡萄糖酸钙含量时，为什么要加入少量的 $Mg^{2+}$ 试液？它会影响测定的结果吗？

（2）本实验能否在配制 EDTA 标准溶液时加入 $Mg^{2+}$ 作辅助指示剂？

（陈志琼）

# 实验课题三
# 分光光度法实验

## 实验八　可见分光光度法测定水样中铁含量

### 【目的要求】

（1）熟悉可见分光光度法测定水样中铁含量的原理和方法。
（2）掌握722型可见分光光度计的使用方法。

### 【预习作业】

（1）Lambert-Beer定律的基本内容是什么？怎样用两种不同的关系式表示？
（2）分光光度法最常用的定量分析方法有哪些？它们具体的方法是什么？
（3）如何绘制吸收光谱曲线？为什么要找最大吸收波长？
（4）分光光度计的基本部件有哪些？各起什么作用？
（5）邻二氮菲法、硫氰酸盐法和磺基水杨酸法测定水样中铁含量的基本原理是什么？试比较三者的相同点和不同点。
（6）请用流程图阐释本实验是如何设计实验步骤，以测定水样中铁含量？请在流程图中标出每个实验步骤要有哪些注意事项，并说明原因。

### 【实验原理】

分光光度法是一种现代仪器分析方法，它的理论基础是物质的吸收光谱和光的吸收定律。根据所用光源波长不同，分光光度法可分为：可见分光光度法（380~760nm）；紫外分光光度法（200~380nm）；红外分光光度法（780~3×10$^5$nm）。Lambert-Beer定律是吸收光谱法的基本定律，它描述物质对单色光吸收的强度与吸光物质的浓度和厚度的关系。根据Lambert-Beer定律，当吸光物质的种类、溶剂、溶液温度一定时，具有一定波长的单色光通过一定厚度（$b$）的有色物质溶液时，有色物质对光的吸收程度（用吸光度$A$表示）与有色物质的浓度（$c$）呈线性关系

$$A = \varepsilon b c \tag{8-1}$$

式（8-1）中，$c$为物质的量浓度，$\varepsilon$为摩尔吸光系数，单位为L·mol$^{-1}$·cm$^{-1}$，它是各种有色物质在一定波长下的特征常数。若溶液组成量度以质量浓度（g·L$^{-1}$）表示，则$A = abc$，$a$称为质量吸光系数，单位为L·g$^{-1}$·cm$^{-1}$。

在分光光度法中，当吸光物质、入射光波长、温度和溶剂一定时，$\varepsilon$，$a$为常数，此时溶液的吸光度（$A$）与有色物质的浓度（$c$）及吸收池厚度（$b$）成正比。Lambert-Beer定律成为可见分光光度法定

量分析的基础，可用于测定吸光物质的含量。

Lambert-Beer 定律只适合于单色光，通常要选择合适的波长。对某一溶液在不同波长下测定其吸光度，以吸光度对波长作图，得到吸收光谱，从吸收光谱中找出最大吸收波长($\lambda_{max}$)作为测定波长。

分光光度法仅适合于微量组分的分析，通常以 $A$ 处于 0.2～0.7 为最佳，可使 Lambert-Beer 定律处于线性范围。

分光光度法常用的定量分析方法有标准比较法和标准曲线法。

标准比较法是指在相同实验条件下配制出标准溶液和试样溶液，分别测出标准溶液和未知物的吸光度

$$A_{标} = abc_{标} \tag{8-2}$$

$$A_{未} = abc_{未} \tag{8-3}$$

因同一实验中所使用的吸收池厚度 $b$ 相等，相同的吸光物质与测定条件，因此 $a$ 亦相同，比较式(8-2)与式(8-3)得

$$\frac{A_{标}}{A_{未}} = \frac{abc_{标}}{abc_{未}} \tag{8-4}$$

$$\frac{A_{标}}{A_{未}} = \frac{c_{标}}{c_{未}} \tag{8-5}$$

$$c_{未} = \frac{A_{未}}{A_{标}} \times c_{标} \tag{8-6}$$

上述方法称为标准对照法(比较法)。

为提高测量的准确度，往往同时测定数个标准溶液的吸光度，并以浓度为横坐标，吸光度为纵坐标，绘制标准曲线(或称工作曲线)，在同样条件下，测定被测溶液的吸光度，从标准曲线上查出该吸光度所对应的溶液浓度，这种方法称为标准曲线法，该曲线可用直线方程 $y = ax + b$，$r^2$ 表示，该方程亦称回归方程。

## 1. 邻二氮菲法

邻二氮菲(邻菲罗啉)是目前可见分光光度法测定铁含量的较好试剂。加入试剂的作用分别是：盐酸羟胺是还原剂、邻二氮菲是显色剂、NaAc 溶液是缓冲溶液用来控制溶液的 pH。在 pH 为 3～9 的溶液中，邻二氮菲与 $Fe^{2+}$ 生成稳定的橘红色配合物 $[Fe(phen)_3]^{2+}$($\lg K_s = 21.3$)，$\varepsilon_{508} = 1.1 \times 10^4$ L·mol$^{-1}$·cm$^{-1}$；也能与 $Fe^{3+}$ 生成淡蓝色 $[Fe(phen)_3]^{3+}$($\lg K_s = 14.1$)。因此，在显色前，加入盐酸羟胺把 $Fe^{3+}$ 还原成 $Fe^{2+}$，让其与邻二氮菲形成稳定配合物，其反应如下：

$$2Fe^{3+} + 2NH_2OH \cdot HCl + 2H_2O = 2Fe^{2+} + N_2\uparrow + 4H_3O^+ + 2Cl^-$$

## 【仪器材料及试剂】

**仪器**　722型分光光度计,万分之一分析天平,容量瓶(50mL×7,1 000mL),吸量管(1mL,2mL×2,5mL,10mL),滴管,可调移液枪(5mL,10mL)

**材料与试剂**　8mmol·L$^{-1}$邻菲罗啉(新鲜配制)溶液,1.5mol·L$^{-1}$盐酸羟胺(临用时配制)溶液,1mol·L$^{-1}$ NaAc溶液,2.000mmol·L$^{-1}$标准铁溶液*

## 【实验步骤】

### 1.1　标准溶液和待测溶液的配制

取50mL容量瓶7支,按表8-1所列的量,用吸量管量取各种溶液加入容量瓶中,加蒸馏水稀释至刻度,摇匀。即配成一系列标准溶液及待测溶液(供试液)(表8-1)。

表8-1　邻菲罗啉标准溶液及待测溶液配制及吸光度测定

| 实验序号 | 1(空白) | 2 | 3 | 4 | 5 | 6 | 7 |
|---|---|---|---|---|---|---|---|
| Fe$^{2+}$标准溶液/mL | 0 | 0.40 | 0.80 | 1.20 | 1.60 | 2.00 | — |
| Fe$^{2+}$水样溶液/mL | — | — | — | — | — | — | 10.00 |
| 盐酸羟胺/mL | 1.00 | 1.00 | 1.00 | 1.00 | 1.00 | 1.00 | 1.00 |
| 邻菲罗啉/mL | 2.00 | 2.00 | 2.00 | 2.00 | 2.00 | 2.00 | 2.00 |
| 醋酸钠溶液/mL | 5.00 | 5.00 | 5.00 | 5.00 | 5.00 | 5.00 | 5.00 |
| $V_{总}$(稀释)/mL | 50.00 | 50.00 | 50.00 | 50.00 | 50.00 | 50.00 | 50.00 |
| $c_{稀释}$(Fe$^{2+}$)/μmol·L$^{-1}$ | | | | | | | |

### 1.2　吸收光谱的测定

取表8-1中的4号溶液,按722型分光光度计的使用方法(详见常用测量仪器使用),在450~560nm波长范围内,以试剂空白作为参比溶液,每隔10nm测定一次溶液的吸光度,在最大吸光度的波长附近可每隔5nm再测其吸光度。记录实验数据于表8-2中。

### 1.3　吸光度(A)的测定及标准曲线的绘制

选择$\lambda_{max}$为入射光波长,以试剂空白作为参比溶液,测出系列标准溶液的吸光度,记录实验数据于表8-3。并以浓度为横坐标,吸光度为纵坐标,绘制标准曲线,该曲线可用直线方程即回归方程$y=ax+b$,$r^2$表示。

### 1.4　待测水样中铁离子浓度的测定

将待测水样(供试液)按与标准曲线系列溶液相同的条件下测其吸光度,记录实验数据于表8-3中。

## 【数据记录与结果分析】

日期:　　　　　　温度:　　　　℃　　　　相对湿度:

(1)绘制吸收光谱并确定$\lambda_{max}$:根据表8-2记录的数据,以波长为横坐标,吸光度为纵坐标,在坐标纸上绘制吸收光谱,找出$\lambda_{max}$。

---

* 标准铁溶液(2.000mmol·L$^{-1}$)的配制:准确称取0.784 2g(NH$_4$)$_2$Fe(SO$_4$)$_2$·6H$_2$O置于烧杯中,加入6mol·L$^{-1}$ HCl 120mL和少量蒸馏水,溶解后转入1 000mL容量瓶中,加蒸馏水稀释至刻度,摇匀备用。

表 8-2　邻菲罗啉的吸收光谱测定

| $\lambda$ / nm | 450 | 460 | 470 | 480 | 490 | 500 | 510 | 520 | 530 | 540 | 550 | 560 |
|---|---|---|---|---|---|---|---|---|---|---|---|---|
| $A$ | | | | | | | | | | | | |
| $\lambda_{max}$ / nm | | | | | | | | | | | | |

吸收光谱:(请附图1于实验报告)

(2) 绘制标准曲线:根据表 8-3 记录的数据,以标准 $Fe^{2+}$ 离子的浓度($\mu mol \cdot L^{-1}$)为横坐标,吸光度($A$)为纵坐标绘制标准曲线。

表 8-3　标准溶液和待测溶液的吸光度测定

| 容量瓶编号 | 1(空白) | 2 | 3 | 4 | 5 | 6 | 7(水样) |
|---|---|---|---|---|---|---|---|
| $c_{稀释}(Fe^{2+})$ / $\mu mol \cdot L^{-1}$ | 0.00 | | | | | | |
| 吸光度 $A$ | 0.00 | | | | | | |

(3) 待测溶液中 $Fe^{2+}$ 离子浓度的确定:根据待测溶液的吸光度,利用标准对照法和标准曲线法分别计算出原水样溶液中未知的 $Fe^{2+}$($\mu mol \cdot L^{-1}$)。

标准曲线:(请附图2于实验报告)

1) 标准曲线法

标准曲线上直接读出供试液 $Fe^{2+}$ 的含量为_____($\mu mol \cdot L^{-1}$);

原水样中 $Fe^{2+}$ 含量为_____($\mu mol \cdot L^{-1}$)。

2) 标准对照法

所选取 $Fe^{2+}$ 标准溶液的吸光度_____;

水样的吸光度_____;

原水样中 $Fe^{2+}$ 含量_____($\mu mol \cdot L^{-1}$)。

样品溶液的原浓度 $c_2(\rho_2)$ = 从标准曲线找出待测溶液的浓度 $c_1(\rho_1) \times$ 稀释倍数 $k$。

## 【思考题】

(1) 为什么要控制被测溶液的吸光度最好在 0.2~0.7 的范围内? 如何控制?

(2) 由工作曲线查出的待测铁离子的浓度是否是原始待测溶液铁离子的浓度?

(3) 从实验结果说明标准对照法与标准曲线法的优缺点。

## 2. 硫氰酸盐法

可见分光光度法仅适合于有色物质含量的测定,而 $Fe^{3+}$ 的稀溶液几乎是无色的,当加入显色剂 KSCN 时,即产生血红色的络合物($\lg K_s = 6.4$)。

$$Fe^{3+} + 6SCN^- \rightleftharpoons [Fe(SCN)_6]^{3-}$$

溶液颜色的深度与 $Fe^{3+}$ 浓度成正比。当溶液中 $SCN^-$ 浓度增大时,平衡右移,溶液颜色也随之加深。因此,在测定时,一定要使 $SCN^-$ 浓度大大超过与 $Fe^{3+}$ 反应所需的浓度,且每次的浓度恒定。此外,为防止 $Fe^{3+}$ 的水解,在溶液中应加入一定量的硝酸。$Fe^{3+}$ 离子还能被 $SCN^-$ 离子慢慢还原为 $Fe^{2+}$ 离子,而使红色变浅,所以在比色溶液中应当含有少量强氧化剂过二硫酸铵,即 $(NH_4)_2S_2O_8$,防止 $Fe^{3+}$ 离子的还原。

## 【仪器材料及试剂】

**仪器** 722型分光光度计,容量瓶(50mL×7,1 000mL),吸量管(1mL,5mL,10mL),滴管

**材料与试剂** 0.2mol·L$^{-1}$ KSCN溶液,2mol·L$^{-1}$ HNO$_3$溶液,浓H$_2$SO$_4$,0.100 0g·L$^{-1}$ Fe$^{3+}$标准溶液*,25g·L$^{-1}$(NH$_4$)$_2$S$_2$O$_8$溶液

## 【实验步骤】

### 2.1 标准溶液和待测溶液的配制

取50mL容量瓶7个,按表8-4所列的量,用吸量管量取试液和试剂分别加入容量瓶,加蒸馏水至刻度,摇匀,配成一系列不同浓度的标准溶液以及空白溶液与待测溶液。

表8-4 Fe$^{3+}$标准溶液和待测溶液的配制及吸光度测定

| 容量瓶号 | 1(空白) | 2 | 3 | 4 | 5 | 6 | 7(待测) |
|---|---|---|---|---|---|---|---|
| 0.1g·L$^{-1}$ Fe$^{3+}$ 标准 / mL | 0 | 0.50 | 1.00 | 1.50 | 2.00 | 2.50 | — |
| 水样 / mL | — | — | — | — | — | — | 10 |
| 2mol·L$^{-1}$ HNO$_3$ / mL | 1.00 | 1.00 | 1.00 | 1.00 | 1.00 | 1.00 | 1.0 |
| 0.2mol·L$^{-1}$ KSCN / mL | 5.00 | 5.00 | 5.00 | 5.00 | 5.00 | 5.00 | 5.00 |
| 25g·L$^{-1}$(NH$_4$)$_2$S$_2$O$_8$ / 滴 | 1 | 1 | 1 | 1 | 1 | 1 | 1 |
| $V_{总(稀释)}$ / mL | 50.00 | 50.00 | 50.00 | 50.00 | 50.00 | 50.00 | 50.00 |
| $\rho_{稀释}$(Fe$^{3+}$) / mg·L$^{-1}$ | | | | | | | |

### 2.2 吸收光谱的测定

取表8-4中的4号溶液,按722型分光光度计的使用方法(详见常用测量仪器使用),在450～520nm波长范围内,以试剂空白作为参比溶液,每隔10nm测定一次溶液的吸光度,在最大吸光度的波长附近可每隔5nm再测其吸光度。记录实验数据于表8-5中。

### 2.3 吸光度($A$)的测定及标准曲线的绘制

根据吸收光谱的$\lambda_{max}$(文献记载的$\lambda_{max}$=480nm),以试剂空白作参比溶液,分别测定各标准溶液的吸光度和待测水溶液的吸光度,并记录数据于下表8-6中。并以浓度为横坐标,吸光度为纵坐标,绘制标准曲线,该曲线可用直线回归方程,$y=ax+b$,$r^2$表示。

### 2.4 待测水溶液中Fe$^{3+}$离子浓度的确定

根据所测得的供试液的吸光度,利用标准对照法和标准曲线法分别计算出水样中Fe$^{2+}$浓度(mg·L$^{-1}$)。

## 【数据记录与结果分析】

日期:　　　　　　　　　温度:　　　　℃　　　　　相对湿度:

(1)绘制吸收光谱并确定$\lambda_{max}$:根据表8-5所记录的数据,以波长为横坐标,吸光度为纵坐标在坐标纸上绘制吸收光谱,找出$\lambda_{max}$。

---

*Fe$^{3+}$标准溶液(0.1g·L$^{-1}$)配制:称取0.864 0g硫酸铁铵[NH$_4$Fe(SO$_4$)$_2$·12H$_2$O]溶于少量蒸馏水中加浓H$_2$SO$_4$ 5mL,等冷后,倾入1 000mL容量瓶中,用蒸馏水稀释至刻度(1.00mL溶液含Fe$^{3+}$ 0.1mg)。

表 8-5　吸收光谱测定

| $\lambda$ / nm | 450 | 460 | 470 | 480 | 490 | 500 | 510 | 520 |
|---|---|---|---|---|---|---|---|---|
| 吸光度 $A$ | | | | | | | | |
| $\lambda_{max}$ / nm | | | | | | | | |

吸收光谱(实验报告附图 1)：

(2) 绘制标准曲线：根据表 8-6 所记录的数据，以标准 $Fe^{3+}$ 离子的浓度($mg \cdot L^{-1}$)为横坐标，吸光度($A$)为纵坐标绘制标准曲线。

表 8-6　标准溶液和待测溶液的吸光度数据记录

| 容量瓶编号 | 1(空白) | 2 | 3 | 4 | 5 | 6 | 7(水样) |
|---|---|---|---|---|---|---|---|
| $\rho_{稀释}(Fe^{3+})$ / $mg \cdot L^{-1}$ | 0.00 | | | | | | |
| $A$ | 0.00 | | | | | | |

标准曲线(实验报告附图 2)

(3) 待测溶液中 $Fe^{3+}$ 离子浓度的确定：根据所测的待测溶液的吸光度，利用标准对照法和标准曲线法分别计算出原未知溶液 $Fe^{3+}$ 浓度($mg \cdot L^{-1}$)。

1) 标准曲线法

标准曲线上直接读出供试液 $Fe^{3+}$ 的含量为_____($mg \cdot L^{-1}$)；

原水样中 $Fe^{3+}$ 含量为_____($mg \cdot L^{-1}$)。

2) 标准对照法

所选取标准溶液的吸光度_____；

水样的吸光度_____；

原水样中 $Fe^{3+}$ 含量_____($mg \cdot L^{-1}$)。

## 【思考题】

在 $Fe^{3+}$ 的显色反应时，为什么要加过量的 KSCN 溶液？

## 3. 磺基水杨酸法

磺基水杨酸($H_2Ssal$)是 $Fe^{3+}$ 的显色剂。$Fe^{3+}$ 离子与磺基水杨酸作用能形成多种配合物，在不同的酸度下生成的配合物的组成和颜色也不同。如 pH 1.8～2.5 时生成紫红色$[FeSsal]^+$；pH 4～8 时生成棕褐色$[Fe(Ssal)_2]^-$；pH 8.0～11.5 时生成稳定的黄色$[Fe(Ssal)_3]^{3-}$，当 pH > 12 时 $Fe^{3+}$ 易水解生成 $Fe(OH)_3$ 沉淀，不能用于比色测定，故磺基水杨酸法应在 pH < 12 以下，且需在恒定不变的 pH 条件下进行。本实验采用在醋酸盐缓冲体系条件下进行测定，此时溶液的 pH 约为 5，$Fe^{3+}$ 与其生成稳定的 1∶2 橙色的$[Fe(Ssal)_2]^-$配合物，其显色反应如下：

$$Fe^{3+} + 2\,\underset{HOOC}{\overset{HO}{\bigcirc}}\!\!-\!SO_3^- \xrightleftharpoons{pH=5} \left[Fe\left[\underset{^-OOC}{\overset{HO}{\bigcirc}}\!\!-\!SO_3^-\right]_2\right]^- + 2H^+$$

在实验条件下，先在 420～490nm 波长范围内，绘制吸收光谱，选择 $\lambda_{max}$ 为测量用单色光，然后在此波长下用标准曲线法或标准对照法测出未知样品的铁含量。

## 【仪器材料与试剂】

**仪器** 722型可见分光光度计，容量瓶（25mL×7），吸量管（2mL×2，1mL），滴管

**材料与试剂** 标准 $NH_4Fe(SO_4)_2 \cdot 12H_2O$ 溶液[含 $Fe^{3+}$ 0.1000 $g \cdot L^{-1}$ 即 0.8634 $g \cdot L^{-1}$ 的 $NH_4Fe(SO_4)_2 \cdot 12H_2O$ 溶液]，100 $g \cdot L^{-1}$ 磺基水杨酸溶液，pH 5.0 的 HAc-NaAc 缓冲溶液，待测水样-铁盐溶液，坐标纸（2张）

## 【实验步骤】

### 3.1 标准溶液和待测溶液的配制

取 25mL 容量瓶 7 支，按表 8-7 所列的量，用吸量管量取各种溶液加入容量瓶中，再用 pH 5 的 HAc-NaAc 缓冲溶液稀释至刻度，摇匀，备用。

表 8-7 标准溶液和待测溶液的配制

| 容量瓶编号 | 1（空白） | 2 | 3 | 4 | 5 | 6 | 7（待测） |
|---|---|---|---|---|---|---|---|
| $Fe^{3+}$ 标准溶液体积 / mL | 0.00 | 0.40 | 0.60 | 0.80 | 1.00 | 1.20 | — |
| 待测溶液 / mL | — | — | — | — | — | — | 1.00 |
| 100 $g \cdot L^{-1}$ 磺基水杨酸溶液 / mL | 2.00 | 2.00 | 2.00 | 2.00 | 2.00 | 2.00 | 2.00 |
| $V_{总}$（稀释）/ mL | 25.00 | 25.00 | 25.00 | 25.00 | 25.00 | 25.00 | 25.00 |
| $\rho_{稀释}(Fe^{3+})$ / $mg \cdot L^{-1}$ | | | | | | | |

### 3.2 吸收光谱的测定

在系列标准溶液中，取表 8-7 中的 4 号溶液，按 722 型分光光度计的使用方法（详见常用测量仪器使用），用 1cm 比色皿，在 420～490nm 波长范围内，以试剂空白作为参比溶液，每隔 10nm 测定一次溶液的吸光度，在最大吸光度的波长附近可每隔 5nm 再测其吸光度，记录实验数据（表 8-8）。

### 3.3 吸光度（$A$）的测定及标准曲线的绘制

按 722 型分光光度计的使用方法，选择 $\lambda_{max}$（文献记载 $\lambda_{max}$ = 466nm），以试剂空白作参比溶液，分别测定各标准溶液的吸光度和待测溶液（供试液）的吸光度，并将数据记录在表 8-9 中。并以浓度为横坐标，吸光度为纵坐标，绘制标准曲线。

### 3.4 待测水溶液中 $Fe^{3+}$ 离子浓度的确定

根据测得的水溶液的吸光度，利用标准对照法和标准曲线法分别计算出水样中铁离子浓度（$mg \cdot L^{-1}$）。

## 【数据记录与结果分析】

日期：　　　　　温度：　　　℃　　　　相对湿度：

（1）绘制吸收光谱并确定 $\lambda_{max}$：据表 8-8 所记录的数据，以波长为横坐标，吸光度为纵坐标在坐标纸上绘制吸收光谱，找出 $\lambda_{max}$。

表 8-8 测定吸收光谱数据记录

| 波长 $\lambda$ / nm | 420 | 430 | 440 | 450 | 460 | 470 | 480 | 490 |
|---|---|---|---|---|---|---|---|---|
| 吸光度 $A$ | | | | | | | | |
| $\lambda_{max}$ / nm | | | | | | | | |

吸收光谱(实验报告附图1):

(2) 绘制标准曲线:根据表8-9所记录的数据,以各标准溶液的浓度($mg·L^{-1}$)为横坐标,相应的吸光度值为纵坐标,绘制出标准曲线,该曲线可用直线方程(亦称回归方程,$y = ax + b, r^2$)表示。

表 8-9  标准溶液和待测溶液的吸光度数据记录

| 容量瓶编号 | 1(空白) | 2 | 3 | 4 | 5 | 6 | 7(待测) |
|---|---|---|---|---|---|---|---|
| $Fe^{3+}$ 溶液的含量 / $mg·L^{-1}$ | 0.00 | | | | | | |
| 吸光度 A | 0.00 | | | | | | |

标准曲线法(实验报告附图2):

(3) 待测溶液中 $Fe^{3+}$ 离子浓度的确定

根据待测溶液的吸光度,利用比较法和标准曲线法分别计算出原未知溶液 $Fe^{3+}$ 浓度($mg·L^{-1}$)。

1) 标准曲线法

标准曲线上直接读出供试液 $Fe^{3+}$ 的含量为_____($mg·L^{-1}$);

原未知样品中 $Fe^{3+}$ 含量 = 从标准曲线查得供试液含量 × 25(倍) 为_____($mg·L^{-1}$)。

2) 标准比较法

所选取标准溶液的吸光度_____;

未知样品中 $Fe^{3+}$ 的吸光度_____;

原未知样品中 $Fe^{3+}$ 含量_____($mg·L^{-1}$)。

【思考题】

(1) 实验中为什么要用缓冲溶液定容?

(2) 本实验测定吸光度的空白溶液是什么?

(王金铃)

# 实验九  分光光度法测定阿司匹林药片的含量

【目的要求】

(1) 掌握可见分光光度法测定阿司匹林药片的方法。

(2) 掌握 722(721)型可见分光光度计的操作方法。

【预习作业】

(1) 查阅相关资料指出阿司匹林的结构和理化性质。

(2) 实验中阿司匹林乙醇溶液可否将溶剂更换为水?

(3) 实验中试剂的加样顺序可否颠倒,为什么?

(4) 影响实验准确度的因素有哪些?

(5) 参照本实验步骤,以流程图方式给出本实验的设计思路,请在流程图中实验步骤注明相应的注意事项及缘由。

(6) 通过文献查阅,指出还有何方法测定阿司匹林药片的含量?

## 【实验原理】

阿司匹林($CH_3COOC_6H_4COOH$)又名乙酰水杨酸,以水杨酸为原料进行乙酰化合成得到的白色晶体化合物,常以片剂的方式用于退烧、消炎和减轻疼痛。

阿司匹林药片中的主要成分为乙酰水杨酸,在碱性条件下,分子中的酯基可与羟胺反应生成羟肟酸,后者在酸性条件下与三氯化铁形成红色的羟肟酸铁,此物质的理论最大吸收波长为520nm,且在一定浓度范围内符合Lambert-Beer定律,可采用标准曲线法求出药片中阿司匹林的含量。

## 【仪器材料与试剂】

**仪器** 722(721)型分光光度计,容量瓶(25mL×6),吸量管(1mL×5,2mL)

**材料与试剂** 2.00mol·$L^{-1}$ NaOH,4.00mol·$L^{-1}$ HCl,10% $FeCl_3$,0.500g·$L^{-1}$ 乙酰水杨酸乙醇溶液,7% 盐酸羟胺乙醇溶液,阿司匹林样品溶液(阿司匹林肠溶片,配制成20片·$L^{-1}$ 乙醇溶液)

## 【实验步骤】

### 1. 乙酰水杨酸标准溶液和样品溶液的配制

依表9-1,分别取0.500g·$L^{-1}$ 乙酰水杨酸乙醇溶液0.00mL、0.50mL、1.00mL、1.50mL、2.00mL和阿司匹林样品乙醇溶液1.00mL置于25mL容量瓶中,再各加入7% 盐酸羟胺乙醇溶液1.00mL,2.00mol·$L^{-1}$ NaOH 1.00mL,放置3min后,加入4.00mol·$L^{-1}$ HCl溶液和10% $FeCl_3$ 溶液各1.00mL,加水至刻度摇匀,放置10min后,备用。

表9-1 乙酰水杨酸标准溶液和样品溶液的配制

| 实验序号 | 1(空白) | 2 | 3 | 4 | 5 | 6 | 7(样品) |
|---|---|---|---|---|---|---|---|
| 0.500g·$L^{-1}$ 乙酰水杨酸溶液 / mL | 0.00 | 0.50 | 1.00 | 1.50 | 2.00 | 2.50 | — |
| 阿司匹林样品溶液 / mL | — | — | — | — | — | — | 1.00 |
| 7% 盐酸羟胺乙醇溶液 / mL | 1.00 | 1.00 | 1.00 | 1.00 | 1.00 | 1.00 | 1.00 |
| 2.00mol·$L^{-1}$ NaOH / mL | 1.00 | 1.00 | 1.00 | 1.00 | 1.00 | 1.00 | 1.00 |
| 4.00mol·$L^{-1}$ HCl / mL | 1.00 | 1.00 | 1.00 | 1.00 | 1.00 | 1.00 | 1.00 |
| 10% $FeCl_3$ / mL | 1.00 | 1.00 | 1.00 | 1.00 | 1.00 | 1.00 | 1.00 |
| $V_{总}$(稀释) / mL | 25.00 | 25.00 | 25.00 | 25.00 | 25.00 | 25.00 | 25.00 |
| $c_{标准}$(稀释) / mg·$L^{-1}$ | | | | | | | |

### 2. 吸光度($A$)的测定及标准曲线的绘制

按722(或721)型分光分度计的使用方法(见常用测量仪器使用)选择波长$\lambda_{max}$ = 520nm和相应的灵敏度档,以空白溶液作为参比溶液,分别测定各标准溶液的吸光度和待测样品溶液的吸光度。记录于表9-2中,并以浓度为横坐标,吸光度为纵坐标,绘制标准曲线,给出回归方程及相关系数。

## 【数据记录与结果分析】

日期：　　　　　　温度：　　　　℃　　　　相对湿度：

（1）乙酰水杨酸标准溶液和样品溶液的吸光度（$A$）

表9-2　乙酰水杨酸标准溶液和样品溶液的吸光度

| 实验序号 | 1（空白） | 2 | 3 | 4 | 5 | 6 | 7（样品） |
|---|---|---|---|---|---|---|---|
| 吸光度 $A$ | | | | | | | |
| $c_{标准}$（稀释）/mg·L$^{-1}$ | | | | | | | |

（2）乙酰水杨酸标准曲线的绘制：以吸光度为纵坐标，乙酰水杨酸标准溶液浓度为横坐标作图，绘制出标准曲线，并给出其线性回归方程及相关系数。

（3）待测药片中阿司匹林含量的确定：从标准曲线找出阿司匹林样品溶液的浓度为 $\rho_1 =$ ＿＿＿＿＿＿mg·L$^{-1}$，原阿司匹林样品溶液的浓度为 $\rho_2 = \rho_1 \times 25$（稀释倍数）= ＿＿＿＿＿＿mg·L$^{-1}$，换算出药片中阿司匹林的含量 = ＿＿＿＿＿＿mg·片$^{-1}$。

## 【思考题】

（1）整个测定过程酸度控制是否一致，为什么？
（2）水杨酸与三氯化铁形成的紫色可干扰测定，如何消除？

（尹计秋）

# 实验十　分光光度法测定磺基水杨酸合铁的组成和稳定常数

## 【目的要求】

（1）巩固对可见分光光度法基本原理的理解。
（2）掌握分光光度计的使用方法。
（3）了解分光光度法测定溶液中配合物的组成和稳定常数的原理和方法。

## 【预习作业】

（1）分光光度法测定物质浓度时显色剂是如何选择的？
（2）本实验是如何控制磺基水杨酸合铁配合物组成的？
（3）本实验为什么选择高氯酸控制溶液的酸度？
（4）如何选择测定波长？如何选择合适的参比溶液？本实验的参比溶液是什么？
（5）通过文献查阅，总结测定配合物的组成和稳定常数的方法有哪些？本实验选择哪种方法？
（6）用等摩尔系列法测定配合物的组成时，为什么说溶液的组成与配合物的组成一致时，配合物的浓度最大？
（7）请用流程图阐释本实验是如何设计实验步骤，以确定测定波长和测定配合物的组成及稳定常数？请在流程图中标出每个实验步骤注意事项及依据？

## 【实验原理】

金属离子 M 和配位体 L 形成配位化合物的反应为

$$M + nL \rightleftharpoons ML_n \text{（忽略离子的电荷）}$$

$$K_s = \frac{[ML_n]}{[M][L]^n} \tag{10-1}$$

式(10-1)中的 $n$ 为配合物的配位数，$K_s$ 为配合物的稳定常数。

如果 M 和 L 都是无色的，而 $ML_n$ 有色，根据朗伯-比尔定律 $A = \varepsilon bc$，则此溶液的吸光度与配合物浓度成正比。通过测定溶液的吸光度，可求出该配合物的组成和稳定常数，本实验采用等摩尔系列法进行测定。

所谓等摩尔系列法又称为连续变化法、浓度比递变法，是将相同摩尔浓度的金属离子溶液和配体溶液，在保持金属离子与配位体的摩尔数之和不变的前提下，按照不同的体积比(亦即摩尔浓度之比)混合，使其总体积相等，以配合物最大吸收波长为入射光，测量系列混合溶液的吸光度。当溶液中配合物的浓度最大时，配位数 $n$ 为

$$n = \frac{c(L)}{c(M)} = \frac{1-f}{f} \tag{10-2}$$

式(10-2)中 $c(M)$ 和 $c(L)$ 分别为金属离子和配体的浓度；$f$ 为金属离子在总浓度中所占分数。且有

$$c(M) + c(L) = c = \text{常数} \tag{10-3}$$

$$f = \frac{c(M)}{c} \tag{10-4}$$

以吸光度 $A$ 对 $f$ 作图(图 10-1)。当 $f=0$ 或 $f=1$ 时，没有加入金属离子或没有加入配体，没有配合物生成，即配合物的浓度为零。当溶液中金属离子与配体的摩尔比与配合物组成一致时，配合物浓度才能最大。图中吸光度值最大处的 $f$ 值，即为配合物浓度达最大时的 $f$ 值。例如：1:1 型配合物，吸光度值最大处的 $f$ 值为 0.5，1:2 型的 $f$ 值为 0.34 等。

若配合物为 ML，从图 10-1 可知，测得的最大吸光度为 $A$，它略低于延长线交点 B 的吸光度 $A'$，这是因为配合物有一定程度的离解，$A'$ 为配合物完全不离解时的吸光度值，$A'$ 与 $A$ 之间差别愈小，说明配合物愈稳定。由此可计算出配合物的稳定常数

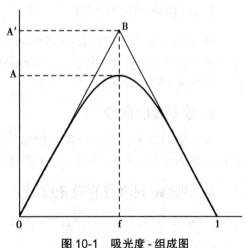

图 10-1 吸光度-组成图

$$K_s = \frac{[ML]}{[M][L]} \tag{10-5}$$

因配合物溶液的吸光度与配合物的浓度成正比，故

$$\frac{A}{A'} = \frac{[ML]}{c'} \tag{10-6}$$

式(10-6)中 $c'$ 为配合物未离解时的浓度

$$c' = c(M) = c(L) \tag{10-7}$$

而

$$[M] = [L] = c' - [ML] = c' - c'\frac{A}{A'} = c'\left[1 - \frac{A}{A'}\right] \tag{10-8}$$

将式(10-6)和式(10-8)代入式(10-5)，整理后得配合物的稳定常数的计算式(10-9)

$$K_s = \frac{\dfrac{A}{A'}}{\left[1 - \dfrac{A}{A'}\right]^2 c'} \tag{10-9}$$

$Fe^{3+}$ 与磺基水杨酸($H_2Ssal$)形成配合物的组成因溶液 pH 不同而不同，在 pH 2～3 时，生成含一个配体的紫红色配合物；pH=4～9 时，生成含有两个配体的红色配合物；pH 9～11.5 时，生成含有三个配体的黄色配合物；pH > 12 时，有色配合物被破坏而生成 $Fe(OH)_3$ 沉淀。由于溶液中磺基水杨酸是无色的，$Fe^{3+}$ 溶液的浓度很稀，也可以认为是无色的，只有磺基水杨酸合铁配离子是有色的。因此溶液的吸光度只与配离子的浓度成正比。对溶液吸光度的测定，可以求出该配离子的组成。本实验是测定 $Fe^{3+}$ 与磺基水杨酸在 pH 2～3 条件下形成配合物的组成和稳定常数，实验中通过加入一定量的 $HClO_4$ 溶液来控制溶液的 pH，其主要优点是 $ClO_4^-$ 不易与金属离子配合，同时用于预防 $Fe^{3+}$ 水解及还原为 $Fe^{2+}$。

## 【仪器材料和试剂】

**仪器**　722 或 721 型分光光度计，容量瓶(50mL×7)，吸量管(10mL×2)，滴管 1 只

**材料与试剂**　0.010 00 mol·$L^{-1}$ 磺基水杨酸溶液，0.010 00 mol·$L^{-1}$ 硫酸铁铵溶液，0.1 mol·$L^{-1}$ $HClO_4$ 溶液

## 【实验步骤】

### 1. 配制系列标准溶液

按表 10-1，用吸量管吸取 0.010 0 mol·$L^{-1}$ 的磺基水杨酸溶液和 0.010 0 mol·$L^{-1}$ 硫酸铁铵溶液分别置于 7 只 50.00mL 容量瓶中，加 0.1 mol·$L^{-1}$ $HClO_4$ 稀释至刻度，摇匀，即得不同浓度的磺基水杨酸合铁溶液。放置 10min。

### 2. 绘制吸收曲线

取 4 号溶液，以蒸馏水为参比溶液，在波长 400～700nm 范围，每隔 20nm 测一次吸光度，峰值附近每隔 5nm 测量一次，数值记录于表 10-1，绘制吸收曲线，找出最大吸收波长，$\lambda_{max}$。

### 3. 测定系列溶液的吸光度

以该配合物最大吸收波长 $\lambda_{max}$ 为入射光，测定步骤 1 配制的系列溶液的吸光度，记录结果于表 10-2。

## 【数据记录与结果分析】

日期：　　　　　　温度：　　　　℃　　　　相对湿度：

(1) 绘制吸收光谱确定测定波长

表 10-1　磺基水杨酸合铁溶液吸收光谱测定

| $\lambda$ / nm | |
|---|---|
| $A$ | |
| $\lambda_{max}$ / nm | |

吸收光谱：

$\lambda_{max}$ = _____ nm。

(2) 配合物组成：以金属离子物质的量浓度与总物质的量浓度之比为横坐标，吸光度为纵坐标作 $A-f$ 图，确定溶液 pH 2～3 时配合物组成，并给出分子式。

表 10-2　磺基水杨酸合铁溶液配制及吸光度 $A$ 测定

| 实验序号 | 1 | 2 | 3 | 4 | 5 | 6 | 7 |
|---|---|---|---|---|---|---|---|
| 0.010 0 mol·L$^{-1}$ H$_2$Ssal 溶液 / mL | 1.00 | 2.00 | 3.00 | 5.00 | 7.00 | 8.00 | 9.00 |
| 0.010 0 mol·L$^{-1}$ NH$_4$Fe(SO$_4$)$_2$ 溶液 / mL | 9.00 | 8.00 | 7.00 | 5.00 | 3.00 | 2.00 | 1.00 |
| 0.1 mol·L$^{-1}$ HClO$_4$ 溶液 / mL | | | | 40 | | | |
| $A$ | | | | | | | |
| $f = \dfrac{c(M)}{c}$ | | | | | | | |

(3) 磺基水杨酸合铁的稳定常数：从 $A-f$ 图中，查得最大吸光度 $A$，延长曲线两边的直线部分，相交于一点，这一点即为配合物未解离时的吸光度 $A'$，求得 $c'$，代入式（10-7）求出磺基水杨酸合铁的稳定常数。

【思考题】

(1) 摩尔系列法测定配合物的稳定常数有什么条件？
(2) 酸度对磺基水杨酸合铁配合物的组成有什么影响？

（王美玲）

# 实验十一　荧光分析法测定维生素 B$_2$ 的含量

【实验目的】

(1) 掌握荧光分析法测定物质含量的原理和方法。
(2) 了解荧光光度计的基本构造。
(3) 学会 930 型荧光光度计的使用。

【预习作业】

(1) 什么是激发光谱？什么是发射光谱？
(2) 在测定过程中，设定的激发光波长为什么比发射光波长短？能否相等？
(3) 物质分子产生荧光的条件是什么？
(4) 在对多种维生素混合物的测量中，能否使用吸收光谱法测定维生素 B$_2$ 的含量？
(5) 请依据荧光分析法原理，参照本实验步骤，用流程图方式阐述本实验的设计思路，并在相应步骤注明操作注意事项及缘由。

【实验原理】

某些物质受紫外光或可见光照射激发后，能发射出比激发光频率低（波长较长）的光，称为荧光。利用荧光现象进行定性、定量分析的方法称为荧光分析法。

荧光物质不同,其特征激发波长和荧光波长不同,这是荧光定性分析的基础。

当荧光物质的浓度极低时,荧光强度 $F$ 与溶液质量浓度 $\rho$ 有如下关系

$$F = 2.3\Phi I_0 ab\rho \tag{11-1}$$

式(11-1)中 $\Phi$ 为荧光效率,$I_0$ 为入射光强度,$a$ 为荧光物质的质量吸光系数,$b$ 为样品池厚度。

对于同一荧光物质,当 $I_0$ 及 $b$ 固定时,荧光强度与该荧光物质的质量浓度成正比,这就是荧光定量分析的基础。

$$F = K'\rho \tag{11-2}$$

维生素 $B_2$ [$C_{17}H_{20}N_4O_6$ ($M_r = 376.37$)] 又称核黄素,是橙黄色结晶性粉末,其结构式如下:

维生素 $B_2$ 的 $0.1 mol \cdot L^{-1}$ HAc 溶液在紫外光照射下,发出黄绿色荧光,可直接进行荧光测定。激发波长可选择 360nm、400nm 或 420nm,发射波长为 530nm。

### 【仪器材料与试剂】

**仪器**　930 型-荧光分光光度计,万分之一分析天平,容量瓶($25mL \times 6$, $50mL$, $1000mL$),吸量管($1mL$, $5mL$),小烧杯,玻璃棒

**材料与试剂**　$0.1 mol \cdot L^{-1}$ HAc 溶液,维生素 $B_2$ 结晶(生化试剂),维生素 $B_2$ 片剂(待测)

### 【实验步骤】

#### 1. 维生素 $B_2$ 标准溶液的配制

准确称取维生素 $B_2$ 约 10.0mg 置于小烧杯中,用少量 $0.1 mol \cdot L^{-1}$ HAc 溶液溶解,转移至 1000mL 容量瓶,用 $0.1 mol \cdot L^{-1}$ HAc 溶液稀释至刻度,摇匀,得到 $10.0 \mu g \cdot mL^{-1}$ 维生素 $B_2$ 的贮备标准溶液,低温、避光保存。

取 6 个 25mL 容量瓶,分别加入 0.00mL,0.50mL,1.00mL,1.50mL,2.00mL,2.50mL 维生素 $B_2$ 的贮备标准溶液,用 $0.1 mol \cdot L^{-1}$ HAc 溶液稀释至刻度,摇匀,得系列维生素 $B_2$ 标准溶液。

#### 2. 维生素 $B_2$ 标准曲线的绘制

按照荧光光度计的使用方法,接通电源,预热 10min。选择 360nm 或 400nm 为激发波长,530nm 为发射波长。用 $0.1 mol \cdot L^{-1}$ HAc 溶液作参比溶液对仪器进行校正,将读数调至零;用浓度最大的标准溶液,调节荧光读数为满刻度。固定条件不变,按由稀至浓的顺序,测定系列标准溶液的荧光强度。根据所记录数据,以溶液浓度为横坐标,荧光强度为纵坐标,绘制标准曲线。

#### 3. 试样测定

取待测维生素 $B_2$ 片剂一片,准确称量(精确至 0.0001g)其质量后用研钵研成粉末,用 $0.1 mol \cdot L^{-1}$ HAc 溶液溶解,定容至 1000.00mL,贮存于棕色试剂瓶中。

取待测溶液 2.50mL 置于 50mL 容量瓶中,用 0.1mol·L$^{-1}$ HAc 溶液稀释至刻度,摇匀。在与测定标准溶液相同条件下,测定待测样品的荧光强度。从标准曲线上查出对应的质量浓度 $\rho_{测}$,得到待测溶液维生素 B$_2$ 的质量浓度 $\rho_{样品}$。由测得溶液浓度,计算片剂中维生素 B$_2$ 的含量。

## 【数据记录及结果分析】

(1) 维生素 B$_2$ 标准曲线的绘制

日期:　　　　　温度:　　　　℃　　　　相对湿度:

表 11-1　维生素 B$_2$ 测定标准曲线的绘制和试样测定

| 实验序号 | 0 | 1 | 2 | 3 | 4 | 5 | 试样 |
|---|---|---|---|---|---|---|---|
| $m$(维生素 B$_2$)/ mg | | | | | | | |
| $\rho_{贮备}$(维生素 B$_2$)/ μg·mL$^{-1}$ | | | | | | | |
| $V_{标}$(维生素 B$_2$)/ mL | 0 | 0.50 | 1.00 | 1.50 | 2.00 | 2.50 | |
| $V_{总}$(0.1mol·L$^{-1}$HAc 稀释后)/ mL | | | 25.00 | | | | 50.00 |
| $\rho_{标}$(维生素 B$_2$)/ μg·mL$^{-1}$ | 0 | 0.20 | 0.40 | 0.60 | 0.80 | 1.00 | |
| $F$ | | | | | | | |
| $m$(待测片剂)/ mg | | | | | | | |

绘制标准曲线:根据表 11-1 记录的数据,以标准维生素 B$_2$ 的质量浓度(μg·mL$^{-1}$)为横坐标,荧光强度($F$)为纵坐标绘制标准曲线。

标准曲线的回归方程为_____,相关系数为_____;

标准曲线上直接读出维生素 B$_2$ 的质量浓度 $\rho_{测}$ 为_____(μg·mL$^{-1}$)。

(2) 试样维生素 B$_2$ 片剂浓度

$$\omega(维生素 B_2)\% = \frac{\rho_{测}(\mu g \cdot mL^{-1}) \times 20(稀释倍数) \times 1\,000mL \times 10^{-3}}{m(待测片剂)} (mg \cdot mg^{-1}) \quad (11-3)$$

## 【思考题】

(1) 为何用 0.1mol·L$^{-1}$ HAc 溶液来配制维生素 B$_2$ 溶液?

(2) 若选择 420nm、440nm 作为激发波长时,对测定结果有无影响?

(周昊霏)

# 实验十二　紫外分光光度法对维生素 B$_{12}$ 的鉴别与含量测定

## 【实验目的】

(1) 掌握紫外分光光度计的使用方法。

(2) 掌握紫外分光光度法定性鉴别维生素 B$_{12}$ 的方法。

## 【预习作业】

(1) 给出维生素 B$_{12}$ 的化学结构,其在体内的主要功能有哪些?

(2) 简要叙述紫外分光光度法的基本原理。

(3) 查阅相关文献，分析哪些因素影响用紫外分光光度法测定维生素 $B_{12}$ 结果的准确性？
(4) 查阅 2015 版药典，指出维生素 $B_{12}$ 的鉴别、检测方法，及紫外分光光度计的要求。
(5) 参照实验步骤，请用流程图描述本实验的设计思路及实验步骤中的注意事项及依据。

## 【实验原理】

紫外分光光度法是通过测定在紫外光区（200～400nm）有特定吸收波长或一定波长范围的被测物质的吸光度，对该物质进行定性和定量分析的方法，且吸收强度与被测物质浓度的关系符合 Lambert-Beer 定律。

维生素 $B_{12}$ 是钴配位化合物（配体为卟啉环，分子式：$C_{63}H_{88}CoN_{14}O_{14}P$，$M_r$ = 1 355.38），为深红色吸湿性结晶，其注射液标示量为 500μg•mL$^{-1}$、250μg•mL$^{-1}$、100μg•mL$^{-1}$、50μg•mL$^{-1}$ 等。维生素 $B_{12}$ 的注射液在 278nm、361nm、550nm 三个波长处有最大吸收，这三个波长的吸光度的比值，可作为维生素 $B_{12}$ 定性鉴别的依据。其比值范围分别为：

$$A_{361nm} / A_{278nm} = 1.70\sim1.88 \qquad A_{361nm} / A_{550nm} = 3.15\sim3.45$$

其在 361nm 波长处的吸收强度最大、干扰较少。测定 361nm 波长处的吸光度，根据维生素 $B_{12}$ 的吸收系数（$E_{1cm}^{1\%}$ 为 207），可计算出维生素 $B_{12}$ 的浓度

$$A_{361nm} = E_{1cm}^{1\%} \cdot b \cdot c \tag{12-1}$$

式（12-1）中，$E_{1cm}^{1\%}$ 指在一定波长时，溶液浓度为 1%（g/100mL），液层厚度为 1cm 时的吸光度数值，$A$ 为溶液的吸光度，$b$ 为液层厚度，$c$ 为溶液浓度。

测定维生素 $B_{12}$ 注射液的含量，浓度单位应表示为 μg•mL$^{-1}$。在计算时，需将浓度单位为 1g/100mL 的吸收系数 $E_{1cm}^{1\%}$ 361 nm = 207 换算成浓度单位为 μg•mL$^{-1}$ 的吸收系数，即

$$E_{1cm}^{1\mu g \cdot L^{-1}} 361nm = \frac{207 \times 100}{10^6} = 207 \times 10^{-4} \tag{12-2}$$

## 【仪器材料与试剂】

**仪器**　754 型-紫外可见分光光度计，容量瓶（10mL），吸量管（1mL）
**材料与试剂**　维生素 $B_{12}$ 注射液（500μg•mL$^{-1}$）

## 【实验步骤】

### 1. 样品溶液的配制

取 3 支维生素 $B_{12}$ 注射液样品，用吸量管各吸取 0.50mL，分别置于 3 个 10mL 容量瓶中，加蒸馏水稀释至刻度，摇匀，配成样品溶液。

### 2. 维生素 $B_{12}$ 定性鉴别及注射液样品含量测定

用 1cm 石英比色皿，以蒸馏水作空白，在仪器上找出样品溶液在 278nm、361nm 及 550nm 处的吸收峰，读取吸光度值。每个样品平行测定 3 次，取其平均值。

## 【注意事项】

(1) 样品溶液必须澄清，不得有浑浊。
(2) 通常样品溶液的吸光度读数，在 0.3～0.7 之间的误差较小。

(3) 在测定不同样品时,应用待测溶液冲洗吸收池 3~4 次,用擦镜纸擦至吸收池的透光面无斑痕(切忌把透光面磨损)。

(4) 测定时不要打开仪器的样品池盖。

## 【数据记录及结果分析】

(1) 维生素 $B_{12}$ 的定性鉴别:计算三个不同波长处的吸光度比值($A_{361nm}/A_{278nm}$, $A_{361nm}/A_{550nm}$),并与药典规定($A_{361nm}/A_{278nm}=1.70\sim1.88$; $A_{361nm}/A_{550nm}=3.15\sim3.45$)进行对照。

(2) 维生素 $B_{12}$ 注射液样品含量:根据 361nm 处测定的吸光度,浓度单位为 $\mu g \cdot mL^{-1}$ 的吸收系数,及注射液稀释倍数,计算注射液待测样品中维生素 $B_{12}$ 的含量($\mu g \cdot mL^{-1}$)。

$$\rho_{样品}(B_{12}) = \frac{A}{E_{1cm}^{1\mu g \cdot L^{-1}} 361nm} = \frac{A}{207 \times 10^{-4}} = A \times 48.31 (\mu g \cdot L^{-1}) \tag{12-3}$$

则原维生素 $B_{12}$ 注射液:$\rho_{原} = A \times 48.31 \times 20$(稀释倍数)($\mu g \cdot mL^{-1}$)

测定值在所用维生素 $B_{12}$ 注射液浓度(材料和试剂中给定的值)标定值的 90%~110% 之间即可视为合格。

## 【思考题】

(1) 测定前应先在仪器上找出三个最大吸收峰的确切位置,意义何在?

(2) 如果取注射液 2mL 用水稀释 15 倍,在 361nm 处测得 $A$ 值为 0.698,试计算注射液每 mL 含维生素 $B_{12}$ 多少 $\mu g$?

(3) 采用吸光系数法直接测定样品含量有何要求?

(武世奎)

# 实验课题四
# 化学原理实验

## 实验十三 稀溶液的依数性及其应用

### 【实验目的】

(1) 掌握用凝固点降低法测定溶质相对分子质量的原理和方法。
(2) 学会使用冰点渗透压计并能准确测定溶液的渗透浓度及渗透压。
(3) 学会使用普通显微镜并了解细胞在不同渗透浓度溶液中的形态。
(4) 学会使用 0.10℃ 分度温度计及巩固分析天平与移液管的使用方法。

### 【预习作业】

(1) 常用测定物质相对分子质量的方法有哪些？实验中常用什么方法？为什么？
(2) 公式(13-5)中 $c_{os} \approx b_B$ 在什么条件下成立？
(3) 为何纯溶剂和稀溶液的冷却曲线不同？理想状态下冷却曲线有哪些特征？
(4) 凝固点降低实验中为什么要在冰水浴中加入较多 NaCl？
(5) 在渗透压测定实验中，溶液为什么会出现过冷现象？实验中应如何解决？
(6) 请用流程图示意完成稀溶液三个依数性实验的主要实验步骤，并注释各步骤的注意事项及目的。
(7) 配制葡萄糖溶液时，尽管葡萄糖溶解较慢，但不能用玻璃棒搅拌溶液，为什么？

### 【实验原理】

难挥发的溶质溶于溶剂形成稀溶液时，溶液的物理性质，如凝固点、渗透压等与纯溶剂不同，其性质的改变与所加入溶质的量成正比，而与溶质的本性无关，这些性质统称为稀溶液的依数性。利用稀溶液的依数性能准确地测定溶质的分子质量。

溶液的凝固点($T_f$)低于纯溶剂的凝固点($T_f^0$)，若非电解质稀溶液的凝固点降低值为 $\Delta T_f$，则由

$$\Delta T_f = T_f^0 - T_f = K_f b_B \tag{13-1}$$

得

$$b_B = \frac{m_B/M(B)}{m_A} \times 1\,000 \, (\text{mol} \cdot \text{kg}^{-1}) \tag{13-2}$$

式(13-2)中，$b_B$ 为溶液的质量摩尔浓度(mol·kg$^{-1}$)，$m_B$ 为溶质的质量(g)，$M(B)$(g·mol$^{-1}$)为溶质的分子质量，$m_A$ 为溶剂质量(g)，$K_f$ 为溶剂的质量摩尔凝固点降低常数(K·kg·mol$^{-1}$，可从化学手

册中查出),故

$$M_r(B) = \frac{K_f m_B}{m_A \Delta T_f} \times 1000 \tag{13-3}$$

式(13-3)中各项的数值均可由实验测定,进而求得溶质的相对分子质量 $M(B)$。并按式(13-4)计算测定的相对误差($E_r$)。

$$E_r(\%) = \frac{实验值 - 理论值}{理论值} \times 100\% \tag{13-4}$$

对比图 13-1 溶剂(a)与稀溶液(b)的冷却曲线可以发现,不同之处在于稀溶液逐渐析出冰的过程将导致溶液浓度逐渐增加,稀溶液的温度没有恒定阶段。由于析出冰时释放了凝固热,溶液温度会迅速回升,并使溶液温度保持在短时间内的相对恒定,之后继续降低。因此,可将回升后的最高温度看作是稀溶液的凝固点 $T_f$。

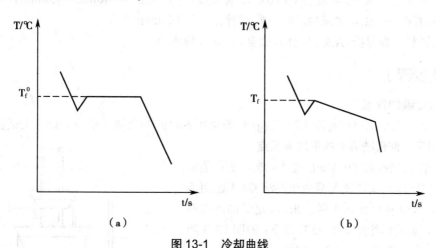

**图 13-1 冷却曲线**
(a)纯溶剂的冷却曲线;(b)稀溶液的冷却曲线

冰点渗透压计是一种利用溶液凝固点降低的性质来测定溶液渗透压和渗透浓度的装置。对非电解质的稀溶液有下列关系

$$c_{os} \approx b_B = \frac{\Delta T_f}{k_f} (mol \cdot L^{-1}) \tag{13-5}$$

$$b_B = \frac{m_B / M(B)}{m_A} \times 1000 (mol \cdot kg^{-1}) \tag{13-6}$$

式(13-5)中,$c_{os}$ 为溶液的渗透浓度($mol \cdot L^{-1}$),通常以水为溶剂且稀溶液的密度约等于 $1g \cdot cm^{-3}$,故

$$M(B) = \frac{m_B}{m_A c_{os}} \times 1000 \tag{13-7}$$

将一定质量的非电解质溶解在一定质量的溶剂中,通过冰点渗透压计测量出该溶液的渗透浓度,根据式(13-7)就可计算出该溶质的相对分子量 $M_r(B)$。

对稀溶液,其渗透压与渗透浓度及温度的关系如下

$$\Pi = c_{os}RT(kPa) \tag{13-8}$$

$\Pi$ 为溶液的渗透压(kPa),$R$ 为理想气体常数($8.134 J \cdot mol^{-1} \cdot k^{-1}$),$T$ 为溶液的绝对温度(K)。

在相同温度下,渗透浓度越大,渗透压就越大。因此,可以直接通过渗透浓度的大小来比较两溶液渗透压的高低。渗透浓度大的溶液称为高渗溶液,渗透浓度小的称为低渗溶液。若用半透膜

将两种渗透浓度不同的溶液分隔开,将发生渗透现象,溶剂从低渗溶液经过半透膜进入高渗溶液。

渗透浓度的测定在临床医学和其他领域都有广泛应用。临床上是以正常人血浆的渗透浓度 (280～320mmol·L$^{-1}$)为标准来确定溶液渗透浓度(或渗透压)的相对高低。若溶液的渗透浓度低于 280mmol·L$^{-1}$ 称为低渗溶液,高 320mmol·L$^{-1}$ 称为高渗溶液。红细胞在临床等渗溶液中形态正常;在高渗或低渗溶液中,形态都会发生变化。

## 1. 凝固点降低法测定溶质的分子量

【仪器材料与试剂】

**仪器**　万分之一天平,温度计(0.10℃),放大镜,干燥大试管(40mm×150mm),套管,烧杯(500mL),移液管(50mL),玻璃棒,洗耳球,搅拌器,量筒(500mL)

**材料与试剂**　葡萄糖(A.R.,固体),食盐,冰块及纯水

【实验步骤】

**1.1　葡萄糖的称取**

在万分之一天平上准确称取 4.2～5.1g(精确至 0.0001g)葡萄糖于洁净干燥的大试管中。

**1.2　测定稀葡萄糖溶液的平均凝固点**

用移液管准确吸取 50.00mL 纯水,放入盛有葡萄糖的大试管中,摇动试管使葡萄糖全部溶解(不能用玻璃棒搅拌)。大烧杯中加入少量自来水、足量的冰块和食盐,混匀,使冰水浴温度在 −5℃ 以下;如图 13-2 所示,将大试管插入套管内,放在大烧杯中。

用搅拌器搅动大试管内的葡萄糖溶液使其慢慢冷却,同时借助放大镜密切注意观察温度计的读数,适时补充适量冰块和食盐,并移去多余的融化冰水直至被测溶液成为过冷溶液。当稀葡萄糖溶液的温度足够低时,溶剂水首先凝结析出冰,溶液的凝固点也进一步降低,当大试管中有冰析出时,停止搅拌。

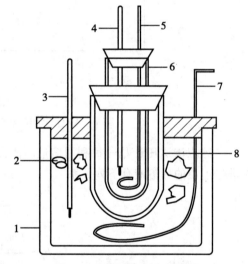

**图 13-2　凝固点测定装置图**
1. 烧杯;2. 冰块;3,4. 温度计;5,7. 搅拌器;6. 大试管;8. 外套管

仔细观察温度的变化,并多次记录回升后的温度,找出相对恒定的最高值(精确到 0.01℃),取平均值。

取出大试管,流水冲洗试管外表面,直至试管内冰屑完全融化,将大试管插入套管内,再重复如上操作,平行测定三次。任意两个测定值之差不得超过 0.05℃,均值即为葡萄糖溶液的凝固点 $T_f$。

**1.3　测定溶剂的凝固点**

将同一大试管洗净,再用纯水洗涤 3 次后,取约 50mL 纯水,放入大试管中,同步骤(2)方法,测定纯水的平均凝固点 $T_f^0$。

【注意事项】

实验中使用大量冰块且温度较低,在搅拌的过程中注意保护玻璃套管、试管和温度计,以免碎裂。

## 【数据记录与结果分析】

将水的摩尔凝固点降低常数 $K_f$(1.86K·kg·mol$^{-1}$)及各测定值代入式(13-3),计算所测定的葡萄糖分子量。按式(13-4)计算测定的相对误差。葡萄糖的分子量的理论值为180。将数据和结果记录在表 13-1 中。

日期：　　　　　　　温度：　　　　℃　　　　　相对湿度：

**表 13-1　凝固点降低法测定葡萄糖的分子量**

| 实验序号 | 1 | 2 | 3 |
|---|---|---|---|
| $m_A$ / g | | | |
| $m_B$ / g | | | |
| $T_f^0$ /℃ | | | |
| $T_f$ /℃ | | | |
| $\Delta T_f$ /℃ | | | |
| $M(B)$/g·mol$^{-1}$ | | | |
| $E_r$ / % | | | |

## 【思考题】

(1) 血清、尿液等生物样品的凝固点能用本实验方法测定吗？凝固点降低法适合测定哪类物质的相对分子量？

(2) 能否用沸点升高法测定葡萄糖相对分子质量？

## 2. 利用溶液的渗透浓度测定溶质的相对分子质量

### 【仪器材料与试剂】

**仪器**　万分之一天平,冰点渗透压计,渗透压测定专用试管(2mL),吸量管(1mL),移液管(50mL),洗耳球,干燥小烧杯(100mL),容量瓶(100mL),玻璃棒。

**材料与试剂**　葡萄糖(A.R.,固体),渗透浓度 $c_{os}$ 为 0.300mol·kg$^{-1}$ 和 0.800mol·kg$^{-1}$ NaCl 标准溶液

### 【实验步骤】

**2.1　葡萄糖溶液的制备**

在万分之一天平上准确称量 9.1~9.5g(精确至 0.0001g)葡萄糖于 100mL 洁净干燥的小烧杯中,用移液管准确吸取 50.00mL 纯水,放入盛有葡萄糖的小烧杯中,使葡萄糖全部溶解(不用玻璃棒搅拌)。

**2.2　冰点渗透压计的校正**

实验前,仔细阅读冰点渗透压计的使用说明。按照冰点渗透压计的使用方法开启冰点渗透压计,使仪器中的冷却池达到适当的冷却温度(-10~-8℃)。

用吸量管取 0.50mL 渗透浓度为 0.300mol·kg$^{-1}$ 的 NaCl 标准溶液于干燥的渗透压测定专用的 2mL 小试管中,在冰点渗透压计上测定其渗透浓度,反复调整测量值稳定在 0.300mol·kg$^{-1}$ 处。再用吸量管取 0.50mL 渗透浓度为 0.800mol·kg$^{-1}$ 的 NaCl 标准溶液于另一个干燥的专用小试管中,

在冰点渗透压计上测定其渗透浓度,反复调整其值稳定在 0.800mol·kg⁻¹ 处。

#### 2.3 测定葡萄糖溶液的渗透浓度(渗透压)

用吸量管取 0.50mL 已配制好的葡萄糖溶液于干燥的专用小试管中,在冰点渗透压计上测定其渗透浓度,重复测定三次。取平均值作为该葡萄糖溶液的渗透浓度。

## 【数据记录与结果分析】

按式(13-7)计算葡萄糖的相对分子质量,并按式(13-4)计算测定的相对误差。将数据和结果记录在表 13-2 中。

日期:　　　　　　温度:　　　　℃　　　　相对湿度:

表 13-2　利用溶液的渗透浓度测定葡萄糖的分子量

| 实验序号 | 1 | 2 | 3 |
|---|---|---|---|
| $m_A$ / g | | | |
| $m_B$ / g | | | |
| $c_{os}$ / mol·L⁻¹ | | | |
| $M(B)$ / g·mol⁻¹ | | | |
| $E_r$ / % | | | |

## 【思考题】

(1) 如何在室温下用冰点渗透压计测定出的溶液渗透浓度来确定溶液的渗透压?
(2) 用冰点渗透压计测定溶液的渗透浓度时应注意哪些问题?

（傅　迎）

# 3. 冰点渗透压计的应用

## 【仪器材料与试剂】

**仪器**　普通显微镜,载玻片×5,冰点渗透压计,渗透压测定专用小试管(2mL),试管(20mL),吸量管(1.00mL),洗耳球,干燥小烧杯(100mL),胶头滴管×5,玻棒

**材料与试剂**　NaCl 标准溶液($c_{os}$ = 0.300mol·L⁻¹ 和 0.800mol·L⁻¹),NaCl 高渗溶液($c_{os}$ = 0.35mol·L⁻¹),NaCl 低渗溶液($c_{os}$ = 0.05mol·L⁻¹),等渗生理盐水,54g·L⁻¹ 葡萄糖溶液,聚乙烯醇滴眼液,动物血液,抗凝剂

## 【实验步骤】

#### 3.1 冰点渗透压计的校正方法同前

#### 3.2 溶液渗透压的测定

用吸量管移取 0.50mL 高渗溶液(0.35mol·L⁻¹ NaCl)于干燥的渗透压测定专用小试管中,在冰点渗透压计上测定渗透浓度,重复测定三次,取平均值作为该溶液的渗透浓度。分别用 0.50mL 低渗溶液(0.05mol·L⁻¹ NaCl)、等渗溶液(生理盐水、54g·L⁻¹ 葡萄糖溶液)、聚乙烯醇滴眼液代替

0.50mL 高渗溶液重复上述实验,记录各种溶液的渗透浓度于表 13-3 中。

### 3.3 红细胞悬液的制备

取新鲜的动物(小白鼠或大白兔)血液 1.0mL,加入 20.0mL 含抗凝剂(0.05mol·L$^{-1}$ 枸橼酸钠或 EDTA 二钠盐溶液 10 滴)的等渗溶液,轻轻搅拌均匀后备用。

### 3.4 观察红细胞的形态

取五支试管,分别加入 1.0mL 0.35mol·L$^{-1}$ NaCl 溶液、0.05mol·L$^{-1}$ NaCl 溶液、生理盐水、54g·L$^{-1}$ 葡萄糖溶液和聚乙烯醇滴眼液,每支试管中加入 0.1mL 红细胞悬液,轻轻摇匀,15~20min 后,在普通显微镜上观察红细胞的形态。

实验前,仔细阅读普通显微镜的使用方法并调试好显微镜。从五支试管中分别取一滴溶液放在不同的载玻片上,刮去多余的溶液,分别将载玻片放在显微镜上,观察并记录红细胞的形态于表 13-4 中。

## 【数据记录与结果分析】

日期: 温度: ℃ 相对湿度:

表 13-3 溶液渗透浓度的测定

| 实验序号 | 1 | 2 | 3 |
|---|---|---|---|
| 0.35mol·L$^{-1}$ NaCl | | | |
| 0.05mol·L$^{-1}$ NaCl | | | |
| 生理盐水 | | | |
| 54g·L$^{-1}$ 葡萄糖 | | | |
| 聚乙烯醇滴眼液 | | | |

表 13-4 红细胞在不同渗透浓度溶液中的形态

| 溶液 | 0.35mol·L$^{-1}$ NaCl | 0.05mol·L$^{-1}$ NaCl | 生理盐水 | 54g·L$^{-1}$ 葡萄糖 | 聚乙烯醇滴眼液 |
|---|---|---|---|---|---|
| 红细胞形态 | | | | | |

## 【思考题】

(1) 红细胞在低渗、等渗和高渗溶液中各呈什么形态?如何解释这些现象?

(2) 为什么在淡水中游泳眼睛会感到胀痛,而在海水中游泳感觉会好得多?

(于 昆)

# 实验十四 置换法测定镁的原子量

## 【实验目的】

(1) 掌握电子分析天平的使用。
(2) 掌握气体体积的测量方法。
(3) 掌握镁的原子量测定原理与方法。

## 【预习作业】

（1）置换法测定镁的原子量的原理是什么？
（2）分析可能导致本实验测定误差的原因。
（3）测定原子量或分子量还有哪些方法，原理各是什么？如给定葡萄糖，或大分子如多糖、蛋白质等，应如何设计实验进行测定？通过查阅文献资料，至少写出其中一种测定方法，并给出设计思路。
（4）以流程图方式，写出本实验测定镁原子量的设计步骤，并注明各步骤的注意事项及原因。

## 【实验原理】

镁是活泼金属，它可置换稀硫酸中的氢离子并放出氢气。反应式如下

$$H_2SO_4 + Mg = H_2\uparrow + MgSO_4$$

准确称取一定质量的金属镁，与过量的稀硫酸反应，在一定温度（$t$）和压力下测出被置换出来的氢气的体积 $V(H_2)$，根据化学计量关系：$n(H_2)=n(Mg)$ 和理想气体状态方程：$pV=nRT$，可计算出 $n(Mg)$。

$$n(Mg)=n(H_2)=\frac{p(H_2)\times V(H_2)}{R\times T}(mol) \tag{14-1}$$

式（14-1）中 $p(H_2)$ 为氢气的分压（单位：kPa），$V(H_2)$ 为收集到的反应气体的体积（单位：L），$R$ 为气体常数（8.314 J·mol$^{-1}$·K$^{-1}$），$T$ 为绝对温度。

由于量气管内收集的氢气被水蒸气所饱和，根据分压定律，若量气管内气压等于大气压时，其压力 $p$ 是氢气分压 $p(H_2)$ 与饱和水蒸汽压 $p(H_2O)$ 的总和，即

$$p_{大气压}=p(H_2)+p_{饱和}(H_2O)(kPa) \tag{14-2}$$

$$p(H_2)=p_{大气压}-p_{饱和}(H_2O)$$

镁的原子量 $A_r(Mg)$ 可以通过式（14-3）计算

$$A_r(Mg)=\frac{m(Mg)}{n(Mg)}=\frac{m(Mg)\times R\times T}{[p_{大气压}-p_{饱和}(H_2O)]\times V(H_2)} \tag{14-3}$$

式（14-3）中，$m(Mg)$ 为镁条的质量（g），$n(Mg)$ 为镁条的物质的量。

## 【仪器材料与试剂】

**仪器** 万分之一分析天平，温度计，气压计，铁架台，量气管（50mL 碱式滴定管），大试管（20mL），长颈漏斗（×2），配塞导管，橡皮管，砂纸，量筒（10mL）

**材料与试剂** 镁条（约 25～35mg），2mol·L$^{-1}$ H$_2$SO$_4$ 溶液，甘油

## 【实验步骤】

### 1. 称量

用万分之一分析天平准确称取三份已用砂纸擦去表面氧化膜的干净镁条，每份重在 0.030 0g 左右（0.025 0～0.035 0g），分别用纸包好，并用铅笔在纸包外标上 1、2 或 3 数字。

### 2. 安装仪器

按图 14-1 装配好仪器，向量气管（50mL 碱式滴定管）内注水至略低于刻度"0"的位置。上下移动漏斗，以赶尽附着在胶管和量气管内壁的气泡。

### 3. 检查装置的气密性

将连接量气管和大试管的塞子塞紧,把漏斗缓慢下移一段距离,并固定在一定位置上。量气管中的液面稍有下降,待液面稳定后读数,3min 后再次读数,若两数据相同,说明装置不漏气;否则说明装置漏气,必须检查并塞紧各接口处。重复气密性试验,直至系统不漏气为止。

### 4. 装硫酸、贴镁条

取下大试管,用一漏斗小心将 5mL 2mol·L$^{-1}$ H$_2$SO$_4$ 溶液注入试管底部(切勿使酸沾在试管壁上)。稍稍倾斜试管,用少量甘油或水将镁条湿润一下,贴在试管壁内上部,确保镁条不与硫酸接触。调整量气管液面,使之尽量接近 0 刻度,并装好试管,塞紧橡皮塞。重复气密性试验,确保装置不漏气。

### 5. 记录初始液面位置

把漏斗移至量气管右侧,使两者的液面保持在同一水平面上,记下量气管中初读数 $V_{初}$。

### 6. 化学反应

适当倾斜铁架台,让镁条与稀硫酸接触(不让硫酸冲出试管进入量气管),反应产生的氢气进入量气管中。为避免管内压力过大,在管内液面下降时,漏斗也相应地向下移动,使两者的液面大体上保持在同一水平面上。

### 7. 记录终读数

反应完全后,用自来水淋洗试管外壁,待整个系统冷却至室温,使漏斗与量气管的液面处于同一水平面上,记下液面读数于表 14-1 中。稍等 1~2min,再次读数,如两次读数相等,表明反应体系温度已与室温一致,否则继续冷却后读数,直至连续两次读数相等为止,记录终读数 $V_{终}$。取下大试管,倒出溶液,将试管冲洗干净。

### 8. 记录室温和反应进行时所对应的大气压

平行测定 3 次。根据表 14-1 中测量数据计算结果,并根据镁的理论原子量计算镁原子量测定的相对误差。

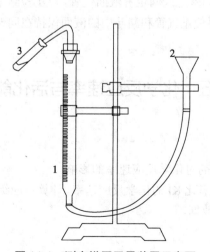

图 14-1 测定镁原子量装置示意图
1. 量气管;2. 漏斗;3. 大试管

## 【数据记录与结果分析】

日期：　　　　　　温度：　　　　℃　　　　相对湿度：

表 14-1　置换法测定镁的原子量

| 实验序号 | 1 | 2 | 3 |
|---|---|---|---|
| $m(Mg)$ / g | | | |
| $V_{终}$ / mL | | | |
| $V_{初}$ / mL | | | |
| $V(H_2)$ / mL | | | |
| $T$ / K | | | |
| $p_{大气压}$ / kPa | | | |
| $p_{饱和}(H_2O)$ / kPa | | | |
| $p(H_2) = p_{大气压} - p_{饱和}(H_2O)$ / kPa | | | |
| $A_r(Mg)$ | | | |
| $\overline{A_r(Mg)}$ | | | |
| $A_{r\,理论}(Mg)$ | | | |
| 相对平均误差 $\overline{E_r}$ / % | | | |

（1atm = 760.15mmHg = 101.325kPa，水的饱和蒸汽压可从附录Ⅱ中查到）

$$\overline{E_r} = \frac{A_r(Mg)_{测定值} - A_r(Mg)_{理论}}{A_r(Mg)_{理论}} \times 100\% \tag{14-4}$$

## 【思考题】

(1) 如果镁条表面上的氧化膜未擦除干净，对实验结果有何影响？
(2) 如果量气管内的气泡没有赶尽，对实验结果有何影响？
(3) 检查漏气与否的操作原理是什么？
(4) 反应后如没有完全冷却就记录滴定管液面位置，对实验结果有何影响？
(5) 读取读数时，为什么要使量气管和漏斗中的液面保持在同一水平面上？

（宋　慧）

# 实验十五　化学反应速率与活化能的测定

## 【实验目的】

(1) 掌握浓度、温度、催化剂对化学反应速率的影响。
(2) 通过测定$(NH_4)_2S_2O_8$氧化 KI 的化学反应速率，计算反应级数、反应速率常数及活化能。
(3) 掌握作图法处理实验数据。

## 【预习作业】

(1) 影响化学反应速率的因素有哪些？如何影响？

(2) 升高温度可以加快化学反应速率，为什么？

(3) 加入催化剂可以改变化学反应速率，为什么？

(4) 请用流程图的形式阐释本实验是如何设计实验步骤来验证浓度、温度、催化剂对化学反应速率的影响，如何测定反应级数、速率常数和活化能？

【实验原理】

在水溶液中，$(NH_4)_2S_2O_8$ 与 KI 的氧化还原反应如下：

$$(NH_4)_2S_2O_8 + 3KI =\!=\!= (NH_4)_2SO_4 + K_2SO_4 + KI_3$$

或

$$S_2O_8^{2-} + 3I^- =\!=\!= 2SO_4^{2-} + I_3^-$$

其反应速率为

$$v = k\, c(S_2O_8^{2-})^m\, c(I^-)^n \tag{15-1}$$

式(15-1)中，$v$ 是反应速率。若 $c(S_2O_8^{2-})$、$c(I^-)$ 是起始浓度，则 $v$ 表示起始速率；$k$ 是速率常数；$m$ 是 $S_2O_8^{2-}$ 的反应级数，$n$ 是 $I^-$ 的反应级数，$m$ 与 $n$ 之和为反应总级数。

为了测定反应速率，必须知道在 $\Delta t$ 时间内 $S_2O_8^{2-}$ 离子浓度的变化。其反应的平均速率可以表示为

$$\bar{v} = -\frac{\Delta c(S_2O_8^{2-})}{\Delta t} \tag{15-2}$$

实验中用平均速率代替起始速率，即

$$\bar{v} = -\frac{\Delta c(S_2O_8^{2-})}{\Delta t} \approx k\, c(S_2O_8^{2-})^m\, c(I^-)^n \tag{15-3}$$

为了测定出在 $\Delta t$ 时间内 $S_2O_8^{2-}$ 离子浓度的变化，在 $(NH_4)_2S_2O_8$ 与 KI 两溶液混合前，先加入一定体积已知浓度的 $Na_2S_2O_3$ 和淀粉溶液，由反应生成的 $I_3^-$ 能很快与 $Na_2S_2O_3$ 反应生成无色的 $S_4O_6^{2-}$ 和 $I^-$

$$2S_2O_3^{2-} + I_3^- =\!=\!= S_4O_6^{2-} + 3I^-$$

由于该反应的速率较快，所以实验开始时看不到蓝色，当 $Na_2S_2O_3$ 耗尽时，反应生成的 $I_3^-$ 立即与淀粉作用，使溶液变为蓝色。

由反应关系可知，$S_2O_8^{2-}$ 减少的量为 $S_2O_3^{2-}$ 减少量的一半，所以

$$\Delta c(S_2O_8^{2-}) = -\frac{\Delta c(S_2O_3^{2-})}{2} \tag{15-4}$$

实验中加入少量的 $Na_2S_2O_3$，因此，在 $\Delta t$ 内 $Na_2S_2O_3$ 完全耗尽，所以 $\Delta c(S_2O_3^{2-})$ 就是 $Na_2S_2O_3$ 的起始浓度。记录从反应开始到溶液出现蓝色所用的时间 $\Delta t$，由式(15-3)求得反应速率

$$\bar{v} = -\frac{\Delta c(S_2O_8^{2-})}{\Delta t} = \frac{c(S_2O_3^{2-})}{2\Delta t} \; (\text{mol·L}^{-1}\cdot\text{s}^{-1}) \tag{15-5}$$

即

$$\bar{v} = -\frac{\Delta c(S_2O_8^{2-})}{\Delta t} = \frac{c(S_2O_3^{2-})}{2\Delta t} = k\, c(S_2O_8^{2-})^m\, c(I^-)^n \tag{15-6}$$

对式(15-6) $v = k\, c(S_2O_8^{2-})^m\, c(I^-)^n$ 两边取对数，可得

$$\lg v = \lg k + m\lg c(S_2O_8^{2-}) + n\lg c(I^-) \tag{15-7}$$

固定 $I^-$ 离子的浓度不变，测定 $S_2O_8^{2-}$ 不同浓度条件下的反应速率，以 $\lg v \sim \lg c(S_2O_8^{2-})$ 作图，得一直线，斜率为 $S_2O_8^{2-}$ 的反应级数 $m$。同理，当固定 $S_2O_8^{2-}$ 离子的浓度不变，改变 $I^-$ 离子的浓度，以 $\lg v \sim \lg c(I^-)$ 作图，得一直线，其斜率为 $I^-$ 离子的反应级数 $n$。根据 $m$、$n$ 可计算出反应级数

($m+n$)。同样根据 $m$ 和 $n$，由速率方程式 $v=k\,c(S_2O_8^{2-})^m\,c(I^-)^n$ 可求得一定温度下反应的速率常数 $k$。

根据 Arrhenius 公式，反应速率常数 $k$ 与反应温度 $T$ 有如下关系

$$\lg k = -\frac{E_a}{2.303RT} + \lg A \tag{15-8}$$

式（15-8）中 $A$ 为反应的特征常数，$R$ 为气体常数（$8.314 \text{J·mol}^{-1}\text{·K}^{-1}$），$T$ 为绝对温度，$E_a$ 为反应的活化能（$\text{J·mol}^{-1}$，或 $\text{kJ·mol}^{-1}$）。

测得不同温度下的速率常数 $k$ 值，以 $\lg k \sim \dfrac{1}{T}$ 作图，可以得一直线，由直线的斜率（$-\dfrac{E_a}{2.303R}$）可求出反应的活化能 $E_a$。

【仪器材料与试剂】

**仪器** 锥形瓶（100mL×1），吸量管（10mL×5），大试管（10mL×1），量筒（10mL×1）或移液枪（1-10mL×1），温度计，秒表，恒温水浴锅（或槽），搅拌器

**材料与试剂** $0.2\text{mol·L}^{-1}$ $(NH_4)_2S_2O_8$ 溶液，$0.2\text{mol·L}^{-1}$ KI 溶液，$0.01\text{mol·L}^{-1}$ $Na_2S_2O_3$ 溶液，$0.2\text{mol·L}^{-1}$ $KNO_3$ 溶液，$0.02\text{mol·L}^{-1}$ $Cu(NO_3)_2$ 溶液，$0.2\text{mol·L}^{-1}$ $(NH_4)_2SO_4$ 溶液，$0.2\%$ 淀粉溶液

【实验步骤】

### 1. 浓度对化学反应速率的影响

在室温条件下，按表 15-1 中实验组 1～9 的用量，取五支 10mL 吸量管分别移取相应体积的 KI、$KNO_3$、$(NH_4)_2SO_4$、$Na_2S_2O_3$、淀粉溶液置于 100mL 干燥锥形瓶中，混合均匀，用量筒量（或移液枪）取相应体积的 $(NH_4)_2S_2O_8$ 溶液，快速将 $(NH_4)_2S_2O_8$ 溶液加到锥形瓶里，同时启动秒表，并不断摇动锥形瓶，当溶液开始出现蓝色时，立即停止计时，记录溶液变蓝所需的时间及反应温度于表 15-1 中。

为了保证溶液离子强度和总体积维持不变，KI 和 $(NH_4)_2S_2O_8$ 的不足量用 $KNO_3$ 和 $(NH_4)_2SO_4$ 补充。

表 15-1 浓度对化学反应速率的影响

| 实验序号 | | 1 | 2 | 3 | 4 | 5 | 6 | 7 | 8 | 9 |
|---|---|---|---|---|---|---|---|---|---|---|
| 试剂用量/mL | $(NH_4)_2S_2O_8$ 溶液 | 2.0 | 4.0 | 6.0 | 8.0 | 10.0 | 10.0 | 10.0 | 10.0 | 10.0 |
| | $Na_2S_2O_3$ 溶液 | 4.00 | 4.00 | 4.00 | 4.00 | 4.00 | 4.00 | 4.00 | 4.00 | 4.00 |
| | 淀粉溶液 | 2.00 | 2.00 | 2.00 | 2.00 | 2.00 | 2.00 | 2.00 | 2.00 | 2.00 |
| | KI 溶液 | 10.00 | 10.00 | 10.00 | 10.00 | 10.00 | 8.00 | 6.00 | 4.00 | 2.00 |
| | $KNO_3$ 溶液 | 0 | 0 | 0 | 0 | 0 | 2.00 | 4.00 | 6.00 | 8.00 |
| | $(NH_4)_2SO_4$ 溶液 | 8.00 | 6.00 | 4.00 | 2.00 | 0 | 0 | 0 | 0 | 0 |
| $c(NH_4)_2S_2O_8)_{初}/\text{mol·L}^{-1}$ | | | | | | | | | | |
| $c(KI)_{初}/\text{mol·L}^{-1}$ | | | | | | | | | | |
| $c(Na_2S_2O_3)_{初}/\text{mol·L}^{-1}$ | | | | | | | | | | |
| 反应温度 / K | | | | | | | | | | |
| $\Delta t$ / s | | | | | | | | | | |
| $v$ / $\text{mol·L}^{-1}\text{·s}^{-1}$ | | | | | | | | | | |

根据表 15-1 中 1～5 组的结果进行相互比较，分析 $S_2O_8^{2-}$ 离子浓度的改变对反应速率的影响。根据 5～9 组的结果进行相互比较，分析 $I^-$ 离子浓度的改变对反应速率的影响。

## 2. 温度对化学反应速率的影响

表 15-2　温度对化学反应速率的影响

| 实验序号 | | 10 | 11 | 12 | 13 | 14 |
|---|---|---|---|---|---|---|
| 试剂用量/mL | $(NH_4)_2S_2O_8$ 溶液 | 5.0 | 5.0 | 5.0 | 5.0 | 5.0 |
| | $Na_2S_2O_3$ 溶液 | 4.00 | 4.00 | 4.00 | 4.00 | 4.00 |
| | 淀粉溶液 | 2.00 | 2.00 | 2.00 | 2.00 | 2.00 |
| | KI 溶液 | 10.00 | 10.00 | 10.00 | 10.00 | 10.00 |
| | $(NH_4)_2SO_4$ 溶液 | 5.00 | 5.00 | 5.00 | 5.00 | 5.00 |
| $c(NH_4)_2S_2O_8)_{初}$ / mol·L$^{-1}$ | | | | | | |
| $c(KI)_{初}$ / mol·L$^{-1}$ | | | | | | |
| $c(Na_2S_2O_3)_{初}$ / mol·L$^{-1}$ | | | | | | |
| 反应温度 / K | | | | | | |
| $\Delta t$ / s | | | | | | |
| $v$/mol·L$^{-1}$·s$^{-1}$ | | | | | | |

按表 15-2 中实验组 10 的用量，在室温条件下完成该组实验。

按表 15-2 中实验组 11 的用量，取四支 10mL 吸量管分别移取相应体积的 KI、$(NH_4)_2SO_4$、$Na_2S_2O_3$、淀粉溶液置于 100mL 锥形瓶中，用量筒（或移液枪）取 5.0mL 0.2mol·L$^{-1}$ $(NH_4)_2S_2O_8$ 溶液于干燥大试管中，将含溶液的锥形瓶和大试管分别置于恒温水浴（高出室温 5℃）中加热，待锥形瓶和试管的温度升到高出室温 5℃时，将大试管中的溶液加到锥形瓶中，同时启动秒表，并不断摇动锥形瓶，待溶液刚呈现蓝色，立即停表，记录所需的时间和温度于表 15-2 实验组 11 中。

按照同样方法分别在高于室温 10℃、15℃、20℃条件下，重复上述实验，记录反应时间和温度于表 15-2 实验组 12～14 中。

根据表 15-2 中 10～14 组的结果进行相互比较，分析温度的改变对化学反应速率的影响。

## 3. 催化剂对化学反应速率的影响

$Cu^{2+}$ 可以催化本反应，加入微量 $Cu^{2+}$ 可以使反应速率大大加快。按表 15-2 中实验 10 的试剂用量在室温下实验，在加 $(NH_4)_2S_2O_8$ 溶液前，先加 2 滴 0.02mol·L$^{-1}$ $Cu(NO_3)_2$ 溶液于锥形瓶中，记录反应时间，并与实验组 10 结果比较，分析催化剂 $Cu(NO_3)_2$ 对化学反应速率的影响。

## 【数据记录与结果分析】

日期：　　　　　　温度：　　　℃　　　　相对湿度：

（1）反应级数和反应速率常数的计算：通过表 15-1 的实验数据可计算出反应的平均速率。根据组 1～5 的数据作出 $\lg v$ 对 $\lg c(S_2O_8^{2-})$ 的曲线，从而求出相对于 $S_2O_8^{2-}$ 的反应级数 $m$，同理以组 5～9 的数据作出 $\lg v$ 对 $\lg c(I^-)$ 的曲线，得出 $I^-$ 的反应级数 $n$。最后从以上的数据处理结果可求出反应速率常数 $k$，记录于表 15-3 中。

表 15-3　反应级数和速率常数

| 实验序号 | 1 | 2 | 3 | 4 | 5 | 6 | 7 | 8 | 9 |
|---|---|---|---|---|---|---|---|---|---|
| $\lg v$ | | | | | | | | | |
| $\lg c(S_2O_8^{2-})$ | | | | | | | | | |
| $\lg c(I^-)$ | | | | | | | | | |
| $m$ | | | | | | | | | |
| $n$ | | | | | | | | | |
| $k$ | | | | | | | | | |
| $k$（平均） | | | | | | | | | |

（2）活化能的计算：根据表 15-2 中实验组 10～14 的数据，计算不同温度下对应的 $k$ 值，以 $\lg k \sim \dfrac{1}{T}$ 作图，得一直线，由直线的斜率求出反应的活化能 $E_a$，记录于表 15-4 中。

表 15-4　反应活化能

| 实验序号 | 10 | 11 | 12 | 13 | 14 |
|---|---|---|---|---|---|
| $k$ | | | | | |
| $\lg k$ | | | | | |
| $\dfrac{1}{T}$ / K$^{-1}$ | | | | | |
| $E_a$ / kJ·mol$^{-1}$ | | | | | |

根据以上实验结果，请分别讨论浓度、温度和催化剂对反应速率的影响。

## 【注意事项】

$(NH_4)_2S_2O_8$ 溶液需要新配制的，若长时间放置，该溶液易分解。

## 【思考题】

（1）在表 15-1 实验 1～4 及 6～9 中分别添加 $(NH_4)_2SO_4$ 及 $KNO_3$ 溶液，其作用是什么？

（2）实验中加入少量的 $Na_2S_2O_3$ 和淀粉溶液，其目的是什么？$Na_2S_2O_3$ 用量多少对实验结果有什么影响？

（3）反应体系中溶液变蓝后，体系中的反应是否停止了？

（李振泉）

# 实验十六　最大气泡压力法测定乙醇溶液的表面张力

## 【实验目的】

（1）掌握最大气泡压力法测定溶液表面张力的原理和方法。

（2）通过测定系列浓度乙醇溶液的表面张力，深入理解溶液浓度与表面张力、表面吸附量的关系。

（3）学会通过表面张力计算表面活性物质的饱和吸附量及其分子横截面积。

## 【预习作业】

(1) 什么是表面张力？
(2) 影响液体表面张力的因素有哪些？
(3) 测定液体表面张力的方法有哪些？各有何优缺点？
(4) 什么是表面活性物质？评价表面活性物质性能的指标都有哪些？
(5) 表面活性物质在医药学及生活中都有哪些应用？
(6) 以流程图方式，写出本实验测定表面张力的设计步骤，并标出注意事项及原因。

## 【实验原理】

由于液体表层分子受力不均衡，液体表面都有趋于收缩的基本特性。沿着液体表面作用于单位长度线段上的收缩张力称为表面张力，其方向为沿着表面切线并指向表面缩小的方向。

影响液体表面张力的因素包括温度、压力及溶液的组成及浓度等。如，在一定的温度和压力下，水的表面张力随加入溶质的种类和浓度的不同而发生变化。能够降低溶剂表面张力的物质称为表面活性物质。当溶液中表面活性物质浓度增加时，溶液的表面张力随之减小，此时溶液界面的浓度大于溶液内部的浓度，溶液界面浓度与溶液内部浓度的差值，称为表面吸附量。溶质的表面吸附量与溶液的表面张力、组成之间的关系符合吉布斯（Gibbs）等温吸附方程

$$\Gamma = -\frac{c}{RT} \times \frac{d\sigma}{dc} (\text{mol} \cdot \text{m}^{-2}) \tag{16-1}$$

式(16-1)中 $\Gamma$ 为表面吸附量($\text{mol} \cdot \text{m}^{-2}$)，$\sigma$ 为溶液的表面张力($\text{N} \cdot \text{m}^{-1}$)，$T$ 为热力学温度(K)，$c$ 为溶质的浓度($\text{mol} \cdot \text{m}^{-3}$)，$R$ 为气体常数($8.314 \text{J} \cdot \text{mol}^{-1} \cdot \text{K}^{-1}$)，$\frac{d\sigma}{dc}$ 为等温条件下表面张力随浓度的变化率。等温条件下 $\sigma$ 与 $c$ 关系曲线如图 16-1 所示。在 $\sigma$-$c$ 曲线上的任一浓度对应点 K 作切线，切线斜率即为表面张力随浓度的变化率 $\frac{d\sigma}{dc}$ 值，借助式(16-1)即可计算出溶质的表面吸附量。若进一步以浓度 $c$ 为横坐标，以表面吸附量 $\Gamma$ 为纵坐标，可以得到 $\Gamma = f(c)$ 关系曲线(图 16-2)。

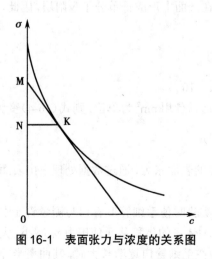

图 16-1 表面张力与浓度的关系图

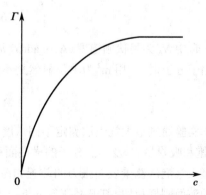

图 16-2 表面吸附量与浓度的关系

表面活性物质分子结构中同时含有亲水和疏水基团。若将表面活性物质分子溶于水，其亲水基团朝向水溶液内部，而疏水基团力图离开水相朝向空气，因此，其易于吸附在水溶液的表面从而

降低了水的表面张力(或表面能)。如图 16-3(a)所示,在浓度较小时,表面活性物质分子零散地排列在水溶液的表面。随着溶质浓度的进一步增加,表面活性物质分子逐渐占据的表面积随之增大。当浓度增加至一定程度时,表面活性物质分子占据了所有的表面,如图 16-3(b),表面活性物质分子在液面上形成单分子吸附层,此时单位面积表面吸附的溶质分子数称为饱和吸附量或最大吸附量,用符号 $\Gamma_\infty$ 表示。若进一步增加溶质的浓度,表面上的分子数不再增加,而是进入溶液中并聚集形成胶束,如图 16-3(c)所示。

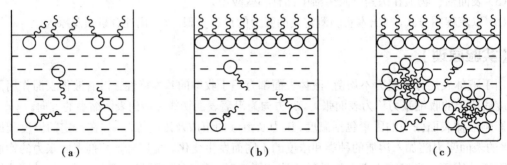

图 16-3　不同浓度情况下表面活性物质分子在溶液中的排布状态
(a) $c < \Gamma_\infty$;(b) $c = \Gamma_\infty$;(c) $c > \Gamma_\infty$

表面吸附量 $\Gamma$、饱和吸附量 $\Gamma_\infty$ 与浓度 $c$ 的关系符合朗格缪尔(Langmuir)提出的等温方程式

$$\Gamma = \Gamma_\infty \times \frac{Kc}{1+Kc} \quad (\text{mol} \cdot \text{m}^{-2}) \tag{16-2}$$

式(16-2)中 $K$ 为一常数。将其稍作变换,得

$$\frac{c}{\Gamma} = \frac{c}{\Gamma_\infty} + \frac{1}{K\Gamma_\infty} \tag{16-3}$$

若以浓度 $c$ 为横坐标,以浓度与表面吸附量的比值 $\frac{c}{\Gamma}$ 为纵坐标作图,可得一条直线,直线斜率的倒数即为饱和吸附量 $\Gamma_\infty$。

当溶液表面吸附的溶质分子数量达到饱和状态时,在液面上形成一单分子吸附层,因此,可以用饱和吸附量 $\Gamma_\infty$ 求出单个溶质分子的横截面积 $S$

$$S = \frac{1}{\Gamma_\infty N_A} \quad (\text{m}^2) \tag{16-4}$$

式(16-4)中 $N_A$ 为阿伏伽德罗(Avogadro)常数(等于 $6.02 \times 10^{23}$)。

由于分子很小,用 $\text{m}^2$ 为单位显然是不合适的,因此通常使用 $\text{nm}^2$ 为单位,则式(16-4)变为

$$S = \frac{10^{18}}{\Gamma_\infty N_A} \quad (\text{nm}^2) \tag{16-5}$$

本实验通过最大气泡法测定不同浓度的乙醇水溶液的表面张力,通过数据处理求出表面吸附量 $\Gamma$、饱和吸附量 $\Gamma_\infty$ 及乙醇分子的横截面积 $S$。

最大气泡法测量表面张力的装置如图 16-4 所示。实验时使毛细管 b 管口与测量管 a 中待测溶液的液面刚好相切,打开减压管 c 下方的活塞放水增大体系的体积从而使压力 $p_0$ 减小,大气压 $p$ 将毛细管内的液面压至毛细管口并形成气泡。当气泡在毛细管口逐渐长大时,其曲率半径先变小后增大,如图 16-5 所示。当气泡的曲率半径刚好等于毛细管的半径时,其曲率半径最小。此时气泡内外的压力差达到最大值,此压力差的数值可由压力计 d 读出。随后,气泡从毛细管口逸出,

测量管中气体得到补偿,压力差数值恢复至较小值。在气泡逸出过程中,最大压力差 $\Delta p$ 及气泡的最小曲率半径 $r'_2$(等于毛细管的半径 $r$)与溶液的表面张力 $\sigma$ 之间的关系满足杨-拉普拉斯公式

$$\Delta p = p - p_0 = \frac{2\sigma}{r} \tag{16-6}$$

由于毛细管的半径难以测定,因此通常使用一种已知表面张力的液体作为标准溶液。用同一根毛细管分别测定标准溶液和待测溶液产生气泡时的最大压力差,代入式(16-6),并将所得到的两个方程式相除,最后可由式(16-7)计算得到待测溶液的表面张力

$$\frac{\sigma}{\sigma'} = \frac{\Delta p}{\Delta p'} \tag{16-7}$$

式(16-7)中 $\sigma$ 和 $\sigma'$ 分别为标准溶液和待测溶液的表面张力($N \cdot m^{-1}$),$\Delta p$ 和 $\Delta p'$ 分别为标准溶液和待测溶液产生气泡时的最大压力差(kPa)。

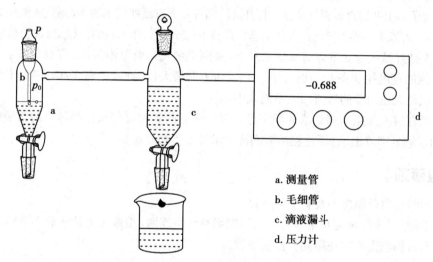

a. 测量管
b. 毛细管
c. 滴液漏斗
d. 压力计

图 16-4　最大气泡法测定溶液表面张力装置

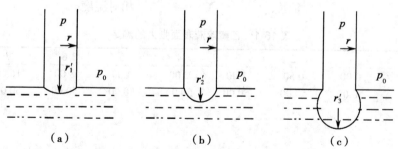

图 16-5　气泡产生过程中曲率半径的变化
(a)$r'_1 > r$;(b)$r'_2 = r$;(c)$r'_3 > r$

## 【仪器材料与试剂】

**仪器**　表面张力测定装置,烧杯(500mL×2),50mL 容量瓶(50mL×8),吸量管(5mL、10mL),移液管(20mL),滴管

**材料与试剂**　无水乙醇(A.R.),纯水

## 【实验步骤】

### 1. 溶液的配制

根据表 16-1 中的试剂用量配制系列浓度的乙醇水溶液待用。

### 2. 表面张力的测定

（1）在滴液漏斗 c（带支管）中加入约三分之二体积的水（勿超过支管），用待测溶液润洗测量管 a 及毛细管 b 3～4 次，将待测溶液加入测量管 a 中，插入毛细管 b 并调整测量管 a 中溶液量，使毛细管管口与待测溶液的液面刚好相切。

（2）打开滴液漏斗 c 上方塞子，接通压力计 d 电源，在系统与大气连通的情况下采零，随后将滴液漏斗 c 上方塞子盖上并检查装置气密性。打开滴液漏斗 c 下方活塞使水滴下，通过滴液增大腔内气体体积，使管内压力低于外界大气压。关闭活塞并维持 1～2min，如压示数数值稳定，说明系统不漏气。

（3）再次打开减压管 c 下方活塞使水滴下，直到毛细管口有气泡冒出，仔细调节活塞，将毛细管口逸出气泡的速度控制在 8～10s 一个。记录压力差最大值，连续记录 3 次，将数据填入表 16-1 中。注意 3 次最大压力差的数值相差不能太大。

（4）将所有溶液按照从稀到浓的顺序进行表面张力的测定。每次更换测定的溶液时，都要将测量管中的溶液全部放出，并用待测溶液润洗测量管 a 及毛细管 b。

## 【注意事项】

（1）测量时尽量使温度不要有明显变化。
（2）润洗测量管时，应避免溶液冲入支管中影响气路连通，造成压力计读数不准确。
（3）所有数据测定必须使用同一根毛细管。

## 【数据记录及结果分析】

日期：　　　　　　温度：　　　℃　　　　　相对湿度：

表 16-1　乙醇溶液表面张力的测定

| 编号 | 1 | 2 | 3 | 4 | 5 | 6 | 7 | 8 |
|---|---|---|---|---|---|---|---|---|
| $V_{乙醇}$ / mL | 0.00 | 0.50 | 1.00 | 2.00 | 4.00 | 8.00 | 16.00 | 20.00 |
| $V_{总}$ / mL | 50.00 | 50.00 | 50.00 | 50.00 | 50.00 | 50.00 | 50.00 | 50.00 |
| $\Delta p_1$ / kPa | | | | | | | | |
| $\Delta p_2$ / kPa | | | | | | | | |
| $\Delta p_3$ / kPa | | | | | | | | |
| $\Delta \bar{p}$ / kPa | | | | | | | | |
| $c_{乙醇}$ / mol·m$^{-3}$ | | | | | | | | |
| $\sigma$ / N·m$^{-1}$ | | | | | | | | |
| $b = c \times \dfrac{d\sigma}{dc}$ | — | | | | | | | |
| $\Gamma$ / mol·m$^{-2}$ | — | | | | | | | |
| $c / \Gamma$ | — | | | | | | | |
| $\Gamma_\infty$ / mol·m$^{-2}$ | — | | | | | | | |
| $S$ / nm$^2$ | — | | | | | | | |

(1) 以浓度 $c$ 为横坐标，乙醇溶液表面张力 $\sigma$ 为纵坐标作图，得 $\sigma$ 随 $c$ 变化的 $\sigma \sim c$ 关系曲线，根据曲线说明 $\sigma$ 随 $c$ 的变化规律。

(2) 在曲线 $\sigma \sim c$ 上，对应任意浓度对应曲线上点 K 作曲线的切线及 x 轴的平行线，分别交 y 轴于 M、N 两点，见图 16-1。从坐标纸上读出 M、N 两点之间的距离 $b$ ($b = c \times \dfrac{d\sigma}{dc}$)，代入吉布斯公式中计算不同浓度下的表面吸附量 $\Gamma$。

(3) 以浓度 $c$ 为横坐标，以表面吸附量 $\Gamma$ 为纵坐标作图，得到 $\Gamma \sim c$ 关系曲线图，根据曲线说明 $\Gamma$ 与 $c$ 的关系。

(4) 以浓度 $c$ 为横坐标，以 $\dfrac{c}{\Gamma}$ 为纵坐标作图，得到 $\dfrac{c}{\Gamma} \sim c$ 关系曲线图，计算乙醇的饱和吸附量 $\Gamma_\infty$ 及单分子所占的横截面积 $S$。

**【思考题】**

(1) 如何对系统进行检漏？若系统漏气，对实验有何影响？
(2) 毛细管管口为何要刚好与液面相切？
(3) 为何表面张力的测定要按溶液浓度从稀到浓的顺序？
(4) 实验是否需要在恒温条件下进行，为什么？

（廖传安）

# 实验课题五

# 化合物制备实验

## 实验十七　氯化钠的精制

### 【实验目的】

（1）掌握溶解、过滤、蒸发、浓缩、结晶、干燥等基本操作。
（2）掌握提纯食盐的原理、方法以及检验其纯度的方法。
（3）了解沉淀溶解平衡在无机物纯化中的应用。

### 【预习作业】

（1）氯化钠在临床上有哪些应用？
（2）固体物质的提纯方法有哪些？
（3）何谓重结晶？氯化钠是否可以用重结晶法来提纯？
（4）固液分离的常用方法有哪些？应如何选择？减压过滤时应该注意哪些细节？
（5）为什么本实验除杂时先除去 $SO_4^{2-}$ 离子，再除 $Ca^{2+}$、$Mg^{2+}$、$Fe^{3+}$ 离子？是否可以改变沉淀剂加入顺序？
（6）蒸发浓缩时为什么不能蒸干？
（7）请用流程图说明本实验是如何设计实验步骤，对氯化钠进行精制的？每个步骤有哪些注意事项并注明原因？

### 【实验原理】

粗食盐中有 $Ca^{2+}$、$Mg^{2+}$、$Fe^{3+}$、$K^+$ 和 $SO_4^{2-}$ 等可溶性杂质和泥沙等不溶性杂质，在制成试剂级 NaCl 之前，必须除去这些杂质且避免引入新的杂质。不溶性杂质可通过溶解后过滤的方法除去。而可溶性杂质则可选择适当的试剂（如 $BaCl_2$、$Na_2CO_3$、$NaOH$ 等）使其生成难溶化合物，再过滤除去。

首先，在粗食盐溶液中加入过量的 $BaCl_2$ 溶液，以除去 $SO_4^{2-}$，

$$Ba^{2+} + SO_4^{2-} = BaSO_4\downarrow$$

过滤除去 $BaSO_4$ 沉淀。然后在滤液中加入 $NaOH$ 和 $Na_2CO_3$ 溶液，以除去 $Ca^{2+}$、$Mg^{2+}$、$Fe^{3+}$ 和过量的 $Ba^{2+}$。

$$Mg^{2+} + 2OH^- = Mg(OH)_2\downarrow$$

$$2Mg^{2+} + 2OH^- + CO_3^{2-} = Mg_2(OH)_2CO_3\downarrow$$

$$Ca^{2+} + CO_3^{2-} = CaCO_3\downarrow$$
$$Fe^{3+} + 3OH^- = Fe(OH)_3\downarrow$$
$$Ba^{2+} + CO_3^{2-} = BaCO_3\downarrow$$

过滤除去沉淀，用稀 HCl 溶液调节溶液的 pH 至 2~3，除去溶液中过量的 NaOH 和 $Na_2CO_3$。

$$H_3O^+ + OH^- = 2H_2O$$
$$2H_3O^+ + CO_3^{2-} = 3H_2O + CO_2\uparrow$$

易溶的 $K^+$ 不能用上述方法除去，但可在蒸发和结晶过程中留在滤液中，趁热减压过滤除去。残留的 HCl 在干燥过程中可被除去。

## 【仪器材料与试剂】

**仪器** 台秤，玻璃漏斗，漏斗架，布氏漏斗，吸滤瓶，电热套，蒸发皿，量筒（10mL，50mL），试管（10mL×6），玻璃棒，烧杯（50mL，100mL），药匙，滴管

**材料与试剂** 粗食盐，6mol·$L^{-1}$ HCl 溶液，6mol·$L^{-1}$ HAc 溶液，6mol·$L^{-1}$ NaOH 溶液，1mol·$L^{-1}$ $BaCl_2$ 溶液，1mol·$L^{-1}$ $Na_2CO_3$ 溶液，$(NH_4)_2C_2O_4$ 饱和溶液，镁试剂，滤纸，广泛 pH 试纸，95% 乙醇

## 【实验步骤】

### 1. 粗食盐的提纯

#### 1.1 称量和溶解

称取 5.0g 粗食盐于 100mL 烧杯中，加入 25.0mL 水，搅拌并加热使其溶解，溶解后如有泥沙等不溶性杂质，可过滤除去。

#### 1.2 除去 $SO_4^{2-}$

当溶液微沸时，边搅拌边逐滴加入 2.0mL 1mol·$L^{-1}$ $BaCl_2$ 溶液（注意：$BaCl_2$ 溶液有毒，废液应回收），继续加热 5min，静置，取少量上清液于小试管中，滴加 2 滴 6mol·$L^{-1}$ HCl 溶液，再滴加 1mol·$L^{-1}$ $BaCl_2$ 溶液 2~3 滴。如果混浊，需向原液继续滴加 $BaCl_2$ 溶液，直至 $SO_4^{2-}$ 沉淀完全。继续加热 5min，静置冷却后过滤，收集滤液于洁净小烧杯中。

#### 1.3 除去 $Ca^{2+}$、$Mg^{2+}$、$Fe^{3+}$ 和过量的 $Ba^{2+}$ 离子

继续加热滤液至微沸，边搅拌边向其中加入 1.0mL 6mol·$L^{-1}$ NaOH 和 2.0mL 饱和 $Na_2CO_3$ 溶液，继续加热 5min，静置冷却后，取少量上层清液放在试管中，滴加几滴 $Na_2CO_3$ 溶液，检查有无沉淀生成。若无沉淀生成，过滤，收集滤液于蒸发皿中。

#### 1.4 除去剩余的 $OH^-$ 和 $CO_3^{2-}$ 离子

向滤液中逐滴加入 6mol·$L^{-1}$ HCl 溶液，至滤液 pH 2~3（用广泛 pH 试纸测定）为止。

#### 1.5 蒸发干燥

小火加热蒸发上述滤液，浓缩至稀糊状（注意：不可将溶液蒸干！）。趁热减压过滤，尽量将结晶抽干，将晶体转移至洁净蒸发皿中。小火加热炒干，冷却至室温，称重，计算产率。

### 2. 产品纯度的检验

取粗盐和精盐各 1.0g，分别置于 25mL 烧杯中，用 5.0mL 去离子水溶解（若粗盐溶液浑浊则须过滤）。将粗盐和精盐澄清溶液分别盛于 3 支小试管中（每支试管取澄清溶液 1.0mL），组成对照组以检验它们的纯度。

### 2.1 $SO_4^{2-}$ 的检验

在第一对照组溶液中分别加入 2 滴 $6mol·L^{-1}$ HCl 溶液,使溶液呈酸性。再加入 2～3 滴 $1mol·L^{-1}$ $BaCl_2$ 溶液。如有白色沉淀,即证明有 $SO_4^{2-}$ 存在。记录结果于表 17-1。

### 2.2 $Ca^{2+}$ 的检验

在第二对照组溶液中分别加入 2 滴 $6mol·L^{-1}$ HAc 溶液,使溶液呈酸性。再加入 3～5 滴饱和 $(NH_4)_2C_2O_4$ 溶液,如有白色沉淀,即证明有 $Ca^{2+}$ 存在*。记录结果于表 17-1。

### 2.3 $Mg^{2+}$ 的检验

在第三对照组溶液中分别加入 3～5 滴 $6mol·L^{-1}$ NaOH 溶液,使溶液呈碱性。再加入 1 滴镁试剂**,如有天蓝色沉淀,即证明有 $Mg^{2+}$ 存在。记录结果于表 17-1。

## 【注意事项】

(1) 加入沉淀剂煮沸过程中,如果溶剂蒸发过多,应适当补充溶剂,避免氯化钠结晶析出。

(2) 加入沉淀剂后,应煮沸并静置后再过滤。

## 【数据记录与结果处理】

日期:　　　　　　温度:　　　　℃　　　　相对湿度:

(1) 粗食盐的提纯

粗食盐 $m(NaCl)=$ _____ g,精制食盐 $m(NaCl)=$ _____ g。

$$产率(\%)=\frac{精制食盐的质量}{粗食盐的质量}\times 100\%=\underline{\qquad\qquad}。$$

(2) 产品纯度的检验

表 17-1　产品纯度检验结果

| 项目 | 实验方法 | 粗食盐溶液 | 精制食盐溶液 |
|---|---|---|---|
| $SO_4^{2-}$ 的检验 | 加入 $BaCl_2$ 溶液 | 现象: <br> 结论: | 现象: <br> 结论: |
| $Ca^{2+}$ 的检验 | 加入饱和 $(NH_4)_2C_2O_4$ 溶液 | 现象: <br> 结论: | 现象: <br> 结论: |
| $Mg^{2+}$ 的检验 | 加入"镁试剂" | 现象: <br> 结论: | 现象: <br> 结论: |

## 【思考题】

(1) 本实验中,能否用其他酸来除去多余的 $CO_3^{2-}$?

(2) 在检验 $SO_4^{2-}$ 时,为什么要加入盐酸溶液?

(乔秀文)

---

\* 用 $(NH_4)_2C_2O_4$ 检验 $Ca^{2+}$ 时,溶液中的 $Mg^{2+}$ 亦可与之作用产生 $MgC_2O_4$ 白色沉淀,但 $MgC_2O_4$ 溶于 HAc,所以加入的 HAc 可排除 $Mg^{2+}$ 的干扰。

\*\* 镁试剂 I 学名为对硝基偶氮间苯二酚,是一种有机染料,在碱性溶液中呈红色或紫色,但被 $Mg(OH)_2$ 沉淀吸附后,则呈天蓝色。

镁试剂的配制:称取 0.01g 镁试剂于 100mL $2mol·L^{-1}$ NaOH 溶液中,摇匀即可。

## 实验十八　硫酸亚铁铵的制备

### 【实验目的】

(1) 了解复盐制备的基本原理。
(2) 练习水浴加热、常压过滤、减压过滤、蒸发、浓缩、结晶和干燥等基本操作。
(3) 学习用目视比色法检验产品质量。

### 【预习作业】

(1) 复盐制备的基本原理是什么？
(2) 如何证明所制备的化合物就是硫酸亚铁铵？
(3) 鉴定等级所用的目视比色法的原理是什么？等级应如何确定？
(4) 根据本实验的实验原理及步骤，以流程图方式写出本实验的实验步骤及设计思路，并在制备流程图中标出反应条件、实验操作条件及保证硫酸亚铁铵成功制备所应采取的措施和所依据的原理。

### 【实验原理】

硫酸亚铁铵$[(NH_4)_2SO_4 \cdot FeSO_4 \cdot 6H_2O]$又称莫尔盐，是浅蓝色透明晶体，比一般的亚铁盐稳定。它在空气中不易被氧化，溶于水但不溶于乙醇。在定量分析中，常用硫酸亚铁铵来配制亚铁离子的标准溶液。

硫酸亚铁铵在水中的溶解度比组成它的组分$FeSO_4$或$(NH_4)_2SO_4$的溶解度都要小，因此只需要将浓度较高的$FeSO_4$溶液与$(NH_4)_2SO_4$溶液混合即得硫酸亚铁铵晶体。

本实验先将铁屑溶于稀$H_2SO_4$，制得$FeSO_4$溶液，再加入$(NH_4)_2SO_4$固体并使其全部溶解，经浓缩、冷却即得溶解度小的硫酸亚铁铵晶体。

$$Fe + H_2SO_4 = FeSO_4 + H_2 \uparrow$$
$$FeSO_4 + (NH_4)_2SO_4 + 6H_2O = (NH_4)_2SO_4 \cdot FeSO_4 \cdot 6H_2O$$

由于$Fe^{2+}$容易被氧化成$Fe^{3+}$，因此在实验过程中要加入强酸$H_2SO_4$，防止$Fe^{2+}$氧化成$Fe^{3+}$。

产品中的杂质$Fe^{3+}$含量可用比色法来鉴定。因$Fe^{3+}$能与过量的$SCN^-$生成血红色的$[Fe(SCN)_6]^{3-}$。在产品溶液中加入$SCN^-$后，若溶液呈较深的红色，则表明产品中含$Fe^{3+}$较多，反之表明产品中含$Fe^{3+}$较少。将它所呈现的红色与$[Fe(SCN)_6]^{3-}$标准溶液颜色进行比较，找出与之深浅程度一致的标准溶液，则该标准溶液所示$Fe^{3+}$含量为产品的杂质$Fe^{3+}$含量，依此便可确定产品的等级。一、二、三级的1g硫酸亚铁铵的含$Fe^{3+}$限量分别为0.05mg、0.10mg及0.20mg。

### 【仪器材料与试剂】

**仪器**　台秤，锥形瓶(250mL)，烧杯(50mL)，水浴锅(内径16cm)，量筒(100mL，10mL)，减压过滤装置，蒸发皿，滤纸，目视比色管(25mL)。

**材料与试剂**　铁屑，$(NH_4)_2SO_4$固体，10% $Na_2CO_3$溶液，3mol·L$^{-1}$ $H_2SO_4$溶液，3mol·L$^{-1}$ HCl溶液，25%乙醇溶液，25% KSCN溶液，$Fe^{3+}$标准溶液，一级试剂(含$Fe^{3+}$0.05mg)，二级试剂(含$Fe^{3+}$0.10mg)，三级试剂(含$Fe^{3+}$0.20mg)。

## 【实验步骤】

### 1. 铁屑的净化

称取 4g 铁屑于锥形瓶中,加入 20mL 10% $Na_2CO_3$ 溶液,水浴加热约 10min,倾去碱液,并用去离子水洗净铁屑。

### 2. 硫酸亚铁的制备

往盛有铁屑的锥形瓶中加入 25mL 3mol·$L^{-1}$ $H_2SO_4$ 溶液,水浴加热约 30min(在通风橱中进行)。在加热过程中,应经常摇动锥形瓶以加速反应,并适当添加少量水分。待反应至无气泡产生后,趁热减压过滤。将滤液转移到小烧杯中,分别用 1mL 3mol·$L^{-1}$ $H_2SO_4$ 和新鲜去离子水洗涤锥形瓶和铁屑残渣($Fe^{2+}$ 在强酸性介质中较稳定,加入硫酸是为了防止滤液中的 $Fe^{2+}$ 转化为 $Fe^{3+}$)。再次减压过滤,合并滤液于蒸发皿中。

收集铁屑残渣,用滤纸吸干后称重,计算参与反应的铁屑量。

### 3. 硫酸亚铁铵的制备

向上述硫酸亚铁溶液中加入约 9.5g$(NH_4)_2SO_4$。水浴加热,充分搅拌使其完全溶解。继续蒸发浓缩至溶液表面出现晶膜,停止加热,静置,冷却至室温,即得硫酸亚铁铵晶体。减压过滤,用少量乙醇洗涤晶体两次,取出晶体置于两张滤纸之间,轻压以吸干母液。晾干,称重,计算理论产量与产率。

### 4. 产品检验

不同级别 $Fe^{3+}$ 标准溶液的配制

$Fe^{3+}$ 标准溶液配制:准确称取 0.4317g $NH_4Fe(SO_4)_2·12H_2O$ 溶解,并定容至 250mL。

一级试剂(含 $Fe^{3+}$ 0.05mg)配制:取 $Fe^{3+}$ 标准溶液 0.25mL,加 2mL 2mol·$L^{-1}$ HCl 溶液和 1mL 1mol·$L^{-1}$ KSCN 溶液,加蒸馏水定容至 25.00mL。

二级试剂(含 $Fe^{3+}$ 0.10mg)配制:取 $Fe^{3+}$ 标准溶液 0.5mL,加 2mL 2mol·$L^{-1}$ HCl 和 1mL 1mol·$L^{-1}$ KSCN 溶液,加蒸馏水定容至 25.00mL。

三级试剂(含 $Fe^{3+}$ 0.20mg)配制:取 $Fe^{3+}$ 标准溶液 1mL,加 2mL 2mol·$L^{-1}$ HCl 和 1mL 1mol·$L^{-1}$ KSCN 溶液,加蒸馏水定容至 25.00mL。

称取约 1g 上述硫酸亚铁铵样品于 25mL 目视比色管中,用 15mL 新鲜去离子水溶解,加 2mL 3mol·$L^{-1}$ HCl 和 1mL 25% KSCN 溶液,加蒸馏水稀释至 25mL,摇匀。与标准溶液进行目视比色,确定产品的等级。

## 【注意事项】

(1)铁屑在使用之前必须净化。

(2)铁屑与 $H_2SO_4$ 溶液的反应须在通风橱中进行。

(3)反应完成后,别用 1mL 3mol·$L^{-1}$ $H_2SO_4$ 和不含氧的蒸馏水洗涤锥形瓶和铁屑残渣,以防止滤液中的 $Fe^{2+}$ 转化为 $Fe^{3+}$。

## 【数据记录及结果分析】

日期：　　　　　　　室温：　　　　℃　　　　相对湿度：

(1) 硫酸亚铁铵的制备

铁屑的质量 $m(Fe)_1=$ _____g；

铁屑残渣重 $m(Fe)_2=$ _____g；

参加反应的铁屑质量 = _____g；

硫酸亚铁铵产量 $m[(NH_4)_2SO_4 \cdot FeSO_4 \cdot 6H_2O] =$ _____g；

硫酸亚铁铵产率(%) = $\dfrac{产品质量(g)}{理论产量(g)} \times 100\% =$

(2) 产品纯度检验

| $Fe^{3+}$ 标准溶液 | 一级($Fe^{3+}$ 0.05mg) | 二级($Fe^{3+}$ 0.10mg) | 三级($Fe^{3+}$ 0.20mg) |
| --- | --- | --- | --- |
| 硫酸亚铁铵产品级别 | | | |

## 【思考题】

(1) 什么叫复盐？复盐与形成它的简单盐相比有什么特点？

(2) 在蒸发及浓缩过程中，若发现溶液变为黄色，可能是什么原因？应如何处理？

(3) 如何计算硫酸亚铁铵的产率？应根据铁的用量还是硫酸铵的用量？

（母昭德）

# 实验十九　转化法制备硝酸钾

## 【实验目的】

(1) 了解复分解反应制备盐类及利用温度对物质溶解度的影响进行分离的方法。

(2) 熟练掌握重结晶法的基本原理和操作方法，进一步巩固溶解、加热、结晶、过滤等操作。

## 【预习作业】

(1) 无机物常用的制备方法有哪些？

(2) $KNO_3$ 的制备方法有哪些？请指出各方法的优缺点。

(3) 什么是重结晶法？操作时注意事项主要有哪些？

(4) 本实验中如何确定各原料的加入量？硝酸钾的理论产量是多少？

(5) 请用流程图阐释本实验的实验步骤，并标明实验过程的注意事项及依据。

## 【实验原理】

工业上，通常采用转化法制备硝酸钾晶体，其反应为：

$$NaNO_3 + KCl \rightleftharpoons NaCl + KNO_3$$

硝酸钾等四种盐类在不同温度下的溶解度见表19-1。由于反应式是可逆的，利用温度对物质溶解度的不同影响，可促使反应向生成产物的方向移动，并将产物进行分离，从而提高硝酸钾晶体的产率和纯度。

表 19-1　不同温度下硝酸钾等四种盐类的溶解度（单位：g/100g $H_2O$）

| 温度/℃ | 0 | 10 | 20 | 30 | 40 | 60 | 80 | 100 |
|---|---|---|---|---|---|---|---|---|
| $KNO_3$ | 13.3 | 20.9 | 31.6 | 45.8 | 63.9 | 110.0 | 169.0 | 246.0 |
| KCl | 27.6 | 31.0 | 34.0 | 37.0 | 40.0 | 45.5 | 51.1 | 56.7 |
| $NaNO_3$ | 73.0 | 80.0 | 88.0 | 96.0 | 104.0 | 124.0 | 148.0 | 180.0 |
| NaCl | 35.7 | 35.8 | 36.0 | 36.3 | 36.6 | 37.3 | 38.4 | 39.8 |

从表 19-1 中数据可看出，氯化钠的溶解度随温度变化不大，氯化钾、硝酸钠和硝酸钾在较高温下具有很大的溶解度，而硝酸钾在低温下溶解度明显减少。根据上述物质溶解度随温度变化的差异大小，可将硝酸钠和氯化钾的混合液进行加热浓缩，使氯化钠达到饱和并析出大量晶体。在 120℃左右，由于硝酸钾的溶解度较大，高温时处于不饱和状态，趁热过滤可除去大部分氯化钠，再将滤液自然冷却至室温。此时，硝酸钾溶解度随温度下降而急剧下降，在室温时以晶体形式大量析出，从而得到硝酸钾粗产品。再经重结晶，即可得到硝酸钾纯品。

## 【仪器材料与试剂】

**仪器**　硬质试管（300mL），小试管（10mL×2），量筒（50mL），烧杯（500mL，100mL），温度计（200℃），铁架台，台秤，玻璃棒，砂芯漏斗，吸滤瓶，循环水真空泵，电陶炉，蒸发皿，表面皿

**材料与试剂**　硝酸钠（化学纯，CP），氯化钾（化学纯，CP），$AgNO_3$ 溶液（0.1mol·$L^{-1}$），$HNO_3$ 溶液（5mol·$L^{-1}$）

## 【实验步骤】

### 1. 硝酸钾的制备

称取 22.0g 硝酸钠和 15.0g 氯化钾于 300mL 硬质试管中，加入 35mL 去离子水。将试管置于甘油浴中加热（试管用铁夹垂直地固定在铁架台上；用 500mL 烧杯盛约 300mL 的甘油作为甘油浴，插入温度计，烧杯放在电陶炉上进行加热，油浴烧杯外与试管内液相水平处画一标记线，油浴温度控制在 140～160℃，搅拌使固体完全溶解，继续加热搅拌，使溶液蒸发浓缩至原体积的 2/3。这时试管内有大量晶体析出，趁热减压过滤，并迅速将滤液倒出。滤液盛于小烧杯中，自然冷却至室温，此时析出的晶体为硝酸钾粗产品，减压过滤，晶体称重，计算理论产量和粗产率，记录结果于表 19-2。

### 2. 硝酸钾的纯化

预留少量（约 0.03g）粗产品供纯度检验外，将粗产品转移入蒸发皿中，按粗产品：水 = 2∶1（质量比）的比例，加入一定量的去离子水，小火加热，并搅拌，待晶体全部溶解后，停止加热。将溶液自然冷却至室温，待充分析出晶体后减压过滤。所得晶体置于表面皿上，放入 120℃烘箱中烘干，冷却，称重，计算产率，记录结果于表 19-2。

### 3. 产品纯度检验

分别取 0.03g 粗产品和一次重结晶得到的产品放入两支小试管中，各加入 3mL 去离子水配成

溶液。在溶液中分别滴加 1 滴 5mol·L$^{-1}$ HNO$_3$ 溶液和 2 滴 0.1mol·L$^{-1}$ AgNO$_3$ 溶液,观察现象,进行对比,记录结果于表 19-2。

## 【注意事项】

(1) 加热反应物时,由于采用油浴及其温度超过 100℃,硬质试管外不能挂水,否则易引起爆沸。

(2) 热过滤后,滤液不能骤冷,以防晶体过细。

(3) 重结晶时,若溶液沸腾,晶体还未全部溶解,可再加入少量蒸馏水继续搅拌至溶解。

## 【数据记录及结果分析】

日期:　　　　　　　温度:　　　　℃　　　　相对湿度:

表 19-2　实验数据及结果处理

| 物质 | 质量/g | 产率/% | 纯度检验现象 |
|---|---|---|---|
| NaNO$_3$ | | — | — |
| KCl | | — | — |
| KNO$_3$(理论值) | | — | — |
| KNO$_3$ 粗产品 | | | |
| KNO$_3$ 纯品 | | | |

## 【思考题】

(1) 硝酸钾的制备过程中,为何第二次减压过滤后得到的晶体是粗产品?

(2) 根据溶解度数据,计算本实验应分别有多少 NaCl 和 KNO$_3$ 晶体析出(不考虑其他盐存在时对溶解度的影响)?

(3) 实验中为何要趁热过滤去 NaCl 晶体?为何要小火加热?

(4) 将粗产品重结晶时,粗产品和水的比例如何确定?

(5) KNO$_3$ 晶体中混有 KCl 和 NaNO$_3$ 晶体时,应如何提纯?

(黄　静)

# 实验课题六
# 综合及研究性实验

## 实验二十 醋酸解离常数的测定与食醋中HAc含量的测定

### 【实验目的】

（1）掌握测定解离平衡常数及解离度的方法。
（2）掌握用pH计测定溶液的pH的方法。
（3）掌握酸碱滴定法测定食醋中HAc的含量的方法。

### 【预习作业】

（1）弱酸的解离常数的意义是什么？
（2）可以用哪些方法测定弱酸的解离常数？
（3）配制不同浓度的HAc溶液时所用容量瓶是否需要干燥？测定不同浓度的HAc溶液pH时所用小烧杯是否需要干燥？
（4）测定不同浓度的HAc溶液pH时为何溶液要按由稀到浓的顺序测定？
（5）请用流程图阐释本实验是如何设计实验步骤，以测定HAc的解离常数与食醋中HAc的含量，并在流程图中标出测定步骤中注意事项及依据？

### 【实验原理】

醋酸是一元弱酸，$K_a = 1.76 \times 10^{-5}$，在水溶液中存在如下解离平衡：

$$HAc + H_2O \rightleftharpoons H_3O^+ + Ac^-$$

一定温度下，解离达到平衡时

$$K_a = \frac{[H_3O^+][Ac^-]}{[HAc]} \approx \frac{[H_3O^+]^2}{c} \tag{20-1}$$

$$\alpha = \frac{[H_3O^+]}{c} \tag{20-2}$$

式（20-1）、（20-2）中，$[H_3O^+]$、$[Ac^-]$、$[HAc]$为平衡体系中各物质平衡浓度，$K_a$、$\alpha$分别为HAc的解离常数和解离度，HAc溶液的原始浓度为$c$，$c$可用NaOH标准溶液滴定测得。通过测定HAc溶液的pH可知$[H_3O^+]$，从而计算该HAc溶液的解离度和解离常数。

醋酸是食醋中的主要成分，其含量约为3.5~5.0g/100mL。可以用NaOH标准溶液直接滴定。由于滴定突跃范围在碱性区域，故常用酚酞作为指示剂。

## 【仪器材料与试剂】

**仪器** 酸度计，碱式滴定管（25mL）或聚四氟乙烯酸碱两用滴定管（25mL），锥形瓶（250mL×3），容量瓶（50mL×4，100mL），移液管（20mL，25mL），吸量管（1mL，5mL），小烧杯（100mL×4），温度计

**材料与试剂** $0.1\text{mol}\cdot\text{L}^{-1}$ NaOH 标准溶液（精确到 0.0001），$0.1\text{mol}\cdot\text{L}^{-1}$ HAc 溶液，市售食醋，酚酞指示剂，标准缓冲溶液（pH 6.86 $KH_2PO_4$-$Na_2HPO_4$，pH 4.00 邻苯二甲酸氢钾）

## 【实验步骤】

### 1. HAc 解离常数的测定

#### 1.1 HAc 溶液浓度的测定

用 20mL 移液管吸取 $0.1\text{mol}\cdot\text{L}^{-1}$ HAc 20.00mL 于锥形瓶中，加入 2 滴酚酞指示剂，用已知准确浓度的 NaOH 标准溶液滴定至溶液呈微红色且 30s 内不褪色，即为滴定终点。记录所消耗 NaOH 标准溶液体积于表 20-1。平行测定 3 次，计算 HAc 溶液的准确浓度。

#### 1.2 配制不同浓度的醋酸溶液

取 4 个 50mL 的容量瓶，按表 20-2 分别加入 2.50mL、5.00mL、25.00mL、50.00mL 已知准确浓度的 HAc 溶液，配制系列不同浓度的 HAc 溶液，并计算稀释后的 HAc 溶液的浓度。

#### 1.3 测定 HAc 溶液的 pH

将上述四种不同浓度的 HAc 溶液 30mL 分别转移至四只干燥小烧杯中，按溶液由稀到浓的顺序分别在酸度计上测定它们的 pH。（pH 计的用法详见第二章：基本操作及实验结果处理）。

根据实验测得的数据和 HAc 溶液的不同浓度，分别计算出 HAc 的解离度和解离常数。

### 2. 食醋中 HAc 含量的测定

食醋待测溶液的配制：用移液管吸取市售食醋 10.00mL，置于 100mL 容量瓶中，用蒸馏水稀释至刻度，摇匀。

食醋中 HAc 含量的测定：用移液管吸取食醋待测溶液 20.00mL，置于锥形瓶中，加酚酞指示剂 2 滴，用已知准确浓度的 NaOH 标准溶液进行滴定，滴定至溶液呈微红色并在 30s 内不褪色即为滴定终点。平行测定 3 次，数据记录于表 20-3，按式（20-3）计算食醋中 HAc 含量

$$\rho_{\text{HAc}} = \frac{c(\text{NaOH}) \times V(\text{NaOH}) \times M_r(\text{HAc})}{20.00 \times \frac{10.00}{100.00} \times 1000} \times 100 (\text{g}\cdot\text{mL}^{-1}) \quad (20\text{-}3)$$

$$M_r(\text{HAc}) = 60.05$$

## 【数据记录及结果分析】

日期：　　　　　　温度：　　　℃　　　　相对湿度：

表 20-1　HAc 溶液浓度的测定

| 实验序号 | 1 | 2 | 3 |
| --- | --- | --- | --- |
| 指示剂 | | | |
| 终点颜色变化 | | | |

| 实验序号 | 1 | 2 | 3 |
|---|---|---|---|
| $V(\text{HAc})/\text{mL}$ | | | |
| $V_{终}(\text{NaOH})/\text{mL}$ | | | |
| $V_{初}(\text{NaOH})/\text{mL}$ | | | |
| $\Delta V(\text{NaOH})/\text{mL}$ | | | |
| $\overline{V}(\text{NaOH})/\text{mL}$ | | | |
| $c(\text{NaOH})/\text{mol·L}^{-1}$ | | | |
| $c(\text{HAc})/\text{mol·L}^{-1}$ | | | |
| $\overline{c}(\text{HAc})/\text{mol·L}^{-1}$ | | | |
| 相对平均偏差 $\overline{d_r}/\%$ | | | |

表 20-2  HAc 溶液离解度和离解平衡常数的测定

| 实验序号 | 1 | 2 | 3 | 4 |
|---|---|---|---|---|
| $c_{原始}(\text{HAc})/\text{mol·L}^{-1}$ | | | | |
| $V_{原始}(\text{HAc})/\text{mL}$ | 2.50 | 5.00 | 25.00 | 50.00 |
| $V_{总}(\text{HAc 稀释后})/\text{mL}$ | 50.00 | 50.00 | 50.00 | 50.00 |
| $c_{稀释}(\text{HAc})/\text{mol·L}^{-1}$ | | | | |
| pH | | | | |
| $[\text{H}_3\text{O}^+]/\text{mol·L}^{-1}$ | | | | |
| $\alpha$ | | | | |
| $K_a$ | | | | |
| $\overline{K_a}$ | | | | |
| 相对平均偏差 $\overline{d_r}/\%$ | | | | |

表 20-3  食醋中 HAc 含量的测定

| 实验序号 | 1 | 2 | 3 |
|---|---|---|---|
| 指示剂 | | | |
| 终点颜色变化 | | | |
| $V_{原始}(\text{HAc})/\text{mL}$ | | | |
| $V_{总}(\text{稀释后 HAc})/\text{mL}$ | | | |
| $V_{待测液}(\text{稀释后 HAc})/\text{mL}$ | | | |
| $V_{初}(\text{NaOH})/\text{mL}$ | | | |
| $V_{终}(\text{NaOH})/\text{mL}$ | | | |
| $\Delta V(\text{NaOH})/\text{mL}$ | | | |
| $\overline{V}(\text{NaOH})/\text{mL}$ | | | |
| $c(\text{NaOH})/\text{mol·L}^{-1}$ | | | |
| $\rho(\text{HAc})/\text{g·mL}^{-1}$ | | | |
| $\overline{\rho}(\text{HAc})\%/\text{g·mL}^{-1}$ | | | |
| 相对平均偏差 $\overline{d_r}/\%$ | | | |

## 【思考题】

（1）解离度和解离常数在反映电解质性质时有何区别与联系？同温下不同浓度的 HAc 溶液的解离度是否相同？其解离常数是否相同？

（2）如果改变测定温度，则 HAc 溶液的解离度和解离常数有无变化？

（3）在 NaOH 溶液滴定 HAc 溶液的实验中能否选用甲基红作为指示剂？若选用其作为指示剂，滴定结果是偏高还是偏低？

（李 蓉）

# 实验二十一　茶叶中钙、镁及铁含量的测定

## 【实验目的】

（1）掌握配位滴定法测定茶叶中钙、镁含量的原理和方法。

（2）掌握可见分光光度法测定茶叶中微量铁的原理和方法。

（3）学习天然产物灰化的处理方法。

## 【预习作业】

（1）请给出 EDTA 的结构和主要理化性质。

（2）通过文献查阅，说明钙、镁及铁对人体的生物学重要性。

（3）如何分别测量钙、镁的含量？

（4）邻二氮菲法的专属性如何？

（5）根据配位滴定法和分光光度法的原理，参照本实验步骤以流程图方式给出实验的设计思路，并在流程图中标注上滴定反应条件、操作条件及保证实验结果的准确度所采取的可行性实验措施及依据。

## 【实验原理】

茶叶中除了主要含有 C、H、O 和 N 等元素外，还含有人体必需的 Ca、Mg、Fe 等多种微量元素。将茶叶在空气中置于敞口的蒸发皿或坩埚中加热，经氧化分解而烧成灰烬，再用酸溶解，可对处理后样品中的钙、镁及铁等多种元素含量进行分析。

借助配位滴定法可测得钙和镁总量，在 pH 10 的条件下，以铬黑 T 为指示剂，用 EDTA 标准溶液进行滴定。用此法时 $Fe^{3+}$、$Al^{3+}$ 离子的存在会干扰钙、镁离子的测定，可用三乙醇胺掩蔽。

茶叶中铁含量较低，可用分光光度法测定。在 pH 2～9 的条件下，$Fe^{2+}$ 与邻二氮菲生成稳定的橙红色的配合物，反应式如下：

该配合物 20℃时的 $\lg K_稳 = 21.3$，摩尔吸收系数 $\varepsilon_{510} = 1.10 \times 10^4$。需要注意的是，溶液的 pH<2 时该显色反应速度较慢。

可用盐酸羟胺将 $Fe^{3+}$ 还原成 $Fe^{2+}$，反应为：

$$2Fe^{3+} + 2NH_2OH \cdot HCl = 2Fe^{2+} + 4H^+ + N_2\uparrow + 2Cl^- + 2H_2O$$

## 【仪器材料与试剂】

**仪器** （紫外）可见分光光度计（如 722、UV800、TU1810 型），万分之一分析天平，容量瓶（50mL×8，250mL×2），吸量管（25mL，10mL），量筒（10mL，25mL），碱式滴定管或酸碱两用型滴定管（25mL 或 50mL），锥形瓶（250mL×5），小烧杯（100mL），蒸发皿或坩埚，漏斗，定量滤纸，玻璃棒等

**材料与试剂** $6mol \cdot L^{-1}$ $NH_3 \cdot H_2O$ 溶液，$6mol \cdot L^{-1}$ HCl 溶液，$2mol \cdot L^{-1}$ HAc 溶液，$6mol \cdot L^{-1}$ NaOH 溶液，$0.0100mol \cdot L^{-1}$ EDTA 标准溶液，$0.010g \cdot L^{-1}$ $NH_4Fe(SO_4)_2 \cdot 12H_2O$ 铁标准溶液，25%($g \cdot g^{-1}$) 三乙醇胺溶液，pH 10 $NH_3$-$NH_4Cl$ 缓冲液，pH 4.6 HAc-NaAc 缓冲液，0.1%($g \cdot g^{-1}$) 邻二氮菲溶液，1%($g \cdot g^{-1}$) 盐酸羟胺溶液，1%($g \cdot g^{-1}$) 铬黑 T，市售茶叶，去离子水等

## 【实验步骤】

### 1. 茶叶灰化及试液的制备

茶叶干燥研磨粉碎后，准确称量茶叶细末 6～8g（精确至 0.0001g），倒入蒸发皿或坩埚中，加热，使茶叶干灰化，冷却后通风橱中用 10.0mL 的 $6mol \cdot L^{-1}$ HCl 溶液溶解。

将溶液转移至小烧杯中，用去离子水约 20.0mL 分三次洗涤蒸发皿，洗涤液全部并入小烧杯，$6mol \cdot L^{-1}$ $NH_3 \cdot H_2O$ 溶液调节至 pH 为 6～7，使其产生沉淀。在沸水浴中加热 30min，过滤，用去离子水洗涤烧杯和滤纸。滤液直接转移入 250mL 容量瓶，并稀释至刻度，摇匀，贴上标签，标明为 $Ca^{2+}$、$Mg^{2+}$ 离子待测试液 1#。

另取 250mL 容量瓶，置于玻璃漏斗下，用 10.0mL 的 $6mol \cdot L^{-1}$ HCl 溶液重新溶解滤纸上的沉淀，用去离子水洗涤滤纸 2 次，滤液全部进入容量瓶后用水稀释滤液至刻度线，摇匀，贴上标签，标明为 $Fe^{3+}$ 离子待测试液 2#。

### 2. 茶叶中钙镁总含量的测定

从 1# 容量瓶中准确吸取 25.00mL 试液置于 250mL 锥形瓶中，加入三乙醇胺 5.0mL，再加 $NH_3 \cdot H_2O$-$NH_4Cl$ 缓冲溶液约 15mL，摇匀。加铬黑 T 指示剂 2 滴，用 $0.0100mol \cdot L^{-1}$ EDTA 标准溶液滴定，记录结果于表 21-1。根据 EDTA 溶液的体积，计算茶叶中钙镁总量，以 CaO 的质量分数表示。采用相同方法平行测定三次。

### 3. 茶叶中铁含量的测定

#### 3.1 铁标准曲线的绘制

分别准确移取 0、1.00mL、2.00mL、3.00mL、4.00mL、5.00mL、6.00mL 铁标准溶液，于 7 只 50mL 容量瓶中，依次加入 5.00mL 盐酸羟胺溶液，5.00mL 的 HAc-NaAc 缓冲溶液，5.00mL 邻二氮菲溶液，用去离子水稀释至刻度，摇匀。放置 10min 后在可见分光光度计上，用空白液作对照于 510nm 波长处分别测定吸光度。以标准铁溶液中铁含量为横坐标，相应的吸光度为纵坐标，绘制

邻二氮菲亚铁的标准曲线。

### 3.2 茶叶中铁含量的测定

从 $2^\#$ 容量瓶中准确移取 2.50mL 待测试液，于 50mL 容量瓶中，操作同 3.1，测定吸光度，数据记录于表 21-2 并从标准曲线上求出待测试液中 $Fe^{2+}$ 的含量 $c(Fe)(g·L^{-1})$，以 $Fe_2O_3$ 质量分数 $\omega(Fe_2O_3)$ 表示。采用相同方法平行测定三次。

### 3.3 计算该茶叶样品中的钙、镁及铁含量

## 【数据记录及结果分析】

日期： 温度： ℃ 相对湿度：

茶叶粉末样品的质量：_____ g

（1）茶叶中总 Ca、Mg 质量分数的测定

表 21-1 茶叶中总 Ca、Mg 质量分数 $\omega$ 的测定

| 实验序号 | 1 | 2 | 3 |
|---|---|---|---|
| 指示剂 | | | |
| 终点前后颜色变化 | | | |
| $V(1^\#$ 水样$)$ / mL | | | |
| 三乙醇胺 / mL | | | |
| $NH_3$-$NH_4Cl$ 缓冲溶液 / mL | | | |
| $V_{终}$(EDTA) / mL | | | |
| $V_{初}$(EDTA) / mL | | | |
| $\Delta V_{消耗}$(EDTA) / mL | | | |
| $c(Ca^{2+}$ 和 $Mg^{2+})$ / μmol·$L^{-1}$ | | | |
| 茶叶中总 Ca、Mg 质量分数 / % | | | |
| 相对平均偏差 $\overline{d_r}$ / % | | | |

$$\omega(CaO)\% = \frac{c(EDTA) \times V(EDTA) \times M_r(CaO)}{m(样品) \times \frac{25.00}{250.00} \times 1\,000} (g·g^{-1}) \quad (21\text{-}1)$$

$$M_r(CaO) = 56.077\,4$$

（2）茶叶中 Fe 的质量分数测定

表 21-2 茶叶中 Fe 的质量分数测定

| 实验序号 | 1(空白) | 2 | 3 | 4 | 5 | 6 | 7 | 8/9/10 |
|---|---|---|---|---|---|---|---|---|
| 标准 $Fe^{2+}$ 溶液 / mL | 0 | 1.00 | 2.00 | 3.00 | 4.00 | 5.00 | 6.00 | - |
| 茶叶样品溶液 / mL | 0 | 0 | 0 | 0 | 0 | 0 | 0 | 2.5 |
| 盐酸羟胺 / mL | 5.00 | 5.00 | 5.00 | 5.00 | 5.00 | 5.00 | 5.00 | 5.00 |
| HAc-NaAc 缓冲溶液 / mL | 5.00 | 5.00 | 5.00 | 5.00 | 5.00 | 5.00 | 5.00 | 5.00 |
| 邻二氮菲 / mL | 5.00 | 5.00 | 5.00 | 5.00 | 5.00 | 5.00 | 5.00 | 5.00 |
| $V_{总}$(稀释) / mL | 50.00 | 50.00 | 50.00 | 50.00 | 50.00 | 50.00 | 50.00 | 50.00 |
| $c_{稀释}(Fe^{2+})$ / g·$L^{-1}$ | | | | | | | | |
| 测定波长 $\lambda$ / nm | | | | | | | | |

| 实验序号 | 1(空白) | 2 | 3 | 4 | 5 | 6 | 7 | 8/9/10 |
|---|---|---|---|---|---|---|---|---|
| 吸光度 $A$ | | | | | | | | |
| $Fe^{2+}$ 待测液的浓度 / $\mu mol \cdot L^{-1}$ | | | | | | | | |
| 茶叶中铁的质量分数 $\omega(Fe_2O_3)\%/(g \cdot g^{-1})$ | | | | | | | | |
| 相对平均偏差 $\overline{d_r}$ / % | | | | | | | | |

绘制标准曲线及给出回归方程及相关系数,再根据待测溶液的吸光度,借助标准曲线法及下列式子计算茶叶中铁含量。

$$\omega(Fe)\% = \frac{c(Fe) \times 50}{m(样品) \times \frac{2.50}{250.00} \times 1\,000} (g \cdot g^{-1}) \tag{21-2}$$

$$\omega(Fe_2O_3)\% = \omega(Fe) \times \frac{M_r(Fe_2O_3)}{M_r(Fe)} (g \cdot g^{-1}) \tag{21-3}$$

$M_r(Fe_2O_3) = 159.688$,$M_r(Fe_2O_3) = 55.845$

【思考题】

(1) 简述分光光度法测定铁的原理,用该法测得的铁含量是否为茶叶中亚铁含量?为什么?

(2) 为什么 pH 6~7 时,能将 $Fe^{3+}$ 离子与 $Ca^{2+}$,$Mg^{2+}$ 离子完全分离?

(3) 分析本实验的误差主要来自哪些过程,比较配位滴定法和分光光度法对实验相对误差的要求。

(胡 新)

# 实验课题七
# 自行设计实验

## 实验二十二 实验设计及研究

### 【实验项目】

（1）$NH_3$ 的解离平衡常数 $K_b$ 测定。
（2）小苏打样品中 $NaHCO_3$ 含量测定。
（3）明矾样品中 $Al^{3+}$ 离子含量测定。
（4）混合物中 $CuSO_4$ 含量测定。
（5）蛋壳中 $Ca^{2+}$ 含量测定。

### 【实验目的】

（1）查阅资料，设计实验方案，依方案测定结果，培养综合实验技能。
（2）掌握定量测定某些物质含量的分析方法。
（3）掌握化学实验规范化基本操作（定性、定量玻璃仪器及仪器设备）。
（4）掌握实验数据处理方法。

### 【仪器材料与试剂】

**仪器** 分析天平（精确至 0.000 1g），台秤，可见光分光光度计，酸度计，电炉，水浴加热锅，滴定台，酒精灯，三脚架，石棉网，吸量管（1mL，2mL，5mL，10mL），移液管（20mL×2，25mL×2），量筒（5mL，10mL，25mL），酸式滴定管（25mL），碱式滴定管（25mL）或两用滴定管（25mL），比色管（25mL×6），烧杯（50mL×6，100mL，250mL），锥形瓶（250mL×3），容量瓶（50mL×3，100mL×3，250mL），玻棒，胶头滴管，称量瓶，漏斗，滤纸等

**材料与试剂**

（1）待测物质：$NH_3·H_2O$（1.0mol·$L^{-1}$）溶液，明矾样品，小苏打样品，硫酸铜样品，蛋壳样品
（2）一级标准物质：无水 $Na_2CO_3$（AR），$KHC_8H_4O_4$（AR），$CuSO_4·5H_2O$（AR），$MgSO_4·7H_2O$（AR），$ZnSO_4·7H_2O$（AR）
（3）其他试剂：6mol·$L^{-1}$ $H_2SO_4$ 溶液，6mol·$L^{-1}$ HCl 溶液，0.1mol·$L^{-1}$ HCl 溶液，6mol·$L^{-1}$ $HNO_3$ 溶液，蒸馏水，pH 10 的 $NH_3$-$NH_4Cl$ 缓冲溶液，1.0mol·$L^{-1}$ $NH_3$-$NH_4Cl$ 缓冲溶液，0.1mol·$L^{-1}$ NaOH 溶液，1.0mol·$L^{-1}$ HAc-NaAc 缓冲溶液，6.0mol·$L^{-1}$ $NH_3·H_2O$ 溶液，0.01mol·$L^{-1}$ EDTA 溶液等

(4) 指示剂: 酚酞, 甲基橙, 铬黑 T(s), 二甲酚橙, 钙红指示剂(s), 广泛 pH 试纸等

# 【具体要求】

## 1. 项目实施(考核)流程

(1) 组队: 学生以 4～5 的人数为限, 自行提前 1～2 周组成小组。

(2) 准备实验方案: 小组根据以上提供的主要仪器材料及试剂, 提前 1～2 周, 通过查阅文献资料, 依表 22-1 设计 5 个项目的最佳实验方案。方案包括: 实验目的与要求、实验原理、仪器材料与试剂、详细的实验步骤、实验现象、实验数据及结果处理结论、讨论及参考文献(其中实验现象、数据记录、结果处理、结论及讨论预留空栏)。

(3) 上交实验方案: 考前 20min, 每个小组将 5 个实验项目的设计方案统一交教师检查, 面对面更正方案错误, 并评分。

(4) 确定考核项目: 每个实验小组要实施的项目, 由教师在考前 5min 从 5 个项目中指定, 不同小组项目不同。

(5) 确定实验操作的考核内容: 教师根据小组人数, 将实验方案分成与实验人数对应的步骤, 组内成员在规定时间内, 根据自己设计的实验方案以"接力"方式, 分别完成其中的一个步骤, 小组团队最终完成整个实验。

(6) 实施操作考核: 教师确定小组内每个成员需完成的实验步骤, 并记录实验过程的不规范化操作点及评分。

(7) 完成实验报告: 实验结束, 小组成员在指定的时间内进行数据处理, 现场完成实验报告, 教师对实验报告评分。

(8) 答辩: 由组内成员口头汇报实验结果, 并根据设计的方案和实验结果回答教师或其他组同学提出的问题, 教师根据答辩标准评分。

## 2. 成绩构成

成绩由操作、五个实验方案、实验报告(数据处理、结果与结论)和答辩的分数构成, 详见表 22-2。

表 22-1 实验项目

| 序号 | 实验项目 | 检测试样用量范围 / g (精确至 ±0.0001g) |
|---|---|---|
| 1 | 0.1mol·L$^{-1}$ NH$_3$ 溶液的解离平衡常数的测定 | , |
| 2 | 小苏打样品中 NaHCO$_3$ 含量测定 | 1.0～1.2 |
| 3 | 明矾中样品中 Al$^{3+}$ 离子含量测定 | 0.3～0.4 |
| 4 | 混合物中硫酸铜含量的测定 | 0.7～0.8 |
| 5 | 蛋壳中钙含量的测定 | 0.15～0.2 |

注: 方案可参考表 22-1 中检测样品的用量范围, 计算出一级标准物质及其他试剂所需的量, 并根据仪器精度及分析的误差要求, 尽可能提高检测结果精度。

表 22-2 实验考核评分表

| 考核时间: | 年 | 月 | 日 | 星期 | 节次 | |
|---|---|---|---|---|---|---|
| 课程 | | | | 教师 | | |
| 年级 | | | | 专业 | | |
| 实验题目 | | | | 实验室 | | |
| 考核编号 | 1 | 2 | 3 | 4 | 5 | 6 |
| 姓名 | | | | | | |
| 班组 | | | | | | |
| 学号 | | | | | | |
| 操作项目 | | | | | | |
| 操作时间 | | | | | | |
| 不规范操作点 | | | | | | |
| 操作得分(50分) | | | | | | |
| 方案得分(25分) | | | | | | |
| 实验报告(数据处理、结果与结论)(20分) | | | | | | |
| 答辩(5分) | | | | | | |
| 合计得分(100分) | | | | | | |

【提示】

(1) 在一定吸收波长范围内, $[Cu(NH_3)_4]^{2+}$ 浓度与吸光值的关系参见表 22-3。

表 22-3 $[Cu(NH_3)_4]^{2+}$ 浓度与吸光值的关系

| $[Cu(NH_3)_4]^{2+}$ 浓度 $c$/ mol·L$^{-1}$ | 0.002~0.01 |
|---|---|
| 吸光度 $A$ | 0.1~0.5 |

(2) 制备 $[Cu(NH_3)_4]^{2+}$ 溶液时,$Cu^{2+}$ 浓度与 $NH_3$ 溶液浓度的参考比:
$$c_{(Cu^{2+})}:c_{(NH_3)} = 1:200 \sim 1:40$$

(3) 0.15~0.2g 蛋壳的溶解与消化大概需要 3~4mL 的 6mol·L$^{-1}$ HNO$_3$,可加热助溶。

(4) 一级标准物质的称量范围一般为量的 ±10%。

(5) 锥形瓶中溶液用量为 20mL 左右,对 25mL 规格的滴定管,溶液消耗量控制在 20mL 左右,滴定的溶液浓度控制在 0.01~0.1mol·L$^{-1}$ 范围。

(6) 建议设计实验方案时,把固体物质样品配成 100.00mL 溶液,每次只取 20.00mL(混合物中硫酸铜含量的测定为 10.00mL,蛋壳中钙含量的测定为 10.00mL)样品溶液进行测定。

【思考题】

(1) 用酸度计测定不同酸强度的液 pH 时,为何要以由稀到浓的顺序进行?

(2) 在用滴定法测定小苏打中 NaHCO$_3$ 含量测定时,为何接近终点时剧烈振摇溶液?

(李福森)

# Part I   Experiment Essentials in Basic Chemistry

# Chapter 1
# Laboratory Rules and Safety Information

### 1. The Purpose of Experiments in Basic Chemistry

The experiment course is an important part in Basic Chemistry. Its purposes, beyond to confirm chemical theories and chemical phenomena so that for students to understand and to grasp the contents of Basic Chemistry, are to study the scientific experimental methods, to learn to combine theory with practice, to apply theory to scientific experimental methods in practice, to cultivate comprehensive experimental skills and rigorous scientific attitude.

Through the rigid discipline in experiments, students should master the standard chemical experiment operation, learn to record and deal with the experiment data; express experiment result correctly, and to analyze experiment phenomena or problems correctly to get the conclusion; learn to summarize the experimental rules, and give the experimental conclusions. By the exercises of experiment design and completion, cultivate original capacity of thinking and solving problems. Meanwhile, foster the strict scientific attitude, master the research methods and apply the experimental skills.

### 2. General Rules in Laboratory

(1) Read the Experiment Instruction Book, and refer to the textbook and references before performing experiments to understand the purposes, principles, methods and procedures of the experiments. Get ready with the preliminary report for the experiment.

(2) Read the instrumental operating manual carefully. Follow instructions carefully and do not perform unauthorized experiments.

(3) Wear white coat in laboratory and are not allowed to wear slippers. Be serious and correct in your experiments. Observe the experiment phenomena and record experiment data carefully. Complete experiment reports based on the source recordings in time.

(4) Never work alone in the laboratory. Report all accidents, no matter how minor, to the instructor

in a timely manner.

(5) Common instruments and sharing reagents are only allowed to use at their original places but not allowed to move anywhere.

(6) Never return chemicals to bottles of their origin. Always read the label twice before using a chemical reagent. Be sure the concentration, as well as the name of the reagent is correct.

(7) Don't lay reagent bottle stoppers down in any way that the part, which goes into the bottle, comes in contact with any surface. Switching reagent bottle stoppers will invariably contaminate the reagent. To avoid this, never have more than one bottle unstoppered at a time. If the stopper is the penny head type, hold it between the fingers of the hand you are pouring with, while pouring. If you do this, you can be certain you are not mixing up stoppers or contaminating the reagent.

(8) Do not take food into the laboratory. Don't taste or eat anything in the laboratory.

(9) Make sure of the safety in experiments. All the flammable chemicals should be away from any flame source. Keep the lab and the table top clean and tidy. Do not dispose of insoluble items on the ground. Used test paper, match etc should be placed in a beaker and be deposed of in designated containers after the experiment finished. Corrosive and toxic chemicals must be reclaimed. Never depose of solid waste, caustic liquid and toxic reagent in sinks.

(10) Do not leave the lab before the ending of experiment without authorization. Only allowed to leave after the experiment finished, the work site and the instruments been cleaned, the reagents been tidy up, the power and water source been turned off, and windows and doors been closed.

## 3. Safety Information

(1) Protection against water, electricity, fire and toxic gases: When entering the laboratory for the first time, you must understand all the escape routes on the floor. Every time you enter the laboratory, first open the laboratory window and ensure good ventilation. Confirm the installation of water, electricity and gas in the laboratory, the storage location of the fire extinguishing equipment and the method of use for emergency use; when using toxic or harmful gases or volatile toxic substances, operate in a fume hood.

(2) Hazardous substance protection: Be cautious to flammable, explosive, toxic, corrosive flammable or highly toxic substances. When using or dealing with it, perform at a well ventilated place far from any flame.

(3) Chemical operation protection: When heating liquids in test tubes, dry the outside wall at first, never point the tube toward yourself or anyone else. Never heat the test tube directly at the bottom but tilt the tube and heat it gently between the bottom of the tube and the top of the liquid.

(4) Chemical reagents protection: When opening reagent bottles of hydrochloric acid, nitric acid, ammonia or hydrogen peroxide, be careful that the reagent dashes out. Do not smell directly near the bottle but fan with a hand then smell slightly. When using concentrated acid, concentrated base, or cleaning solution, avoid the chemical spilling on skin or cloth. Especially protect eyes from chemical injuring (wearing protective glasses).

(5) Personal safety protection: Wear a lab apron to protect your skin and clothing whenever you are

working with hot or corrosive liquids. Leather shoes are preferred to canvas or open-toed shoes.

(6) Chemicals treatment: Keep tabletops clean. Wipe up acid and base spills promptly. If you spill a chemical on your skin, flush the area immediately with plenty of water, then wash the area with soap and water.

(7) Use of electrical equipment: When using electrical equipments, be sure of the voltage-current-power matching. Never touch electric connector with wet hand.

(8) Laboratory spare first aid kit: The first-aid kit is required for the student's laboratory. Common first-aid kits include: Medical gauze block, medical elastic bandage, band-aid, medical breathable tape, iodophor disinfectant, medical applicator, medical alcohol cotton film, clean wipes, burns, first aid manuals, etc., for emergency use.

## 4. The Treatment of Accidents in Chemical Experiments

(1) Cut: Coat the injured area with adhesive bandages after applying iodine.

(2) Scald: Daub the wound with scald medicament, or wet the wound with a concentrated $KMnO_4$ solution till the skin turn to brown, then apply vaseline or scalding ointment.

(3) Acid or base injury: rinse with a great deal of water at once. For acid injury, after rinsing, treat with saturated sodium bicarbonate solution, or diluted ammonia, or soapy water. In case of scald of concentrated sulfuric acid, it is necessary to wipe with cotton wool before rinsing with water. For base injury, after rinsing, treat with 2%~5% acetic acid or 3% boric acid solutions. If the eye is spattered in by acid, wash with water and then treat with 1%~3% sodium bicarbonate solution, followed by rinse with water; and if the eye is spattered in by base, wash with water and then treat with 3% boric acid solutions, followed by rinse with water. After the above treatments, send to the hospital for further therapy.

(4) Toxic gas inhalation: If inhaling toxic gases such as bromine, chlorine or hydrogen chloride, immediately leave the toxic atmosphere and breathe fresh air outside. A small amount of alcohol mixed with ether can be inhaled to detoxify. If inhaling hydrogen sulfide or carbon monoxide gas, immediately take a breath of fresh air outside.

(5) Poison entry: Drink a cup of diluted cupric sulfate solution, touch the throat with a finger to bring anabole, and then go to the hospital.

(6) Electric shock accident: The electric appliance is isolated from the water and the experimental tabletop is kept dry and tidy. When using electrical equipment, do not touch the electric appliance and the plug with wet hands. If the electric appliance is on fire, immediately turn off the power to prevent electric shock and then extinguish the fire.

## 5. Fire Prevention, Anti-electricity and Extinguishing Knowledge

(1) Some main causes of fire in lab

1) Flammable materials near the fire. Laboratory safety requirements: Do not use open flame heating.

2) Wire aging, connector badness and electric equipment trouble.

3) Improper chemical reaction, such as unclear chemical properties and improper operation.

4) Mixing the following together may cause fire:
- Acticarbon and ammonium nitrate;
- Clothing contaminated by strong oxidants such as potassium chlorate;
- Rag and concentrated sulfuric acid;
- Inflammable, such as lignum and fiber, and concentrated nitric acid;
- Organics and liquid oxygen;
- Aluminum and organic chloride;
- Hydrogen phosphide, silane, metal alkyl or white phosphorus contacting air.

(2) Extinguishing methods: Once a fire burns in lab, do not panic, but extinguish following the methods in Table1 and call the fire brigade in time.

Table 1  Extinguishing Methods for Inflammable

| Inflammable | Extinguishing method | Note |
| --- | --- | --- |
| Paper, textile or wood | Sand, water, fire extinguisher | Need cooling and isolated from air |
| Organic solvents such as oil, benzene etc | $CO_2$ or dry-chemical fire extinguisher, asbestos cloth, dry sand etc | Applicable for valuable instruments. Do not use a foam fire extinguisher to extinguish oil, gas, etc. |
| Alcohol, ether etc | Water | Need to dilute, cool and be apart from air |
| Wire, electric meters or instruments | $CCl_4$、$CO_2$ extinguishers | Extinguish material unable to conduct electricity. Do not use water and foam extinguisher |
| Flammable gas | Close the gas, fire extinguisher | Do everything possible to cut off the gas source |
| Active metals such as potassium and sodium, or phosphide contacting water | Dry sand, dry-chemical fire extinguisher | Do not use water and foam or $CO_2$ extinguishers |
| Clothing | Roll on the ground to put out the fire or put off clothing, cover the clothing with fireproofing material | Do not run, otherwise it aggravates burning |

(3) Protection against electric shock: First cut off the power supply, use insulators such as dry wooden sticks or bamboo poles as soon as possible to make the electric shocker out of the power supply. Breathing artificially when necessary and immediately sent to the hospital for rescue.

(籍雪平)

# Chapter 2
# Ordinary Instruments in Chemical Experiments

## 1. Introduction of Ordinary Instruments in Chemical Laboratory

The ordinary instruments in laboratory refer to glasswares, such as volumetric glasswares and non-glasswares.

(1) Container: A container is any receptacle or enclosure for holding a product used in storage, packaging, and shipping at room temperature or under heating conditions, including test tube, beaker, flask, funnel and so on, as shown in Figure 1. Choose the containers of different types and specifications according to their usage and dosage of objects.

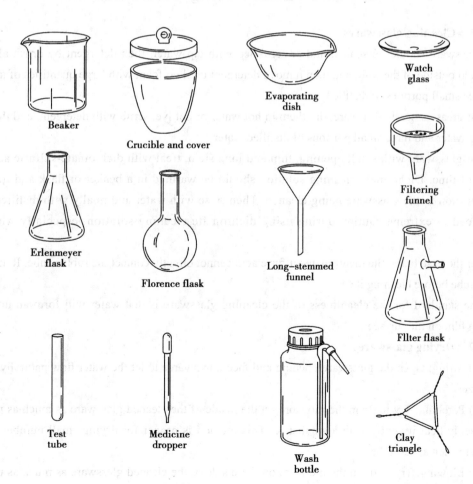

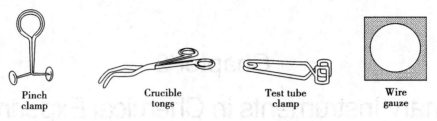

Figure 1　Ordinary glassware in laboratory

(2) Volumetric Glassware: Volumetric glassware refers to glass container that is designed to measure the volume of liquids, take and move solution, mainly including graduated cylinder, buret, volumetric flask, pipette and volumetric pipette. They have different degree of accuracy. The volumetric glassware can not be used as an experimental container, for example, for dissolving and diluting operations, not for taking hot solution, heating, long-term storage of solution. Volumetric glasswares are marked as to the manner of calibration. Items marked "E" are designed to contain the specified volume when filled liquid reaches the calibration line. Items marked "A" are designed to deliver the specified volume when liquid is delivered from the calibration line.

## 2. Cleaning and Drying Glasswares

### 2.1　Cleaning glasswares

Glassware should be soaked and/or scrubbed with tap water, then detergent by brush along the inside and outside of the wall, and then remove detergent until no foam with large quantities of tap water and three small portions of distilled water.

For grease smeary glassware, dip them in hot water or hot lye, scrub with detergent, and then rinse with tap water and three small portions of distilled water.

For glassware with small opening, thin and long stem, treat with dichromate-sulfuric acid for a period of time. Dichromate cleaning solution should be warmed in a beaker or flask and transfer a small portion to the glassware being cleaned. Then rinse with water and finally with distilled water. (You need to extreme caution during using dichromate cleaning solution, especially when it is warm.)

For the base buret, the dichromate-sulfuric acid cannot directly contact the rubber tube. Remove the rubber tube before cleaning it.

The standard for the cleanliness of the cleaning glassware is that water will form an unbroken, uniform film on the surface.

### 2.2　Drying glasswares

(1) Airing: Open the glassware opening and face downwards to let the water flow naturally out and volatilize.

(2) Parching: Try to drain the moisture on the inside of the cleaned glassware as much as possible, then evenly heat it gently with a small fire. This method is proper for drying small number of little glassware such as test tube.

(3) Blowing: Try to drain the moisture on the inside of the cleaned glassware as much as possible,

then dry it with an electric blower, as drying a beaker.

(4) Baking: Try to drain the moisture on the inside of the cleaned glassware as much as possible, then dry it in an oven at 105℃. This method is suitable for drying large numbers of thin and long glasswares.

(5) Volatilizing with organic solvents: Try to drain the moisture on the inside of the cleaned glassware as much as possible, then rinse it with volatile solvent such as acetone or alcohol, recover the solvent, rinse it again with ether. Finally dry it by airing or blowing.

Glassware expands and contracts with rising and falling temperatures, and an object that has expanded on heating does not always return to the same volume on cooling, volumetric glassware should not be heated much above its calibration temperature.

The method of cleaning base buret is just as that of acid buret, however, the chromic acid lotion cannot directly contact the rubber tube, remove the rubber tube before cleaning it.

# 3. Operation of Instrument

## 3.1 Operation of basic volumetric glasswares
### 3.1.1 Burets

A burette (also buret) is a graduated glass tube with a tap at one end, for delivering known volumes of a liquid, especially in titrations. There are acid, base, and polytetrafluoroethylene (PTFE) buret with different structure. Acid buret consists of a long tube of highly uniform diameter marked with volume graduations and fitted with a ground-glass stopcock for the sake of containing acidic or oxidative reagents, as shown in Figure 2(a). While the base buret fitted with a rubber tube with a glass ball in it for containing alkali liquid, and a short delivery tip, as shown in Figure 2(b).

An PTFE buret, which is the same as the structure of acid buret. As the stopcock of the PTFE buret is made of polytetrafluoroethylene, the buret has the function of acid and alkali resistance, which possesses the same operating mode to the acid buret.

A conventional buret is marked in 0.1 mL increments from 0 to 25 mL or 50 mL, on which a precision of 0.01 mL can be estimated. However, a semimicroburette or a microburette has the volume of 10, 5, 2 or 1 mL with the minimum calibration of 0.05, 0.01 or 0.005 mL.

(1) Checking of liquid-tightness and assemblage

1) Before using a buret, dry the stopcock with soft tissue and grease it with a little vaseline. A thin layer of stopcock grease is applied uniformly to the stopcock, using very little grease near the hole and taking care not to get any grease in the hole (Figure 3). Insert the stopcock and rotate it. There should be a uniform and transparent layer of grease, and the stopcock should not leak.

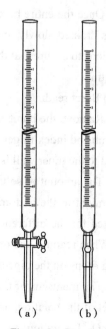

(a)　　(b)

Figure 2　Burets

(a) Acid (PTFE) buret; (b) Base buret

2) Base buret need not be greased, but the glass ball should be selected to fit the rubber tube to avoid any leak.

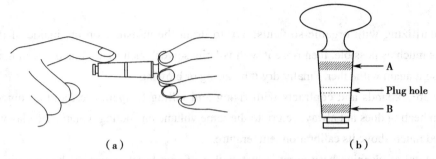

Figure 3  Stopcock of acid burette and greasing
(a) Grease the stopcock; (b) All parts of the stopcock

(2) Rinsing: Before filling the operation solution, rinse the buret several times with 10 mL solution to remove traces of water (ensure the correct concentration). Hold the buret in a horizontal position and rotate it to coat the entire surface with the rinse solution. The rinse solution should be expelled through the stopcock by opening the stopcock of the acid buret or pinching the glass ball of the base buret.

(3) Filling the operation solution and removing air bubbles

1) It is common practice to overfill a buret, then open the stopcock fully, or tilt the acid buret to $15°\sim30°$ off-horizontal plane and open the stopcock slowly, the bubbles will be removed from the buret. For base buret, bend the rubber tube, pinch the glass ball, then the bubbles will be removed by ejecting the liquid (Figure 4).

2) Once the entire buret is free of bubbles, the liquid is drained slowly until the level reaches the graduations to ensure that there is enough titrant for the titration.

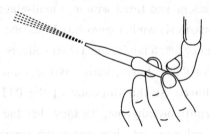

Figure 4  Remove the air bubbles in a base buret

(4) Buret reading

1) Remove the buret from the buret clamp, and the buret should be holded upright naturally using the thumb and index finger or middle finger. Read the scale $1\sim2$ min after the liquid is stable (the liquid attached to the inner wall is filled or delivered).

2) The amount of solution added in or drained out needed to be read correctly by observing at eye level straight to the bottom of "Meniscus" for titration. The initial and final volumes collected will be calculated for the difference in volume which equal to the total volume of solution drained out of the buret. When reading, the droplet should not be hung on the pipe wall. There should be no air bubbles or hanging drops on the pipe tip. The reading should be accurate to the nearest 0.01 mL.

3) The meniscus of the colorless or light solution is clearer, the bottom of the meniscus is taken as the level of the solution, as shown in Figure 5(a); but the meniscus of the colored solution is dim, such as $KMnO_4$, $I_2$ solution, etc., the eye should be level with the top of the meniscus. The reading in Figure 5(a) is 23.55 mL.

4) Using the buret with a colorless solution is sometimes difficult to observe the bottom of the meniscus so "Blue Strip" technique can help to accurately observe and measure the number on the scale, as shown in Figure 5(b), the eye should be level straight to the intersection where two meniscuses meet. For colored solution, the eye should be level straight to the top of the meniscus. The reading in Figure 5(b) is 25.00 mL.

5) The reading is best taken by using a meniscus illuminator, as shown in Figure 5(c). The meniscus illuminator has a white and a black field, the black field is positioned just below the meniscus about 1 mm, and the meniscus turns to black. Then the bottom of the meniscus is taken as the level of the solution. The reading in Figure 5(c) is 32.45 mL.

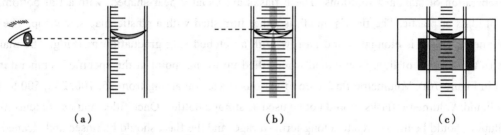

Figure 5  Read the buret
(a) The way of reading volume for buret; (b) The way of reading volume for blue strip buret; (c) The way of reading volume by black background for buret

(5) Titration

1) The buret should be clamped on the stand upright. The droplet on the pipe tip of the buret should be knocked off by the outer wall of container. The titration is performed with the sample solution in an Erlenmeyer flask. The flask is placed on a white color paper as the background, and the buret tip is positioned about 1 cm from the neck of the flask.

To acid burets, the reverse grip method is used for titration. The stopcock is manipulated with left hand. The grip on the buret should be pointed to the right and maintains a slight inward pressure on the stopcock with the thumb, index finger and middle finger to insure that leakage will not occur. The flask is swirled with the right hand, and the buret tip is positioned into the conical flask by about 1～2 cm by lifting the flask. Then the flask is rotated in the same direction circularly while titration, as shown in Figure 6 (do not vibrate back and forth, otherwise solution will be spilled out). The rate of delivery is about 10 mL·min$^{-1}$, ie, 3～4 drops per second. As the end point of a titration is approached, drop stepwise, rinse the droplets adhering to the inner wall of the conical flask with a small amount of distilled water; when the color halo disappears slowly, the end point of a titration will be reached at this time, and it can be added half a drop until the end (ie, the stopcock is slightly opened, the droplet is suspended at the pipe tip, the droplet should be knocked off by the inner wall of the conical flask, then the droplet on the inner wall of the flask should be cleaned with distilled water). At the end, the volume value should be eliminated if the

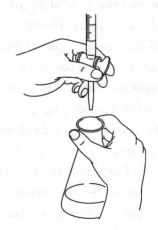

Figure 6  Titration operation

droplet is hanged on the pipe tip.

2) The titration operation of PTFE burets is similar to that of acid buret.

3) To base burets, pinch the rubber tube just on the glass ball or upper part with thumb and index finger to form a gap, and the solution can flow out of the tip of the buret. The flow rate of the solution can be controlled according to the size of the gap by pinching. Don't pinch the lower part, otherwise bubbles will form.

### 3.1.2 Volumetric Flasks

A volumetric flask (measuring flask or graduated flask) is a piece of laboratory apparatus, calibrated to contain a precise volume at a certain temperature. Volumetric flasks are used for precise dilutions and preparation of standard solutions. These flasks are usually pear-shaped, with a flat bottom, and made of glass or plastic. The flask's mouth is either furnished with a plastic snap/screw cap. The neck of volumetric flasks is elongated and narrow with an etched ring graduation marking. The marking indicates the volume of liquid contained when filled up to that point at the specified temperature, as shown in Figure 7(a). Volumetric flasks are of various sizes, containing from 25, 100, 250, 500 to 1 000 mL of liquid. Volumetric flasks should not be used as storage bottles. Once filled and its contents mixed, the solution should be transferred to a long-term storage, and the flask should be rinsed and cleaned.

(1) Checking of liquid-tightness: Fill the flask with water to two-thirds of the volume. Keeping the stopper on securely by using the index finger, and hold the flask body with another hand, invert the flask and swirl or shake it vigorously. If there is no leakage, turn the bottle upright and then turn the stopper 180°, invert the flask and check it to ensure liquid-tight.

(2) Transfer and constant volume of solution

1) Volumetric flasks only contain dissolved samples. A weighed sample is dissolved with a 10%~20% of the solvent in a beaker, transferred to the appropriate size flask, and diluted to the mark with solvent. If the solute is insoluble, it can be heated to help dissolving, and it should be transferred at room temperature to avoid volume error.

2) The method of transferring solution into a flask is shown in Figure 7(b). The lower part of the glass rod extends into the flask neck and rests under the calibration mark (other parts of the glass rod should not be touched the flask to avoid flowing to the outer). The beaker mouth is close to upper part of the glass rod and the beaker is tilted, then the solution flows into the beaker along the glass rod. The beaker is gently lifted along the glass rod and slowly stood upright, so that the droplets attached between the glass rod and the beaker mouth flow back to the beaker.

The beaker and the glass rod should be rinsed 3 times with a small amount of solvent, and then the rinsed mixture is added to the flask about a half of the volume. Swirl the solution before to obtain most of the mixing.

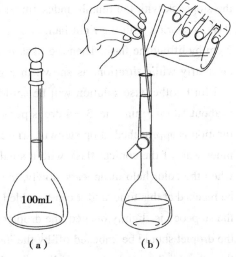

Figure 7  Volumetric flask and its operation
(a) Volumetric flask; (b) Operation

Finally, solvent is carefully added with a dropper so that the bottom of the meniscus is even with the middle of the calibration mark (at eye level). The solution is finally thoroughly mixed by swirling or turn upside down flask vigorously.

3) Don't add solvent to fill the flask if the meniscus of the liquid is below the calibration mark caused by the solution attached to the inner wall of flask neck, which does not affect the concentration of the solution prepared, otherwise, the concentration will be reduced. If the meniscus of the liquid is above the calibration mark, repreparation is needed.

3.1.3 Volumetric pipette and Pipette

(1) Volumetric pipette: These pipettes have a large bulb with a long narrow portion above with a single graduation mark as it is calibrated for a single volume (like a volumetric flask). Typical volumes are 1, 2, 5, 10, 20, 25, 50 and 100 mL. Volumetric pipettes are commonly used in analytical chemistry to make laboratory solutions from a base stock as well as to prepare solutions for titration.

A volumetric pipette or bulb pipette allows extremely accurate measurement of the volume of a solution. It is calibrated to deliver accurately a fixed volume of liquid at a certain temperature (indicated by TD, usually 25℃) with an accuracy of 0.01 mL. Typical volumes are 5, 10, 25, and 50 mL. These pipettes have a large bulb with a long narrow portion above with a single graduation mark as it is calibrated for a single volume, as shown in Figure 8(a). Measuring pipettes are straight bore pipettes with different graduation marks, as shown in Figure 8(b).

There are five steps on operation of volumetric pipettes: "reading" "rinsing" "drawing" "holding" and "draining".

1) Reading: Before operation, it is necessary to confirm the type, range, the direction of the calibration mark, the scale between graduated lines, and the word "吹" or "快" on the nozzle as well.

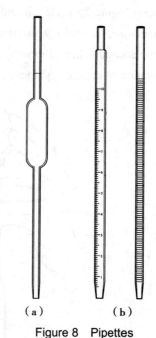

Figure 8　Pipettes
(a) A large bulb transfer pipettes; (b) Measuring pipettes

2) Rinsing: Before transferring the solution, volumetric pipette should be cleaned and rinsed 2～3 times with small portions of distilled water, and then water is completely drained. Squeeze out the air inside the aurilave, the tip of the aurilave is close to the volumetric pipette. The volumetric pipette extends vertically and rests under the liquid surface 1.5 cm. Loosen the aurilave slowly and draw a solution of about 1/3. Rotate tile pipette while holding it in a horizontal position to ensure that the rinse solution contacts the entire inside surface, and drain through either the tip or the top opening. Repeatedly, volumetric pipette should be rinsed 2～3 times with small portions of the liquid to be transferred to ensure a constant concentration.

3) Drawing and Holding: Hold over the top of the calibration mark on the volumetric pipette with the thumb and middle finger, extends vertically and rests under the solution surface 1.5~2 cm (do not touch the bottom), as shown in Figure 9(a). Pipettes are filled by drawing liquid in through the tip by the sucking action of an evacuated rubber bulb held over the top opening with the other hand. Liquid is drawn in above the calibration mark and the bulb is replaced quickly with the index finger.

4) Draining: The pipette is allowed to drain slowly until the level reaches the calibration mark. This is accomplished best by reducing the pressure of the index finger and rotating the pipette back and forth between the thumb and middle finger. The tip of the pipet should be touched to the side of a waste container to remove any partial drops that may have formed and are clinging to the outside. Finally, the index finger is removed, the pipette is held vertically and the tip is touched onto the side of the receiving vessel to allow smooth delivery freely without splashing, and so that the proper volume will be left in the tip. Most volumetric pipettes are calibrated to deliver with certain small volume remaining in the tip. Wait for 10~15 s, the forces of attraction of the liquid on the wall of the vessel will draw out a part of this, as shown in Figure 9(b). This should not be shaken or blown out if the pipette is marked with "吹" or "快"; Otherwise, do not blow out remaining in the tip, as the small volume remaining in the tip has been taken into account when calibrating the pipette.

The operation of measuring pipettes is the same as the transfer pipettes.

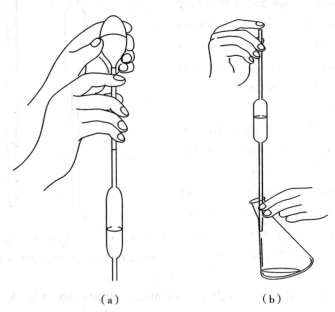

(a)          (b)

Figure 9　Operation of pipettes
(a) Drawing; (b) Draining operating Pipettes

(2) Pipette: A pipette, known as eppendoff, is used to transfer a measured volume of liquid. The common type of pipette is single piece glass pipettes (Figure 10). Pipettes are available in sizes from 0.1 μL to 5 mL. It works by creating a partial vacuum above the liquid-holding chamber and selectively releasing this vacuum to draw up and dispense liquid.

## Operation of pipettes

1) Installing of the pipette tip: The pipette tip (also known as a tip) can be installed with a rotary mounting method. Press the pipette vertically into the tip to make it tightly combined.

2) Range setting: Before operation, it is necessary to confirm its capacity specification and set volume. The range of pipettes can be set from large to small by rotating the knob counterclockwise. On the contrary, the scale should be 1/4 exceeded the set range value by over-rotating the knob clockwise and then to the set value, mechanical clearance can be excluded to ensure that the set range is accurate. Do not over-rotate the knob out of the range, otherwise, the internal mechanism will be jammed and the pipette will be damaged.

3) Drawing: The appropriate pipette tip is connected, then the control button is pressed to the first stop with the thumb, the pipette tip should be 1~6 mm below the liquid surface vertically: 1~2 mm below the liquid surface with pipette tip range of 0.1~10 μL, 2~3 mm with tip range of 2~200 μL, 3~6 mm with tip range of 1~5 mL, then the

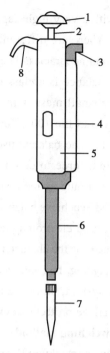

Figure 10 Structure of pipettes
1. Push button; 2. Push button rod; 3. Tip Ejector; 4. Display cover; 5. Handle; 6. Tip Ejector collar; 7. Pipette tip; 8. Grip cover

liquid is drawn by reducing the pressure of the thumb. Wait for 1~3 s, the pipette tip should be knocked off by the inner wall of container to remove any partial drops that may have formed and are clinging to the outside. When drawing organic solvent or high-volatile liquid, a negative pressure will formed by volatile gas in the white sleeve chamber, resulting in leakage. Rinsing for 4~6 times, the negative pressure can be eliminated by saturate gas.

4) Draining: The pipette tip is positioned in the container vertically. Slowly press the control button to the first stop for about 1~3 s. Drain the remaining liquid by pressing the control button to second stop repeatedly to ensure no residual liquid in the pipette tip.

5) The pipette tip can be ejected by pressing the "tip ejector" button. The pipette is set to its maximum range with the aim of keeping the spring loose, then the pipette is placed vertically on the pipette rack. If the liquid is left in the pipette tip, the pipette should not be placed horizontally or upside down to prevent the liquid from flowing back to corrode the piston spring.

In addition, when transferring high viscosity liquid, bio-active liquid, easily bubbled liquid or a trace amount of liquid, press the control button to the second stop directly, draw more than the set amount of liquid, slowly release the button to the original position, and then press the button to the first stop to drain the set amount of liquid. Keep holding the button in the first stop (no longer press to the second stop), remove the pipette tip with residual liquid, and discard it.

### 3.2 Electronic Balance

Electronic balance is a single-pan analytical instrument, which is capable of measuring weight or mass with a very high degree of accuracy and precision. There are different electronic balance with different accuracies. The one ten-thousandth electronic balance is a class of balance designed to measure small mass with accuracy of 0.1 mg (or 0.0 001g) in the 100~200 g range, as shown in Figure 11. The balance uses an electromagnet to generate a force to counter the sample being measured and outputs the result by measuring the force needed to achieve balance. Such measurement device is called electromagnetic force restoration sensor, which has the characteristics of accurate and reliable weighing, fast and clear display, automatic detection system, simple automatic calibration device, overload protection device and so on.

**The weighing methods**

(1) Direct weighing: Direct weighing is proper for samples stable in air and nonhygroscopic, such as metal or ore, etc. When weighing: ① Do not weigh the hot samples; ② Samples can not be placed directly on the pan to prevent the samples from corroding the pan, and placed on a sheet of weighing paper (the four corners of the paper are folded twice into a small carton at a quarter of the diagonal extension, the size of the carton depends on the size of the pan or the mass of the weighing samples), or the receiving vessel.

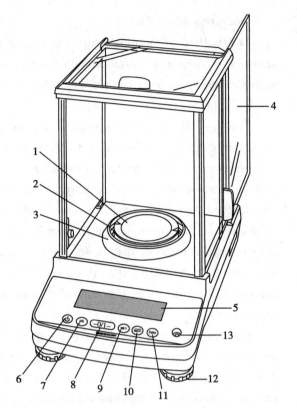

Figure 11　Electronic balance
1. Tray; 2. Tray support; 3. Tray circle; 4. Windshield; 5. Display portion; 6. Start switch; 7. "CAL" sensitivity calibration key; 8. "0/T" Determination (zero/buckle weight) key; 9. Unit switch; 10. Function switch; 11. Data output menu key; 12. Leveling pad; 13. Leveling equipment

1) Adjusting the balance level: Make sure the bubble in the leveling equipment is centered by adjusting the leveling pad. The balance can be rised by adjusting the leveling pad clockwise to elongate the pad, otherwise the balance will be decreased. Clean the pan by a small brush.

2) Warming up: Power on the balance, warm the balance up for at least half an hour or more.

3) Adjusting the sensitivity: Start the sensitivity calibration with the built-in weight automatically by pressing the "CAL" sensitivity calibration key, then return after the [END] is displayed.

4) Adjusting zero or removing tare: The mode of weighing mass is used (mass unit "g", etc.). Open the slide glass doors (the enclosure is often called a draft shield), put the container on the pan, then close them again. After the display is stable, set zero point by pressing the key "0/T".

5) Weighing: Open the draft shields, place the weighing vessel or the weighing carton on the pan and close the slide doors. Press the "TARE" switch after the display is stable, and the balance displays "0.0". Remove the weighing bottle top and place it upward on the table. Hold the weighing bottle with the left hand, place the bottle mouth above the center of the weighing carton or vessel about 5~10 cm (prevent samples from falling to the table). Carefully remove the samples with a special medicine spoon

and gently shake it into the weighing carton or vessel until the desired mass value is shown, then close the two slide doors. Read the mass value after the display is stable.

6) Packaging samples: Samples should be packaged in the way of pharmacy packaging by using weighing paper. The name, mass, time of weighing of the samples, and the name of the weighing person should be written down on the outside of the package.

7) Powering off the balance: Clean the balance immediately by a small brush after using it, adjust zero point, press the on/off switch until "OFF" is displayed, and then unplug power.

(2) Indirect Weighing: Indirect Weighing involves weighing the container, generally the weighing bottle, and its contents, transferring a portion of contents to another vessel, and reweighing the container. The difference equals the weight of the transferred material. This technique is especially useful when weighing hygroscopic or volatile substances because the weighing container can be kept closed.

Most solids absorb atmospheric moisture to some extent, thereby undergoing a change in composition. In such instances, it is necessary to put the solid in a weighing bottle and dry in an oven for 1~2 hours before weighing, as shown in Figure 12(a). Put the solid into the desicator after drying, as shown in Figure 12(b). To remove the samples from the weighing bottle to a vessel, the weighing bottle and its top need to be hooped with a slip of paper for holding, knock off the samples from the bottle slightly with its top (Figure 13). Don't spill any samples out of the vessel.

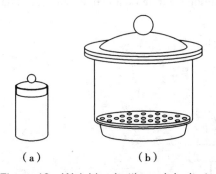

Figure 12  Weighing bottle and desicator
(a) Weighing bottle; (b) Desicator

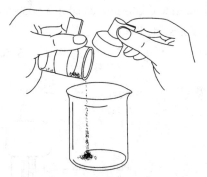

Figure 13  Knock off the samples

**3.3  pH Meter**

A pH meter, the acidometer, is used to measure pH of a solution by potentiometry in which the difference in potential between an indicating electrode and a reference electrode immersed in the solution is determined.

(1) Principle: pH meter consists of a reference electrode, an indicator electrode and an exact potentiometer.

1) Reference Electrodes: A reference electrode must have a potential that is independent of the solution in which it is immersed and that does not change significantly when a small amount of current is passed through it. Most routine measurements are made using a saturated calomel electrode or a silver-silver chloride electrode.

### Saturated calomel electrode

A schematic drawing of a commercial saturated calomel electrode (SCE) is shown in Figure 14(a). The working part of the electrode, consisting of a platinum wire immersed in a slurry of solid mercurous chloride (whose common name is calomel), liquid mercury, and aqueous saturated potassium chloride, is contained in the inner tube. The outer tube is merely a saturated potassium chloride salt bridge that permits the entire assembly to be placed directly in the solution to be measured.

The calomel electrode half-cell may be represented as

$$\text{Pt(s)} \mid \text{Hg(s)} \mid \text{Hg}_2\text{Cl}_2(\text{s}), \text{KCl(sat'd)} \parallel$$

Mercurous chloride is reduced and elemental mercury is oxidized in the reversible electrode reaction

$$\text{Hg}_2\text{Cl}_2(\text{s}) + 2\text{e}^- \rightleftharpoons 2\text{Hg}(\text{l}) + 2\text{Cl}^-(\text{aq})$$

Since the activities of the liquid Hg and solid $\text{Hg}_2\text{Cl}_2$ are both unity, the potential of the electrode is described by the Nernst equation

$$\varphi = \varphi^{\ominus}_{\text{Hg}_2\text{Cl}_2/\text{Hg}} - \frac{0.05916}{2} \lg c(\text{Cl}^-)^2 \tag{1}$$

the concentration of $\text{Cl}^-$ in equation (1) is fixed (saturated KCl is about 4.2 mol·L$^{-1}$), so the electrode remains constant, 0.2415 V at 25 °C, as long as the salt bridge solution is not permitted to mix with the test solution.

### Silver-silver chloride electrode

The silver-silver chloride electrode, shown in Figure 14(b), consists of a silver wire coated with silver chloride that is immersed in a potassium chloride solution saturated with silver chloride. The half-cell is represented as

$$\text{Ag(s)} \mid \text{AgCl(s)}, \text{KCl(c)} \parallel$$

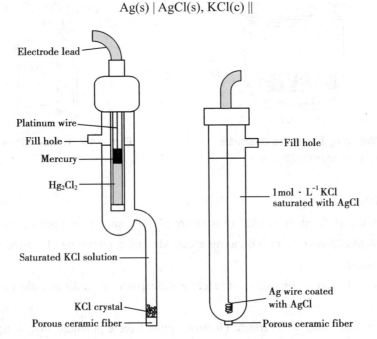

Figure 14  Reference electrode
(a) Saturated calomel electrode; (b) Silver-silver chloride electrode

for which the half-reaction is
$$AgCl(s) + e^- \rightleftharpoons Ag(s) + Cl^-(aq)$$
According to the Nernst equation, the potential of the electrode depends only on the concentration of $Cl^-$

$$\varphi = \varphi_{AgCl/Ag^+}^{\ominus} - \frac{0.05916}{1}\lg c(Cl^-) \qquad (2)$$

This electrode, like the calomel electrode, the potassium chloride solution is saturated. Usually the electrode is small and can be used at higher temperatures.

2) Indicator Electrodes: An indicator electrode has a potential that responses to the certain species in test solution. Glass pH electrode is a kind of membrane electrode that develops a potential determined by difference in concentration of hydrogen ion on two sides of a special glass membrane. When the hydrogen ion concentration on both sides of the special glass film is not equal, a certain potential will be generated.

**Glass pH Electrode**

The glass pH electrode shown in Figure 15 illustrates the construction of glass membrane electrode. The internal element consists of a silver-silver chloride electrode immersed in a pH 7 buffer saturated with silver chloride. The thin, ion-selective glass membrane, consisting mainly of $SiO_2$ (about 70%), and the remaining 30% being a mixture of CaO, BaO, $Li_2O$, and $Na_2O$, is fused to the bottom of a sturdy, nonresponsive glass tube so that the entire membrane can be submerged during measurement. When placed in a solution containing hydrogen ions, this electrode can be represented by the reaction

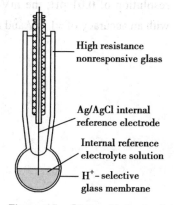

Figure 15  Glass pH electrode

$$Ag(s), AgCl \mid Cl^- \text{ (inside)}, H_3O^+ \text{ (inside)} \mid \text{glass membrane} \mid H^+ \text{ (outside)}$$
whose potential at 25 ℃ is given by
$$\varphi_{glass} = K_{glass} + 0.05916\lg a(H_3O^+) = K_{glass} - 0.05916\text{pH} \qquad (3)$$
where $K_{glass}$ is an unknown constant that includes the potential of the internal reference electrode with a constant concentration of $Cl^-$, a potential at the glass membrane due to the concentration of internal $H_3O^+$, and a term known as the asymmetry potential.

If a cell is constructed at 25 ℃ with a saturated calomel electrode and a glass pH electrode when determining pH of a solution, the electromotive force of the cell is given by
$$E = \varphi_{SCE} - \varphi_{glass} = 0.2415 - (K_{glass} - 0.05916\text{pH}) \qquad (4)$$
therefore, $K_{glass}$ must be determined by calibration with a standard buffer of known pH
$$\text{pH} = \frac{E + K_{glass} - 0.2415}{0.05916} \qquad (5)$$

3) Combination electrode: Nowadays most of pH electrodes are also available physically combined with an external silver-silver chloride reference electrode, all enclosed in a single tube (Figure 16). Glass pH electrode in the combination electrode is located in the middle surrounded by a reference electrode filled with a reference electrolyte. These combination electrodes have the

advantage of being more compact than a separated, two-electrode system. Another convenient electrode is the composite electrode with the temperature probe, so that a simple temperature compensation can be performed, which is also known as a three-in-one electrode.

(2) Operation of the FE28 pH meter: The FE28 pH meter (Figure 17) is a voltmeter with high input impedance, intuitive flexible operation and saving space. The voltage readout is calibrated to read in pH units and millivolts (mV). The pH scale ranges 0~14 with an accuracy of measurements to ± 0.01 pH and resolution of 0.01 pH; the mV scale ranges −2 000~2 000 mV with an accuracy of ±1 mV and resolution of 1 mV.

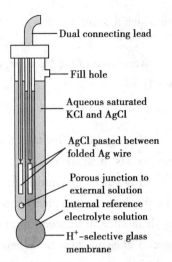

Figure 16  Combination electrode

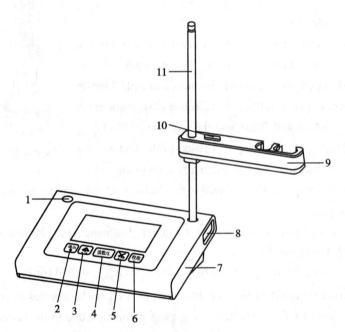

Figure 17  A diagram of FE28 pH meter

1. Left installation location side of the electrode bracket rod; 2. On/off switch; 3. Store/Echo;
4. Read/End way; 5. Mode/Setting; 6. Calibration; 7. shell; 8. Storage space of bracket rod;
9. bracket arm; 10. Fasten button; 11. Electrode bracket rod

1) Installing: Install the electrode rod, connect power cable of pH meter to the 100~240 V power supply, and connect the electrode. Press the on/off switch down to warm the instrument up for 0.5 h.

2) Calibration

① The temperature of ATC and samples can be displayed automatically. Otherwise, the manual temperature mode should be switched from the automatical mode. Press the "模式/设置" key for a long time, the temperature value will flash, then select the temperature value (the default setting is 25 ℃) by pressing the arrow keys, finally set the MTC temperature compensation by pressing "读数

$\sqrt{A}$ " key. Select standard buffer solution by arrow keys, and select calibration method (Linear mode: Lin. or Segment mode: Seg.), resolution (0.1 or 0.01), temperature unit (℃ or ℉), then return to the measurement interface by pressing the on/off switch.

② 1-point calibration or multi-point calibration is need (1 point calibration can adjust offset only; slope and offset can be updated by 2-point calibration; the slope or zero can be updated and shown in the corresponding position of the display by 3- point or more calibration). The pH of standard buffer is related to temperature (Table 2).

Table 2  pH of Standard Buffers at Different Temperature

| Temperature (℃) | Potassium acid phthalate (0.05 mol·L$^{-1}$) | $KH_2PO_4$ + $Na_2HPO_4$ (0.025 mol·L$^{-1}$ + 0.025 mol·L$^{-1}$) | Sodium tetraborate (0.01 mol·L$^{-1}$) |
|---|---|---|---|
| 0 | 4.00 | 6.98 | 9.46 |
| 10 | 4.00 | 6.92 | 9.33 |
| 15 | 4.00 | 6.90 | 9.27 |
| 20 | 4.01 | 6.88 | 9.22 |
| 25 | 4.01 | 6.86 | 9.18 |
| 30 | 4.02 | 6.85 | 9.14 |
| 35 | 4.03 | 6.84 | 9.10 |
| 40 | 4.04 | 6.84 | 9.07 |
| 50 | 4.06 | 6.83 | 9.01 |
| 60 | 4.10 | 6.84 | 8.96 |

③ Clean and wipe pH electrode tip, place the electrode below the standard buffer solution surface by 3～5 cm, and begin the calibration by spressing the "校准" key. Switch the automatic or manual ending mode by pressing "读数$\sqrt{A}$" key for a long tome. When the signal is stable, "$\sqrt{A}$" appeared, the display will lock the reading automatically. If manual ending mode is selected, record pH value of the buffer by pressing "读数$\sqrt{A}$" to after the signal is stable, and "$\sqrt{M}$" appears.

④ For 1-point calibration, complete it by pressing "读数$\sqrt{A}$" key.

⑤ For 2-point calibration, rinse the electrode with distilled water and dry it, place the electrode in the next calibration buffer, and complete the calibration by pressing "校准"key.

⑥ For 3-point calibration, repeat the step ⑤ on the basis of 2-point calibration.

⑦ By analogy, the device can be calibrated by 4-point or 5-point calibration. (Note: Seg line calibration is only meaningful for 3-point or multi-point calibration.) Return to the measurement interface by pressing "退出" key.

3) Measurement of pH

① Set pH measurement mode by pressing the "模式设置" key.

② Clean and wipe pH electrode tip, immerse the electrode below the solution surface by 3～5 cm, and measure the pH of solution by pressing the "读数$\sqrt{A}$" key. Complete the measurement after the signal is stable (automatic end mode, "$\sqrt{A}$" appears) or by pressing "读数$\sqrt{A}$" (manual end mode, "$\sqrt{M}$" appears). Record pH value of the solution at the current temperature.

③ Place the electrode on the bracket arm and remove the sample solution. Clean the electrode with distilled water and dry it. Power off the pH meter by pressing "退出" key for a long time.

4) Measurement of mV: Set mV measurement mode by pressing the "模式设置" key. The operation steps are the same as the measurement of pH on ② and ③ steps of the part "3)".

### 3.4 722-Spectrophotometer

(1) Principle: The absorption energy by ions and molecules provide the basis for both qualitative and quantitative analysis. It is known that substances absorb certain wavelengths of radiation. When a monochromatic light passes through the colored solution (Figure 18), a portion of the light is absorbed by the light obsorbing substance and the rest is transmitted and reflected. The relationship between energy absorption and concentration is of the utmost importance and was first proposed by Beer. Beer's law states that the amount of monochromatic radiant energy absorbed or transmitted by a solution is an exponential function of the concentration of the absorbing substance present in the path of the radiation.. a hundred years earlier Bouguer and Lambert independently had shown a similar relationship to the thickness of the absorbing substance. Thus,

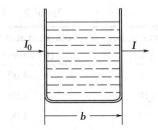

Figure 18  Schematic of light absorption

$$T = \frac{I}{I_0} \tag{6}$$

$$A = \lg\frac{1}{T} = \lg\frac{I_0}{I} \tag{7}$$

$$A = abc \tag{8}$$

where $I$ is the intensity of transmitted radiant power, $I_0$ is the intensity of incident radiant power, $T$ is the transmittance, $A$ is the absorbance, $a$ is the absorptivity, $b$ is the thickness, and $c$ is the concentration. The last equation is named as Lambert-Beer's law. The direct, linear relationship between absorbance and concentration is often used as fundamental test of a system's conformity to Lambert-Beer's law. Deviations from this linear behavior are the result of chemical phenomena such changes in equilibria and solute-solvent or solute-solute interactions or of instrumental factors such as nonmonochromatic radiant energy, stray radiant energy, and multiple-reflection effects.

(2) The optical system of 722-spectrophotometer: 722-spectrophotometer is used for photometric analysis at wavelengths ranging from 360 nm to 800 nm, the wavelengths of near ultraviolet and visible lights. The main advantages of spectrometric method are nice stability, reproducibility, functional and a relatively simple methodology. The optical system of 722-spectrophotometer is shown in Figure 19. White light from light source (Tungsten lamp or halogen lamp) is focused by a mirrored lens and reaches a reflecting mirror. The continuous radiation emitted by the light source lamp (halogen lamp) strikes the condenser lens, converges and then passes through the mirror corner 90° and is reflected to the entrance slit and the collimating mirror. After the incident light is reflected by the collimating mirror, a parallel beam of light is emitted toward the grating (C-T monochromator) and is scattered and reflected by a slight deflection at an angle, converging on the exit slit through the collimating mirror. The light goes through the exit slit, through the condenser lens for focusing, through the sample cell, and then gates to pH otomultiplier tube by the shutter, the resulting photocurrent is inputed to the display, the optical data can be readed directly.

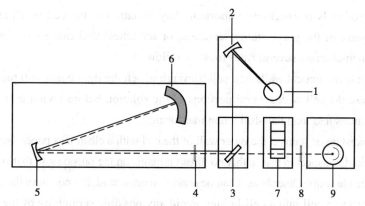

Figure 19  The optical system of 722-spectrophotometer
1. Light source; 2. Mirrored lens; 3. Reflecting mirror; 4. Slit; 5. Collimating mirror;
6. Grating; 7. Sample cell; 8. Shutter; 9. Photomultiplier tube

(3) Structure of 722-Spectrophotometer: The fundamental requirements of a spectrophotometer are illustrated by the simplified schematic presentation (Figure 20). Radiation over a broad range of wavelengths emitted from the source is separated into nearly individual components by the monochromator (prism or grating with slit system). A particular wavelength is selected by proper positioning of the dispersive element in the monochromator and is then passed to the sample where a portion of the energy may be absorbed. The photoelectric detector measures the amount of transmitted radiant energy, converting it to an electrical current that is amplified to give a reading on a meter or a recorder.

Figure 20  Schematic presentation of a spectrophotometer structure

(4) Cells: Sample containers, called cells or cuvettes, are characterized by having two flat, parallel windows through which the radiation passes (Figure 21). The rectangular cell of 1-cm path length is certainly the most widely used of various types. Longer-path-length cells get more analyte in the optical beam and thereby increase the measured absorbance. Glass cells are suitable for measurements with visible radiation but cannot be used in the ultraviolet because of their strong absorption. Quartz cells can be used in either ultraviolet or visible spectrophtometry. Instruments often are designed to accommodate a set of cells simultaneously. The cells in a set different thickness are always made from the same batch of glass or quartz, and therefore have nearly identical optical qualities.

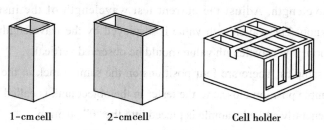

1-cm cell    2-cm cell    Cell holder

Figure 21  Absorption cells and Cell holder

The handling of cells is extremely important. Any variation in the cell (such as a change in cell width or the curvature of the glass, stains, smudges, or scratches) will cause varying results. To avoid significant experimental errors several rules must be followed:

1) Do not handle the optical walls of a cell (through which the light beam will pass).

2) Always rinse the cell with several portions of the solution before taking a measurement. Never fill a cell with solutions less than half-full when measuring.

3) Wipe the moisture or stain on the outer wall of the cell with a clean lens paper. Never wipe cells with towels or handkerchiefs. Inspect to insure that no thread remains on the outside and that small air bubbles are not present on the inside walls. Otherwise, it can be soaked in nitric acid, but not more than 3 hours.

4) When inserting a cell into a cell-holder, avoid any possible scratching of the cell in the optical path, make the optical walls facing toward the windows of the holder. Be sure that the cell is set upright.

5) When using two or more cells simultaneously, use one of them always for the blank solution and the others for the various samples to be measured, and insert the cell for blank solution in the nearest cabin of the holder. If absorption spectra are being run on several different samples at the same time, matched cells must be used.

(5) Operation of 722-spectrophotometer: The exterior of a 722 spectrophotometer is shown in Figure 22.

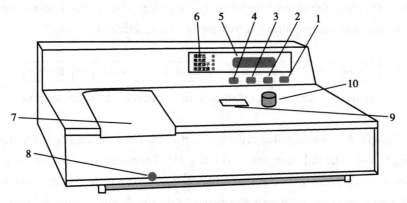

Figure 22　722-spectrophotometer

1. 100%T-adjust key; 2. Zero-adjust key; 3. Function key; 4. Mode key 5. Display; 6. Four scales (" transmittance ", "absorbance", "concentration factor", "concentration direct reading"); 7. Sample chamber; 8. Sample-holder pull; 9. Wavelength scale window; 10. Wavelength selector knob

1) Warming up: Power on the instrument switch, warm the instrument up for 0.5 h.

2) Adjusting wavelength: Adjust the current test wavelength of the instrument by using the wavelength selector knob. The wavelength value is displayed by the wavelength scale window on the left side near to the knob. The wavelength value should be observed vertically.

3) Puting in the samples: There are four positions on the sample rack in the instrument's standard configuration. The sample position closest to the tester in the object tank is called "0" position, followed by "1" "2" and "3", generally a blank sample is placed into the "0" position.

4) Zero-adjust: Open the sample cover or keep out the light path in the sample room using the

opaque materials, and then press the "0%" key.

5) 100%T-adjust: Put the blank solution and samples into the sample rack rod in the sample chamber, close the sample cover (the light door will be opened automatically) and press the "100%" key (If there is an error, press again).

6) Absorbance measure: Set "Absorbance" mode by pressing the mode key. Pull different samples into the optical path by pulling the sample-holder pull. Read out the absorbance values of each sample separately from the display. Readjust 100% T when close the cover of sample chamber each time, and then determine the absorbance.

7) Turn off the switch and unplug the power after the instrument is used up.

### 3.5 FM-9X Osmometer

(1) Principle: Properties of solutions that depend on the amount of solute are colligative properties. These properties include vapor pressure lowering, boiling point elevation, freezing point depression, and osmotic pressure. Among them the freezing-point depression of the solution, $\Delta T_f$, is expressed as

$$\Delta T_f = i K_f b_B \tag{9}$$

where $K_f$ is the freezing-point depression constant of the solvent, which is 1.857 K·kg·mol$^{-1}$ for water, and $b_B$ is the molality of the solute with the unit of mol·kg$^{-1}$. The correction factor $i$ is the amount of particles in solution divided by amount of solute dissolved; that is, $i$ would be 1 for nonelectrolyte solute, 2 for AB type electrolytes such as KCl, CaSO$_4$ and NaHCO$_3$, 3 for AB$_2$ or A$_2$B type electrolytes such as MgCl$_2$, Na$_2$SO$_4$.

The osmotic pressure, $\Pi$, according to van't Hoff's equation, is expressed as

$$\Pi = i c_B RT \tag{10}$$

where $c_B$ is the amount-of-substance concentration of the solute with the unit of mol·L$^{-1}$, and $R$ is the gas constant, and $T$, the absolute temperature.

The amount-of-substance concentration and the molality of the solute in a diluted solution are nearly equal in value at room temperature

$$b_B (\text{mol·kg}^{-1}) \approx c_B (\text{mol·L}^{-1}) \tag{11}$$

therefore,

$$\Pi \approx i b_B RT \text{ (kPa)} \tag{12}$$

The osmotic pressure of body liquid is of important physiologic functions. The freezing-point depression is often used to measure the osmotic pressure in body liquid because it is simple to determine, and the measurement is at lower temperature that biological samples will not be ruined. And also it reaches higher precision because of its larger value of $K_f$. Most instruments measuring the osmotic pressure in the world are designed on the principle of the freezing-point depression.

The freezing point is the temperature at which the ice and the liquid solvent are in equilibrium. During the cooling process the solvent may not solidify when the temperature reaches the freezing point, that is the "undercooling". The liquid solvent is so unstable that it will crystallize at once when some disturbances occur, and its molecules will release the heat called "crystallization heat". The crystallization heat will raise the temperature of the super cooling liquid. Figure 23 shows the cooling curve, in which A is start of cooling, B represents the undercooling, and C refers to the temperature when solid forms, the CD segment is the freezing point temperature stabilization time, and the DE segment is solid state and continues to cool. From C to D the temperature will last for a period, which is the freezing point of the liquid. FM-9X osmometer is designed to adopt a high sensitive semiconductor thermistor to measure the freezing point and to transform the measured current into osmotic pressure display.

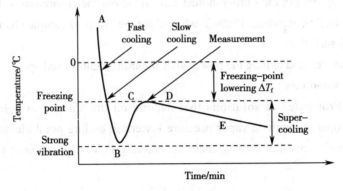

Figure 23  The cooling curve

(2) Operation for FM-9X osmometer: The osmometer is equipped with a semiconductor refrigeration setiing which a nonfreezing solution is used as the cooling media (Figure 24). There is a vibrator for causing crystallization, and a high sensitive measuring system to measure the freezing point and to determine the osmotic pressure of the sample. This osmometer should be calibrated with a 300 mmol·L$^{-1}$ or 800 mmol·L$^{-1}$ standard NaCl solution to match its precision.

1) Preparation and adjustment of osmometer

① Fill in the cooling pond with 60 mL of nonfreezing solution till the liquid flow off from the overflowing tube on the right side of the osmometer.

② Plug in the meter to warming up for 30 min. The osmometer at rest will reach the reference point automatically and its display will show the temperature of the cooling pond. If the display shows "- - - -", the temperature is much higher.

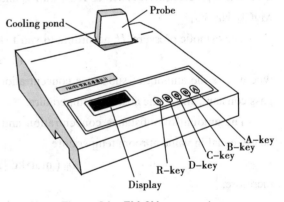

Figure 24  FM-9X osmometer

2) Calibration

① Calibration requirements: Take a clean and dry test tube, extract 0.5 mL of 300 (or 800) mmol·L$^{-1}$ standard solution from the standard solution bottle into the test tube for normal calibration (The results of early or non-freezing cannot be used as normal calibration). The standard solution should not be contaminated by syringes and tubes.

② Calibration method: To calibrate the instrument, be sure that the small tube and the glass injector are dry and clean and transfer 0.5mL of 300 (or 800) mmol·L$^{-1}$ NaCl solution from a storage bottle into the tube with the injector, than set the tube vertically to the probe and put it in the cooling pond.

Push C-key to set the meter in measurement program, the meter shows "300" as the osmotic concentration of the standard NaCl solution. If a 800 mmol·L$^{-1}$ NaCl solution is used push D-key to switch the display. Then push B-key for calibration. The display shows the temperature change of NaCl solution and when the temperature reaches $-6\ ℃$ the vibrator will vibrate strongly, then the display will show the freezing point followed by 300E (or 800E) to end the calibration. Push D-key to store the calibration, and the display shows "300P" (or "800P"). After the finish of calibration the probe must be taken out of the cooling pond to avoid damaging the sensor of the probe.

In the process of the measurement the A-key can be pushed to withdraw. The measuring time (second) can be checked by pushing B-key and the temperature be checked by pushing D-key.

3) Measure the sample solution

① Fill in a dry tube with the sample, clean and dry the probe with soft tissue, and set the tube vertically to the probe and put it in the cooling pond, then push D-key to set the meter in measurement program. The meter shows the temperature change and when the temperature reaches $-6\ ℃$ the vibrator will vibrate strongly, then the display will show the freezing point followed by the osmotic concentration of the sample. After finished take the probe out of the cooling pond.

② In the process of the measurement the D-key can be pushed to check the temperature, the C-key can be pushed to vibrate manually, the B-key can be pushed to check the measuring time and the A-key can be pushed to withdraw.

Manipulating vibration by pushing C-key is applied to interfere the premature freezing unless the freezing is normal.

4) Additional functions of the keys

① Sensing data: push B-button when the meter is at rest state to show the last 12 data, obtaining the time by pushing down the key, and the measuring data by releasing it.

② Setting the time: push R- and D-keys, then release R-key followed by D-key to set the function. The display shows the month and day by four digits. Push D-key to shift the decimal point that indicates the digits being changed, pushing B-key for adding by 1 and A-key for reducing by 1. Then shift decimal point to the right and push C-key to store the date. After that the display shows the hour and minute, operate the keys the same way can set the time.

③ Standardizing the temperature of thermistor and the linearity of voltage: This function is used when the instrument loses the storaged data. Transfer 0.5 mL of nonfreezing solution into the tube and set it vertically to the probe and put them in the cooling pond. Push R-key plus C-key and then release R-key followed by D-key to set the function, the display shows "- - 0 0" (Error! No index). Push the probe to the lower part of the colling pont, the meter shows four digits, in which the first two are for temperature and the last two are the time. Finally a "- - - -" appears on the display indicating the end, then push R-key to finish the reset. It needs about 1 hour to complete the process.

④ Setting the temperature of cooling pond: Generally the temperature of cooling pond is set by $-9.0\ ℃$. Push B-key plus R-key, then release R-key followed by B-key, the meter shows four digits as "L - X X" where the "- X X" gives the original temperature. Push D-key to shift the decimal point that indicates the digits being changed when pushing B-key for adding by 1 and A-key for reducing by

1. Then shift decimal point between two "X"s and push C-key to store the date. After that the display shows the decimal point after "L" as the data accepted, push R-key for ending.

⑤ Examining the date of calibration: Push A-key plus R-key and then release R-key followed by A-key, the meter shows the last date for calibration.

5) Premature freezing and nonfreezing of the sample: Both the premature and the nonfreezing of the sample can not reveal the correct measuring results.

① Premature freezing: Premature freezing may take place when the temperature of the nonfreezing solution in cooling pond is lower than $-9.0℃$, or the osmotic pressure of the sample is too low; or the test tube is not clean or there are some particle impurities such as crystallization in the sample; or there is any vibration to the instrument.

Generally one measurement can finish within 300 seconds. If the measure time is longer than 300 s the premature freezing occurs, the measurement should be sopped. Otherwise the meter will show "- - - -" after vibration to inform of the failure. In this case the sample tube and the thermistor are freezed together, and be attention that never pull the tube off unless the iced sample melts.

② Nonfreezing: Nonfreezing of the sample may take place when the temperature of the nonfreezing solution in cooling pond is higher than $-9.0℃$, or the osmotic pressure of the sample is too high; or the vibrator is too weak to beat the wall of the tube; or there are some bubbles in sample.

If nonfreezing occurs it needs to dilute the sample, or strengthen the vibration.

### 3.6 Centrifuges

(1) Principle: A centrifuge, is a special machine that rotates rapidly to separate solid from a mixture of different phases (l-s, l-l), as shown in Figure 25(a). When a test tube containing a mixture of a liquid and solid is placed in a test tube holder in the centrifuge, the solid can be "packed" into the bottom of the test tube by the rotatory motion of the centrifuge. Then the liquid can be decanted (poured off) into another container leaving the solid behind. Sometimes we use a capillary pipet to withdraw the liquid and leave the solid in the test tube.

Before the centrifuge is turned on, a "balanced tube" must be placed opposite the test tube of interest to prevent excessive vibration, which will destroy the centrifuge otherwise, The balance tube should be the same size as the test tube that contains the mixture to be separated, and it should contain a volume of water equal to the volume in the other test tube.

Centrifugal separation is widely applied in hospitals, food and industrial departments, and research institutes to perform chemical and biochemical examinations, separation of suspension, and so on.

Now many centrifuges are improved for controlling. These improvements will be illustrated by the use of LDZ5-2 centrifuge. It is controlled by a integrated circuit, its control board is shown in Figure 25(b). It will start up tardily, and it has convenience and direct digital displays of timing and speed. It is also designed to equip the over-speed alarm and the self-brake system. The system can protect the main control circuit and overspeed operation caused by wrong operation.

(2) Procedure for LDZ5-2 centrifuge

1) Place the centrifuge on a stable platform.

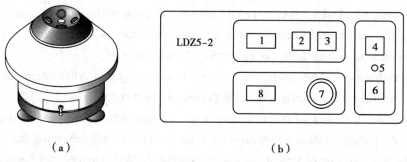

Figure 25  Centrifuge and schematic diagram of low-speed LDZ5-2 centrifuge and its control panel
(a) A centrifuge; (b) Schematic diagram of low-speed LDZ5-2 centrifuge and its control panel
1. Timer (min); 2. Time set; 3. AUTO/MAN timing; 4. Start; 5. Over-speed protection; 6. Power; 7. Speed adjuster; 8. Rotate speed ($\times 100$ r/min)

2) Balance sample: The centrifuge is equipped with an automatic balancing system, the sample can be balanced simply. The amount of solution placed in each test tube is approximated by eyesight, and then placed symmetrically into the centrifugal spacer. The maximum unbalance amount shall not exceed 20 g, and each rotating body shall operate at the corresponding speed, and shall not exceed 5 000 r·min$^{-1}$ (the centrifuge will alarm if exceeds it). It can be balanced with water in a symmetrical tube when centrifuging a small amount of test solution.

Check the balance by eyesight that the test tubes are in the opposite test tube holders and each of test tubes contains the same volume of liquid. If the sample is so small amount that it cannot be divided into two portions, a balance tube containing a volume water equal to the sample in the test tube must be used. Select the maximum rotate speed lesser than the rating in Table 3 for each set of test tubes.

Table 3  Rotate Speed, Centrifugal Force and Volume of LDZ5-2 Centrifuge

| Rotate speed r/min | Centrifugal force g | Volume of test tube mL | Number of test tubes | Overall volume mL | Imbalance < g |
| --- | --- | --- | --- | --- | --- |
| 5 000 | 4 360 | 50 | 4 | 200 | 20 |
| 5 000 | 4 360 | 10 | 12 | 120 | 20 |
| 3 500 | 2 100 | 10 | 32 | 320 | 20 |

3) Examination
① Examine carefully that the speed knob is at "0".
② Test tubes are in right places and tightened.
③ Power connect, plug in, the lid is closed.
④ The centrifuge is plugged in and grounded.

4) Operational program
① Turn on the power switch 4 (POWER) to "ON" position, the indicator lights, and speed display shows "00". If timing is needed, select "AUTO" by pushing the manual manual auto key 3 (MAN-AUTO), the timingdisplay twinkles, and then push the "timed tuning" key 8 (SET) to set the time.
② Push the "Start" key 6 (START), turn the adjustment knob 3 (SPEED) clockwise, the centrifuge

starts to rotate counterclockwise, observe the tachometer, and adjust to the required speed.

③ During the set time, the display table is time subtracted. When the time reaches "00", the centrifuge stops running, and the indicator on the START key turns off.

④ Repeat at this speed if necessary, you can reset the time required. Afterpushing the "Start" key 6 (START), the centrifuge can automatically run at the preset speed and time.

⑤ If timing is not needed, or if centrifuging time is longer than 59 min, switch the AUTO/MAN timing to "MAN". Though random number displayed on the timer is still subtracted, but it is untimed. It needs to stop the centrifugation by turn off the power switch 4 (POWER) to the "OFF" position at the end.

⑥ When it is used repeatedly, first turn on the power switch 4 (POWER) and then press the start key 6 (START), the centrifuge can be automatically accelerated to the set speed (still untimed).

⑦ After the centrifuging is finished, return the speed adjustor to "0" and turn of the power switch 3 (SPEED), move the power switch to "OFF". Open the lid and take the tubes out of the centrifuge. Never open the lid when the machine is working or try to stop the machine with your hand.

（周春艳　李雪华）

# Chapter 3
# Experiment Results and their Expressions

## 1. Experimental Error and Significant Figures

### 1.1 Experimental Error

Experiment error is ubiquitous. Even if the parallel determination is strictly performed, one usually does not obtain precisely the same result. Each measurement is subject to experimental error. Experiment error comes from the following causes.

(1) Systematic error: Systematic errors have a definite value and an assignable cause, and are of the same magnitude for replicate measurements made in the same way. Systematic error results from method error, instrument error, reagent error and operation error. Systematic error produces values that are either all higher or all lower than the actual value, it can be avoided, through calibration of measuring device, improvement of experimental method, changing reagents or by comparing with a known standard.

(2) Random error: Random error results from unpredictable reasons such as the small variation of surrounding temperature, humidity, pressure, performance of measuring device and measurer's treatment. Random error, in the absence of systematic error, produces some values that are higher and some that are lower than the actual value. Random error always occurs, but its effect on the accurate measurement can be eliminated by taking the average value of many parallel repeating measurements.

Occasionally the fault occurs because of the negligences or non-compliance with operating procedures, such as taking a wrong reading, or an incorrect operation, etc., which must be avoided.

### 1.2 Accuracy and Precision

(1) Accuracy: Accuracy refers to the closeness of a single measurement ($x_i$) to its true value ($T$). Accuracy is represented by error, the smaller the error, the higher the accuracy.

This definition of error is also named as absolute error ($E$), which is given by

$$E = x_i - T \tag{13}$$

This definition of error is also named as absolute error. Another representation of accuracy often used is relative error ($E_r$).

$$E_r = \frac{E}{T} \times 100\% \tag{14}$$

Since the relative error reflects the proportion of error in the true value, it is more reasonable to measure the accuracy of the measurement. Therefore, the relative error is usually used to indicate the accuracy of the measurement results.

(2) Precision: Precision refers to the degree of closeness between parallel measured values. Precision is represented by deviation, the smaller the deviation, the higher the precision.

The difference between a certain measured value ($x_i$) and the arithmetic mean ($\bar{x}$) of the multiple measured values is called the absolute deviation ($d$), that is

$$d_i = x_i - \bar{x} \tag{15}$$

The average value of the sum of the absolute values ($|d_i|$) of the absolute deviations of each measurement is called the absolute mean deviation ($\bar{d}$), and the number of measurements is $n$, then

$$\bar{d} = \frac{|d_1| + |d_2| + |d_3| + \cdots + |d_n|}{n} \tag{16}$$

Another method of evaluating precision is to determine the relative average deviation ($\bar{d_r}$), which can be calculated as

$$\bar{d_r} = \frac{\bar{d}}{\bar{x}} \times 100\% \tag{17}$$

When discussing experimental the concept of standard deviation ($s$) and relative standard deviation (RSD) is in common use.

The standard deviation is defined as

$$s = \sqrt{\frac{d_1^2 + d_2^2 + d_3^2 + \cdots + d_n^2}{n-1}} \tag{18}$$

The standard deviation highlights the impact of larger deviations in a single measurement on the measurement, which is an indication of better precision.

The relative standard deviation can be expressed as

$$\text{RSD} = \frac{s}{\bar{x}} \times 100\% \tag{19}$$

In practice, the relative standard deviation is used to indicate the precision of the analytical results.

### 1.3 Significant Figures

To indicate the precision of a measured number or a result of calculations on measured numbers, the concept of significant figures is often used, which are those digits in a measured number or a calculated numbers that include all certain digits plus a final one having some uncertainty. When measuring the volume of a liquid with a cylinder graduated in milliliter, for instance, we have a certain value of 23 mL plus an estimated value, 0.2 mL, thus the volume of the liquid is 34.2 mL. This measured value has 3 significant figures, its error is ± 0.1 mL.

(1) Number of significant figures: To count the number of significant figures in a given measured quantity you observe the following rules:

1) Digit "zero": All digits are significant except zeros at the beginning of the number. Thus, 34.2 mL and 0.0 342 mL all contain three significant figures.

Terminal zeros ending at the right of the decimal point are significant. Each of the following has three significant figures: 9.00 cm, 9.10 cm, and 90.0 cm.

Terminal zeros in a number without an explicit decimal point may or may not be significant. More generally, any uncertainty can be removed in such case by expressing the measurement in scientific notation.

2) Scientific notation: Scientific notation is the representation of a number in the form $A \times 10^n$, where $A$ is a number with a single nonzero digit to the left of the decimal point and n is an integer. In scientific notation, the measurement 900 cm precise to two significant figures is written $9.0 \times 10^2$ cm.

3) Significant figures in a common logarithm: A common logarithm has the number of digits at the left of the decimal point corresponding to the power of a rational number, thus whose digits at the right of the decimal point are significant figures. Therefore a pH, the minus logarithm of the concentration of hydrogen ion, 2.88 has two significant figures, and the concentration of hydrogen ion is $1.3 \times 10^{-3}$ mol·$L^{-1}$.

(2) Exact numbers: So far only numbers that involve uncertainties are discussed. However, exact numbers are also encountered. An exact number is a number that arises when counting items or sometimes when defining a unit. For example, when saying there are 9 reagent bottles on a table, that means exactly 9 not 8.9 or 9.1. Also, the inch is defined to be exactly 2.54 centimeters; the 2.54 here should not be interpreted as a measured number with three significant figures. In effect, the 2.54 has an infinite number of significant figures, but of course it would be impossible to write out all infinite number of digits. The conventions of significant figures do not apply to exact numbers. The number of significant figures in a calculation result depends only on the numbers of significant figures in quantities having uncertainties. For example, the calculation of the total mass of 9 bottles, suppose one reagent bottle has a mass of 20.1 grams, is

$$20.1 \text{ g} \times 9 = 180.9 \text{ g}.$$

(3) Rounding off a number

1) In the actual measurement, often a plurality of measuring instruments are used in combination, and multiple measurements are performed. When multiple measurement data are to be processed, it is necessary to make a trade-off to the extra digits of the measurement data according to the law of error transmission, that is, rounding off.

2) The rounding is usually handled according to the rules as follows. That is: when the digit removed is 4, simply drop it and all digits farther to right, and when it is 6, add 1 to the last digit to be retained and drop all digits farther to the right. Thus rounding off 1.2 143 and 1.2 162, for instance, to three significant figures gives to 1.22 and 1.21 respectively. When the digit removed is 5, and the digit after which is zero or null, the last digit retained is added to by 1 to make it even if it is odd or remains unchanged if it is even. Thus rounding off 1.215, for instance, to three significant figures gives 1.22.

(4) Significant figures in calculations: When measuring data for data processing, the following two rules are commonly used to determine the significant figure of calculation results.

1) Addition and subtraction: When adding or subtracting measured quantities, give as many of decimal digits in answer as there are in the measurement with the least number of decimal digits. Therefore the sum of $34.2 + 0.21$ is 34.4. Now consider the addition of 184.2 g and 2.324 g. On a calculator, you find that $184.2 + 2.324 = 186.524$. But because the quantity of 184.2 g has the least number of decimal places —— one, whereas 2.324 g has three —— the answer is 186.5 g.

2) Multiplication and division: When multiplying or dividing measured quantities, give as many significant figures in answer as there are in the measurement with the least number of significant figures. Therefore the product of 34.2 × 0.21 is 7.2, has two significant figures.

In doing a calculation of two or more steps, it is desirable to retain nonsignificant digits for intermediate answers. This ensures that accumulated small errors from rounding do not appear in the final result. If using a calculator, simply enter numbers one after the other, performing each arithmetic operation and rounding off only the final answer.

## 2. Treatments of Experimental Data

To express the experiments result and to find the rules in experiment, it needs to deal with a great many experimental data by means of tabling, graphing and equation.

### 2.1 Tabling

It is very common for experiments to result in tables full of data. List the experimental data in the order of a dependent variable corresponding to its independent variable(s) to form a three-line table so that the relationship between variables is clear.

Each table should have a laconic title. The table consists of three lines. If it is not particularly needed, there must be no vertical lines. The name and unit of the variable shown in the table should be noted. The data in the table can no longer be taken with the unit. Pay attention to the number of significant figures for each of the data. Usually the simpler varials such as temperature, time and concentration are selected as independent varials.

### 2.2 Graphing

While it is sometimes possible to come to meaningful conclusions from tables, it is often easier to discern relationships between variables visually using appropriate graphs. A graph shows how one variable changes as another is varied. Furthermore the slop, the intercept and the extrapolated value can be obtained by graphing. The guidelines below should help how to prepare readable graphs.

(1) Select logical scales that utilize as much of the graph paper as possible. For many graphs, the two axes will have different scales, and the zero value will not be at origin. Each major division should represent 1, 2, 2.5 or 5 units or some power of ten times one of these units. Never use 3, 7 or another number that makes it difficult to locate points on the graph.

(2) Label the axes and indicate physical quantity units used. Conventionally the horizontal axis is chosed for independent variable, and the vertical axis for dependent variable.

(3) Locate the points and put a dot at the proper locations. Circle the dots with one of such figures as O, △, ×, or □, which area is about equal to the range of measuring error, and each point uses the length of the "Ⅰ" to indicate the standard deviation of the value.

(4) If the points appear to fall on a straight line, use a straight edge to draw the line that best fits (averages the deviations) the data. Do not sketch straight lines and do not connect the dots with straight line segments. If the points seem to fall on a smooth curve draw the best curve possible through the data. Or fit with "excel" software.

(5) Title the graph: Among the graphs, the line graph is an extremely efficient way to present large amount of data and, at the same time, permits important conclusions to be drawn easily and with confidence. Although many different relationships exist between experimental variables, linear relationships are especially attractive, partly because straight lines are easier to draw than curved lines, but more so because the calculations associated with straight lines are easier to perform than those for curved lines.

### 2.3 Computer treatment

Now the technic of computer makes it more convenient for both tabling and graphing dealing with computer, and some data processing softwares such as Excel perform very well.

How to use excel to get a linear regression equation:

(1) Enter the data into the excel table (column 2).

(2) Select all the experimental data, then click "Insert" → "Scatter plot" in the menu to display the dot plot.

(3) Right-click on any data point in the graph and select "Add trendline" from the options that appear, and immediately display the line in the graph.

(4) In the "Trend line options" above, select "Show formula" to display the linear regression equation immediately. The slope and intercept of the line can be obtained from the linear regression equation.

## 3. Experiment Reports

The experiment reports demonstrate for students a personal understanding of the nature of chemistry. By participating in experiments with the chemicals and the equipment students lean to collect data from measurements, to observe changes and to interpret the data and observations. In preparing and performing their experiment reports students learn the skills of planning, carrying out, and reporting on the results of real investigations.

Students should carefully prepare before they come to the laboratory, including carefully reading the textbooks, reviewing the references, simulating the identity of the designer, briefly describing the experimental design ideas, principles, procedures, experimental conditions, and precautions to complete the preparatory work. In their laboratory period they should follow the instruction, observe and record earnestly. At the end of the laboratory period they should hand in the report forms on which you have recorded your data and observations, as well as any calculations and conclusions based on these data and observations.

An experiment report includes a title, student's name and registration code, date; objectives (purposes), principles, apparatus and reagents, procedure, conclusions and discussions, and finally the questions. The common requirement for a experiment report is being scientific, precise and clear, clean and tidy, factual, and correct. According to the aim of the experiment, from the perspective of cultivating students' comprehensive experimental skills and the combination of theory and practice of "Inorganic Chemistry" "Analytical Chemistry" and "Physical Chemistry", and the different levels of teaching, the experiments in this textbook are divided into the following three-stage chemistry experiment training content: basic operation standardization and skill training experiment, comprehensive experiment of cultivating scientific thinking by simulation designer (capacity analysis, instrument analysis, chemical principle, inorganic compound preparation), and comprehensive experimental skill improvement self-designed experiments. The followings are two typical report forms of inorganic compound preparation and quantitative analysis.

Inorganic compound preparation experiment

Experiment X   Preparation of Ammonium Iron(II) Sulfate

Name_____   RD_____   Date_____

1. Objectives

……

2. Pre-Lab Assignments

……

3. Principles

……

4. Apparatus, Material and Reagents

……

5. Procedures

……

6. Data and Results

(Three-line form and data processing items for experimental data recording in advance)

Date: _____    Room temperature:_____ ℃    Relative Humidity _____

(1) Preparation of Ammomium Iron(II) Sulfate

Weigh iron filings $m(Fe)_1 = $ _____ g;

The residue of iron filings $m(Fe)_2 = $ _____ g;

$m[(NH_4)_2SO_4 \cdot FeSO_4 \cdot 6H_2O] = $ _____ g.

Yield of $(NH_4)_2SO_4 \cdot FeSO_4 \cdot 6H_2O$ (%) $= \dfrac{\text{weight of product (g)}}{\text{yeild of theoretical (g)}} \times 100\% = $

(2) Test the purity of the product

| $Fe^{3+}$ standard solution | Grade 1 | Grade 2 | Grade 3 |
|---|---|---|---|
| Grade of the product | | | |

7. Questions

Quantitative analysis

Experiment X   Standardization of HCl Concentration

Name_____   RD_____   Date_____

1. Objectives

……

2. Pre-Lab Assignments

……

3. Principles

……

4. Apparatus, Material and Reagents

……

5. Procedures

……

6. Data and Results

……

Date: _____    Room temperature:_____ ℃    Relative Humidity _____

(1) Weighing Primary Standard, $Na_2CO_3$

Weighing bottle + $Na_2CO_3$ ($W_1$) _____ g;

Weighing bottle + $Na_2CO_3$ retained ($W_2$) _____ g;

Mass of $Na_2CO_3$ ($W_1$-$W_2$) _____ g;

Mass of $Na_2CO_3$ in each titration _____ g.

(2) Solution preparation

……

(3) Data record and result analysis

| Samples | 1 | 2 | 3 |
|---|---|---|---|
| $V(Na_2CO_3)$ / mL | | | |
| $V_{Final}(HCl)$ / mL | | | |
| $V_{Initial}(HCl)$ / mL | | | |
| $\Delta V(HCl)$ / mL | | | |
| $c(HCl)$ / mol·L$^{-1}$ | | | |
| $c_{Average}(HCl)$ / mol·L$^{-1}$ | | | |
| $\overline{d_r}$ /% | | | |

(4) Calculation of NaOH concentration

7. Questions

(籍雪平)

# Part II  Experiments in Basic Chemistry

## Chapter 1
## Experiments of Basic Operation Practice

### Experiment 1  Operation of Volumetric Analysis

【Objectives】

(1) To master laboratory techniques for using volumetric glassware including volumetric flask, volumetric pipette, buret, and Erlenmeyer flask.

(2) To practice general laboratory techniques including cleaning and using of common glassware, solution preparation and transfer, and so on.

(3) To master the principles of acid-base titration analysis.

【Pre-lab Assignments】

(1) Point out the difference between a container and a measuring apparatus, and then explain the operating requirements.

(2) Briefly write down key points for using quantitative analysis glassware including volumetric flask, volumetric pipette, graduated pipettes, buret and Erlenmeyer flask, respectively.

(3) Review literatures on acid-base titration. Briefly describe rules of indicator selection, principle and types of acid-base titration, respectively.

【Apparatus, Reagents, and Materials】

**Apparatus**  Beaker (100 mL, 250 mL), Erlenmeyer flask (250 mL × 3), volumetric flask (100 mL), pipet bulb, volumetric pipet (20 mL × 2), graduated pipette(10 mL), acid buret (25 mL), base buret (25 mL), or PTFE buret (25 mL), buret stand, 0.1mg- electronic balance, stirring rod

**Reagents and Materials**  0.1 000 mol·L$^{-1}$ HCl solution, 0.1 000 mol·L$^{-1}$ NaOH solution, 1.000 mol·L$^{-1}$ NaCl solution, Na$_2$CO$_3$(A.R.), phenolphthalein, methylate orange

## 【Procedures】

### 1. Cleaning of Volumetric Glassware

Clean Erlenmeyer flask, volumetric pipette, and buret according to commonly used measuring apparatus in Chapter 2, Part I.

### 2. Practicing Pipettes

#### 2.1 Practicing a Volumetric pipette

Practice transferring tap water from a 250 mL beaker to a 20 mL pipette and adjusting the meniscus to the calibration mark until you are proficient. Then practice transferring 20.00 mL of water from a 250 mL beaker to a 250 mL Erlenmeyer flask with a 20 mL pipette. Repeat the above practice several times until you are proficient.

#### 2.2 Practicing a Graduated Pipette

Take a 10 mL graduated pipette, practice transferring 10.00 mL of water from a beaker to a 250 mL Erlenmeyer flask. Practice rinsing, filling and draining the pipette with tap water. It is required to deliver liquid of 2.00 mL each time by slightly rolling the index finger and apply pressure to stop the flow until all the 10.00 mL liquid drains out to the flask. Repeat the practice until you are proficient.

### 3. Practicing a Volumetric Flask

#### 3.1 Check liquid-tightness and Transferring Solution

According to the instructions of the volumetric flask in Chapter 2, Part I, practice checking liquid-tightness and transferring solutions.

#### 3.2 Preparing solution

(1) Preparing 0.2 000 mol·L$^{-1}$ NaCl solution: Transfer 20.00 mL of 1.0 000 mol·L$^{-1}$ NaCl solution from a reagent bottle to a 100 mL volumetric flask with a 20 mL pipette. Use the method above to make 100 mL of 0.2 000 mol·L$^{-1}$ NaCl solution. Repeat the practice several times until you are proficient.

(2) Preparing standard solution of $Na_2CO_3$: Weigh anhydrous sodium carbonate 1.0~1.1 g (accurate to 0.0 001 g) with 0.1 mg-electronic balance. Pour the sodium carbonate into a 50 mL beaker. Add 30 mL of distilled water and stir the solution with a stirring rod until all sodium carbonate is dissolved. Then transfer the solution to a 100 mL volumetric flask. Rinse the beaker several times with a small amount of distilled water and transfer it to the volumetric flask. Add water to the calibration mark and shake the flask sufficiently to homogenize the solution.

### 4. Practicing an Erlenmeyer flask

Take a 250 mL Erlenmeyer flask, and then transfer 20.00 mL of tap water into it with a 20 mL pipette. Hold the neck of the Erlenmeyer flask with right hand. Agitate the tap water in the Erlenmeyer flask with a circle way. Do not make water splash out of the Erlenmeyer flask during agitating. Repeat the above practice several times until you are proficient.

## 5. Practicing a Buret

According to the operation techniques of buret in Chapter 2, Part I, check liquid-tightness and assemble the acidic buret and basic buret. Practice rinsing, filling liquid, removing bubbles, and reading the buret.

## 6. Titration

### 6.1 Practice of Titration

Practice titration according to the operation specification of titration in Chapter 2, Part I.

### 6.2 Practice of Acid-base titration

(1) With phenolphthalein as the indicator, titrate 0.1 000 mol·L$^{-1}$ HCl solution with 0.1 000 mol·L$^{-1}$ NaOH solution.

Take a clean base buret or PTFE buret, check liquid-tightness with tap water. If not, rinse the buret with distilled water and rinse it again with about 5 mL to 6 mL 0.1 000 mol·L$^{-1}$ NaOH solution for 2～3 times. Pour 0.1 000 mol·L$^{-1}$ NaOH directly from the reagent bottle into the buret. Free the bubbles at the buret tip, adjust the liquid level, and then read the buret as the initial volume, such as $V_{initial}$ = 0.01 mL.

Rinse three 250 mL Erlenmeyer flasks with distilled water for 2～3 times. With a 20 mL pipette, accurately transfer 20.00 mL of 0.1 000 mol·L$^{-1}$ HCl solution into one Erlenmeyer flasks. Add 2 drops of phenolphthalein and titrate with 0.1 000 mol·L$^{-1}$ NaOH solution. The end of titration reaches when the color of the solution becomes faint pink and does not disappear within 30 seconds. Record the volume of NaOH solution as $V_{final}$ in Table 1-1. Repeat the titration for 3 times.

Calculate the volume ratio of NaOH solution to HCl solution. Calculate the concentration of the HCl solution and the relative deviation.

(2) Using methyl orange as the indicator, titrate 0.1 000 mol·L$^{-1}$ NaOH solution with 0.1 000 mol·L$^{-1}$ HCl solution in the same way. Record the corresponding data in Table 1-2 to calculate the volume ratio of HCl solution to NaOH solution. Calculate the concentration of the NaOH solution and the relative deviation.

## 【Data and Results】

Date:　　　　　　　Temperature:　　　　℃　　　Relative Humidity:

Table 1-1　Use 0.1 000 mol·L$^{-1}$ NaOH solution as titrant to titrate 0.1 000 mol·L$^{-1}$ HCl solution

| Experiment No. | 1 | 2 | 3 |
|---|---|---|---|
| Indicator | phenolphthalein | phenolphthalein | phenolphthalein |
| Color of solution at the end of titration | | | |
| $V$(HCl) / mL | | | |
| $c$(NaOH) / mol·L$^{-1}$ | | | |
| $V_{Final}$(NaOH) / mL | | | |
| $V_{Initial}$(NaOH) / mL | | | |
| $\Delta V_{Consume}$(NaOH) / mL | | | |

续表

| Experiment No. | 1 | 2 | 3 |
|---|---|---|---|
| $\Delta V_{\text{Average}}$ (NaOH) / mL | | | |
| $c$(HCl) / mol·L$^{-1}$ | | | |
| $\bar{c}$(HCl) / mol·L$^{-1}$ | | | |
| $V$(HCl) / $V$(NaOH) | | | |
| $\bar{d_r}$ / % | | | |

Table 1-2  Use 0.1 000 mol·L$^{-1}$ HCl solution as titrant to titrate 0.1 000 mol·L$^{-1}$ NaOH solution

| Experiment No. | 1 | 2 | 3 |
|---|---|---|---|
| Indicator | methyl orange | methyl orange | methyl orange |
| Color of solution at the end of titration | | | |
| $V$(NaOH) / mL | | | |
| $c$(HCl) / mol·L$^{-1}$ | | | |
| $V_{\text{Final}}$(HCl) / mL | | | |
| $V_{\text{Initial}}$(HCl) / mL | | | |
| $\Delta V_{\text{Consume}}$(HCl) / mL | | | |
| $\Delta V_{\text{Average}}$ (HCl) / mL | | | |
| $c$(NaOH) / mol·L$^{-1}$ | | | |
| $\bar{c}$(NaOH) / mol·L$^{-1}$ | | | |
| $V$(HCl) / $V$(NaOH) | | | |
| $\bar{d_r}$ / % | | | |

【Questions】

(1) Before titration, burets and pipettes must be subsequently rinsed with distilled water and the solution to be transferred. Why?

(2) Why each titration starts from a position close to the zero mark?

(3) Why the ratios of HCl and NaOH volumes differ when different indicators such as phenolphthalein and methylate orange are used in titration?

（李献锐）

# Experiment 2　Weighing Practice with an Electronic Balance

【Objectives】

(1) To master the basic operations of electronic balance and common weighing methods.

(2) To understand the usage of weighing bottle and desiccator.

【Pre-lab Assignments】

(1) How many kinds of weighing methods are there? Describe the differences between the direct

weighing method and the indirect method in respect of operation process and scope of application.

(2) Briefly explain the general operation procedures of using an electronic balance.

(3) When using an electronic balance, what should be pay attention to?

(4) According to the experimental contents, use a flow chart to illustrate the operation steps of an electronic balance, and then indicate the specific weighing method and requirements of each step in the flow chart.

## 【Principles】

Refer to Chapter 2 of Part I: electronic balance, desiccator and weighing bottle.

A sample can be weighed in many ways, such as direct weighing method and indirect weighing method. In direct weighing, a receiving vessel or a sheet of weighing paper is placed on the pan and weighed. Once the sample is transferred into the container, reweigh the container and the sample. This method is recommended for weighing samples stable in air and nonhygroscopic, such as metal or ore. Weighing by indirect method involves weighing the container, generally the weighing bottle and its contents, after transferring a portion of contents to another vessel. Then reweigh the container and the remaining sample. The difference in mass is the mass of the sample transferred.

In this experiment, we practice using an electronic balance to accurately weigh samples with both methods.

## 【Apparatus, Reagents and Materials】

**Apparatus**   0.1 mg-Electronic balance, platform balance, desiccator, weighing bottle, beaker (50 mL), spatula, a slip of paper and weighing paper

**Reagents and Materials**   Magnesium carbonate and borax

## 【Procedures】

### 1. Checking the Balance

Check the balance carefully. Make sure that the balance is in horizontal state by checking the water level behind the stand of the balance. Otherwise, adjust the base-screws until the bubble moves to the center of the level indicator.

### 2. Turning on the Balance

Turn on the balance using an on/off switch. Once the balance is on and has gone through its initial calibration check, the digital display will show "0.0 000 g". If not, push the tare button or O/T button. Preheat the balance for at least 30 minutes if it is switched on after disconnected with power for a long time.

### 3. Direct Weighing

Place a container in the center of the pan and close the sliding glass doors. When the green dot on the left of digital screen disappears (indicating that the weight is stable), the weight (g) of the container will appear in the unit area of the display. Then push the tare button or O/T button to deduct the weight of the container. Carefully add 0.20~0.25 g of magnesium carbonate to the container. Record the sample mass $m$ (g) in Table 2-1 after closing the sliding glass doors again. Repeat the weighing for another 2 times.

### 4. Indirect Weighing

Take a weighing bottle containing borax with a slip of paper out of a desiccator and place it in the center of the pan. Push the tare button or O/T button till the reading of 0.0 000 g appear.

After that, transfer 0.4~0.6 g of borax sample into a clean and dry beaker by hitting the rim of the weighing bottle with its lid right above the beaker. Record the sample mass in Table 2-1 when the weighing sample is within the required range. Repeat the weighing for another 2 times for borax sample.

### 5. inishing Weighing

After weighing, remove the weighing container from the pan. Push the on/off button to turn off the balance. Then clean the balance with a brush. Write down the status of the balance in the register book and sign your name.

【Notes】

(1) Never exceed the loading limit of a balance for mass of sample.

(2) Never drop an object onto the balance pan surface. Gently place the object on the balance pan. Make sure the balance door is closed before recording the weight.

(3) Never weigh an object directly on the balance pan. Always use a weighing bottle, weighing paper, or other containers to protect the surface of the balance pan. Corrosive or hygroscopic sample should be weighed in a closed container.

(4) Never weigh hot or cold object. Wait until the object is at room temperature before placing it on the balance.

(5) Always return the balance back to 0.0 000 g before turning off the balance.

【Data and Results】

Date:                    Temperature:            ℃        Relative Humidity:

Table 2-1  Weighing practice with an electronic balance

| Experiment | No. | 1 | 2 |
|---|---|---|---|
| Direct weighing | $m$ (beaker) / g |  |  |
|  | $m$ (sample) / g |  |  |
| Indirect weighing | $m$ (sample) / g |  |  |

【Questions】

(1) How to transfer samples from the weighing bottle when using the indirect weighing method?

(2) What are the factors affecting the reading stability of an electronic balance?

（赵全芹）

# Experiment 3  Preparation, Properties and pH Measurement of Buffer Solutions

## 【Objectives】

(1) To learn how to prepare buffer solutions and isotonic phosphate-buffered saline (PBS).
(2) To understand the properties of buffer solutions.
(3) To consolidate skills of using a volumetric pipette.
(4) To practice the operation of micropipettes.
(5) To practice pH measurements with pH test papers and pH meter, respectively.

## 【Pre-lab Assignments】

(1) What are the differences between common solutions and buffer solutions? Why does a buffer solution have buffer action?

(2) What are the properties of a buffer solution? What are the key factors that determine pH of a buffer solution?

(3) How to measure the buffer capacity? What are the key factors that determine the buffer capacity of a solution?

(4) How to detect pH change of a buffer solution? Is a pH meter necessary?

(5) According to pH testing methods in this experiment, explain whether the experiment is quantitative, qualitative, or half quantitative?

(6) Use flow chart to illustrate how the experimental procedures are designed. Make sure to label notes for each step on the flow chart.

## 【Principles】

A buffer solution is an aqueous solution consisting of two components: a weak acid (HB, called anti-base component) and its conjugate base ($B^-$, called anti-acid component), a weak base and its conjugate acid, or an acid salt of polyprotic acid and its secondary salt. There is a proton-transfering equilibrium between the two components as follows:

$$HB + H_2O \rightleftharpoons H_3O^+ + B^-$$

pH of a buffer solution can be calculated by the Henderson-Hasselbalch equation

$$pH = pK_a + \lg \frac{c(B^-)}{c(HB)} \tag{3-1}$$

where, $K_a$ is the acid ionization constant and $\frac{[B^-]}{[HB]}$ is called the buffer ratio.

On addition of a small amount of a strong acid (or a strong base), it will be comsumed by the anti-acid (or anti-base) component in the buffer solution. Thus pH of the buffer solution can be maintained. When the buffer solution is slightly diluted, both the conjugate acid and base are diluted to the same degree, making both the buffer ratio and pH of the solution constant.

The definition of buffer capacity, $\beta$, is expressed by formular (3-2)

$$\beta = \frac{dn_B}{V|dpH|} \qquad (3\text{-}2)$$

A buffer capacity varies with the total concentration ($c_{total} = [HB] + [B^-]$) and buffer-component ratio ($[B^-]/[HB]$) of the solution. For the same buffer system, if the buffer-component ratio is fixed, the more concentrated the components of a buffer, the greater the buffer capacity. Meanwhile, for a given concentrated buffer solution, the nearer the the buffer-component ratio is to 1, the greater its buffer capacity. A buffer has a highest capacity when the buffer ratio is equal to 1 or $[B^-] = [HB]$.

Buffer solutions are mainly applied in solution environments that require stable pH. For example, biochemical reactions in all kinds of organs of the organisms should maintain proper pH. pH of PBS, a commonly used buffer solution for *in vitro* experiments, is in pH range of healthy human plasma.

In this experiment, you will practice the preparation of buffer solutions and verify the anti-acid/anti-base/anti-dilution properties of buffer solutions.

## 【Apparatus, Reagents and Materials】

**Apparatus**  Beaker (50 mL × 6, 200 mL × 2, 500 mL × 1), test tube (5 mL × 12), colorimetric tube (10 mL × 3), volumetric flask (50 mL × 1), volumetric pipette (5 mL × 5), micropipette (5 mL × 6, tips × 6; 1 mL × 3, tips × 3), colorimetric tube support, test tube rack, medicine dropper, stirring rod, wash bottle, pipette bulb, pH meter

**Reagents and Materials**  1.0 mol·L$^{-1}$ HAc solution, 0.5 mol·L$^{-1}$ HAc solution, 0.1 mol·L$^{-1}$ HAc solution, 1.0 mol·L$^{-1}$ NaAc solution, 0.1 mol·L$^{-1}$ NaAc solution, 1.0 mol·L$^{-1}$ NaOH solution, 1.0 mol·L$^{-1}$ HCl solution, 0.15 mol·L$^{-1}$ NaCl solution, 0.15 mol·L$^{-1}$ KCl solution, 0.15 mol·L$^{-1}$ KH$_2$PO$_4$ solution, 0.1 mol·L$^{-1}$ Na$_2$HPO$_4$ solution, standard buffer solutions (pH 4.00, 6.86 and 9.18), distilled water, universal pH test paper, two sample solutions (volume not less than 50 mL) prepared by students

## 【Procedures】

### 1. Preparation of Buffer Solutions

#### 1.1 Preparation of PBS (pH 7.40)

According to the volumes of each reagent in Table 3-1, use measuring pipettes or micropipettes to transfer the corresponding solutions into a 50 mL volumetric flask. Mix the solution homogeneously.

#### 1.2 Preparation of HAc-NaAc Buffer Solutions

According to the volumes of each reagent in Table 3-2, use measuring pipettes or a micropipettes to transfer the corresponding solutions into 3 colorimetric tubes (10 mL), respectively. Mix the solution homogeneously and mark as buffer solutions A, B and C.

### 2. Properties of Buffer Solutions and Factors Affecting Buffer Capacity

According to Table 3-3, test the anti-acid/anti-base/anti-dilution properties of a common solution (0.5 mol·L$^{-1}$ HAc) and a buffer solution (0.5 mol·L$^{-1}$ HAc −0.5 mol·L$^{-1}$ Ac$^-$). Analyze results in Table 3-3 and try to summarize properties of buffer solutions and factors affecting buffer capacity.

## 3. Measuring pH of Buffer Solutions

Calibrate a pH meter according to the instruction in Chapter 2 "Common Instruments in Basic Chemistry Experiments".

Transfer the prepared PBS and your own sample solutions into three 50 mL beakers, respectively. Measure their pHs using pH test paper and pH meter sequentially. Write down the data in Table 3-4.

According to pH tested with different methods, explain the precision of different pH determination methods.

### 【Data and Results】

Date:　　　　　　　　　Temperature:　　　　　　　　　Relative Humidity:

Table 3-1　Preparation of PBS (pH 7.40)

| Reagents | KCl | $KH_2PO_4$ | $Na_2HPO_4$ | NaCl |
|---|---|---|---|---|
| $c$ /mol·$L^{-1}$ | 0.15 | 0.15 | 0.10 | 0.15 |
| Volume /mL | 0.86 | 0.56 | 4.90 | Fill to 50 mL |

Table 3-2　Preparation of HAc-Ac- buffer solutions

| Solutions No. | A | B | C |
|---|---|---|---|
| 1.0 mol·$L^{-1}$ NaAc / mL | 2.50 | 4.00 | / |
| 1.0 mol·$L^{-1}$ HAc / mL | 2.50 | 1.00 | / |
| 0.1 mol·$L^{-1}$ NaAc / mL | / | / | 2.50 |
| 0.1 mol·$L^{-1}$ HAc / mL | / | / | 2.50 |
| $V_{total}$/mL | | | |
| $c$(NaAc) /mol·$L^{-1}$ | | | |
| $c$(HAc) /mol·$L^{-1}$ | | | |
| Buffer ratio = $\dfrac{c(\text{NaAc})}{c(\text{HAc})}$ | | | |
| $c_{total} = c(\text{NaAc}) + c(\text{HAc})$ / mol·$L^{-1}$ | | | |

Table 3-3　Properties of common solutions and buffer solutions

| Testing items | Anti-acid properties | | | | Anti-base properties | | | | Anti-dilution properties | | | |
|---|---|---|---|---|---|---|---|---|---|---|---|---|
| Reagents | 0.5 M HAc | A | B | C | 0.5 M HAc | A | B | C | 0.5 M HAc | A | B | C |
| Volume / mL | 1.00 | 1.00 | 1.00 | 1.00 | 1.00 | 1.00 | 1.00 | 1.00 | 1.00 | 1.00 | 1.00 | 1.00 |
| $pH_1$ | | | | | | | | | | | | |
| 1.0 mol·$L^{-1}$ HCl /drops | 3 | 3 | 3 | 3 | | | | | | | | |
| 1.0 mol·$L^{-1}$ NaOH/drops | | | | | 3 | 3 | 3 | 3 | | | | |
| $H_2O$ / mL | | | | | | | | | 4.0 | 4.0 | 4.0 | 4.0 |
| $pH_2$ | | | | | | | | | | | | |
| $\Delta pH = |pH_2 - pH_1|$ | | | | | | | | | | | | |

Table 3-4  Measuring of pHs

| Sample | PBS buffer solution | Sample 1 | Sample 2 |
|---|---|---|---|
| pH (pH paper) | | | |
| pH (pH meter) | | | |

【Questions】

(1) How about the accuracies of pH test paper and pH meter, respectively?

(2) Why should the combination electrode be thoroughly cleaned with distilled water and dried with tissues before making another measurement?

(3) Can properties of buffer solutions be determined by HAc and NaOH solutions, HCl and $NH_3 \cdot H_2O$ solutions, or $KH_3PO_4$ and NaOH solutions? If it can do, how do you design the experiment?

（赖泽锋）

## Experiment 4    Preparation and Properties of Colloidal Systems

【Objectives】

(1) To understand the preparation methods of sols.

(2) To verify major properties of colloidal systems.

(3) To understand the protection of macromolecular solution to sols.

【Pre-lab Assignments】

(1) How to classify dispersed systems according to the size of dispersed particles? Can they permeate filter paper or semi-permeable membrane?

(2) How to classify colloid disperse system? What are the properties of sols?

(3) Write the structure of $Fe(OH)_3$ and AgI micelles, respectively. Determine the direction of the colloidal particles move in electrophoresis, respectively.

(4) Why are sols thermodynamically unstable dispersions and kinetically stable systems?

(5) Explain effects of heating and electrolytes on the stabilization of sols.

(6) Please use flow chart to illustrate the experimental steps and mark the notes of each experiment step.

【Principles】

### 1. Preparation of the Colloidal System

The colloid is a dispersion system of matter. When 1~100 nm of particles disperse in the another medium, it will become a dispersed system of colloid. The dispersed system of colloid mainly includes two types: sol and macromolecular solution. Colloids are usually prepared by chemical reaction or physical coagulation.

## Chapter 1  Experiments of Basic Operation Practice

### 1.1  Chemical Reaction Method

Preparation of $Fe(OH)_3$ sol by hydrolyzation of $FeCl_3$: A dark red colloidal suspension of iron (Ⅲ) hydroxide may be prepared by mixing a concentrated solution of iron (Ⅲ) chloride with hot water.

$$FeCl_3 + 3H_2O \longrightarrow Fe(OH)_3 + 3HCl$$
$$Fe(OH)_3 + HCl \longrightarrow FeOCl + 2H_2O$$
$$FeOCl \longrightarrow FeO^+ + Cl^-$$

In this preparation, $Fe(OH)_3$ colloidal nucleus preferentially adsorbs $FeO^+$ ions which are similar to its own composition to form a positive sol.

Preparation of AgI sol by double decomposition: A creamy yellow AgI sol is produced by the reaction of KI solution in excess with $AgNO_3$ solution.

$$AgNO_3 + KI \longrightarrow AgI$$
$$KI \longrightarrow K^+ + I^-$$

In this preparation, because KI is in excess, AgI colloidal nucleus preferentially adsorbs $I^-$ ions to form a negative sol.

### 1.2  Physical Coagulation Method

The sulfur sol is prepared by changing solvent: according to the different solubility of substances in different solvents, the ethanol saturated solution of sulfur is dripped into the water, due to poorly soluble in water, supersaturated sulfur atoms are gathered to form a sulphur sol in the water.

## 2. Optical and Electrical Properties of Colloids

The sol has a strong light scattering phenomenon. When a beam of light is passed through a colloid, an obvious light path can be seen in its vertical direction. This phenomenon is called the *Tyndall* effect. But when a strong beam of light passes through a solution, no *Tyndall* effect is observed because the solution particles are too small to scatter the light. Thus, the *Tyndall* effect is a basic characteristic of the sols which is different from the true solution.

Another important property of sol is that its colloidal particle surface has charge. The reason for the charge of colloidal particles is the selective adsorption of the ions with the similar composition and the dissociation of the molecules on the colloidal surface. The charge of a colloid may be determined by an electrophoretic device, through placing a sol in a U tube with two electrodes. When the current passes through the u-tube, the negative colloidal particles migrate to the positive pole, and the positive colloidal particles migrate to the negative pole. The movement of charged suspended particles toward an oppositely charged electrode is called electrophoresis. Such as ferric hydroxide particles will move towards the negative electrode, thus the positively charged; most of the metal sulfides such as antimony trisulfide will move to the cathode, showed that the negatively charged.

## 3. Purification of Colloids

The electrolyte ions often exist in the prepared sol, which affects the stability of the sol. If these impurity ions are removed, the sol can remain stable for a long time. Water, ions, and small molecule flows can cross a semipermeable membrane, but the colloidal particles cannot, which can be used to separate and purify the colloid. The membrane is called a dialyzing membrane. This process is called dialysis.

## 4. Coagulation of Colloids

The stabilizing factors of the sol include colloidal charge, hydration film on the surface of the colloidal particles and Brown movement. The most important factor is the charge of colloidal particles. There are several ways to cause the coagulation of sol, the most effective method is to add electrolyte. If a certain amount of electrolyte is added into the sol, the charge will be partially or completely neutralized, and the sol will easily aggregate and settle. The electrolyte ions that deposit the sol are mainly ions (counterions) which are oppositely charged with colloidal particles, and the higher the counter valence is, the stronger the aggregation ability of the electrolyte is. For example, ferric hydroxide sol is a positive sol, which can be aggregated by a solution containing high price anions (such as $PO_4^{3-}$). However, AgI sol is a negative sol, and the high-value electrolyte with positive charge is the effective ion of its aggregation. Therefore, adding $AlCl_3$ to AgI sol is better than adding an equal amount of NaCl.

In addition to the coagulation effect of electrolyte, the coagulation of two oppositely charged sols can also happen. Besides, heating is also a commonly used method for sol coagulation.

## 5. Protective of Macromolecules to Sols

If a sufficient amount of macromolecular solution is added into the sol, macromolecular material will form protective layer around the colloidal particles, so that colloidal particles are not easy to coagulate. So macromolecular solution can increase the stability of sol and protect the sol.

## 【Apparatus, Reagents and Materials】

**Apparatus**  Device for electrophoresis, laser pointer, magnetic stirrer, alcohol lamp, beaker (100 mL × 5), graduated cylinder (10 mL × 4, 20 mL × 3), test tube, petri dish, dropper, tripod, glass rod

**Reagents and Materials**  3% $FeCl_3$ solution, 0.05 mol·$L^{-1}$ $AgNO_3$ solution, 0.05 mol·$L^{-1}$ KI solution, saturated sulfide solution in ethanol, saturated NaCl solution, 0.01 mol·$L^{-1}$ NaCl, 0.01 mol·$L^{-1}$ $CaCl_2$ solution, 0.01 mol·$L^{-1}$ $AlCl_3$ solution, 0.1 mol·$L^{-1}$ $NH_3 \cdot H_2O$, 0.05 mol·$L^{-1}$ $I_2$ solution, 0.1 mol·$L^{-1}$ KSCN solution, freshly prepared 3% gelatin solution, 2% $CuSO_4$ solution, dialysis bag, starch solution, pH indicator paper

## 【Procedures】

### 1. Preparation of Colloids

#### 1.1  Preparation of $Fe(OH)_3$ Sol

Add 30 mL of distilled water to 100 mL beaker and heat to boil, then add 3 mL of 3% $FeCl_3$ solution and continue boiling for 2~3 minutes. Observe change of the solution, write micelle structure and keep it for later use. Successful preparation can be proved by detecting the light column with a laser pointer. (Note: if the Tyndall phenomenon of $Fe(OH)_3$ sol is not obvious, the boiling in preparation should be prolonged.)

### 1.2 Preparation of AgI Sol

Transfer 20 mL of 0.05 mol·L$^{-1}$ KI solution into a small beaker. Add 10 mL of 0.05 mol·L$^{-1}$ AgNO$_3$ solution dropwise under stirring to get a creamy yellow AgI sol. Observe change of the solution, write micelle structure and keep it for later use.

### 1.3 Preparation of Sulfur Sol

Add 20 mL of distilled water into a small beaker. Add 2~3 mL of saturated sulfur solution in ethanol under stirring. Observe change of the solution, write micelle structure and keep it for later use.

## 2. Properties of Colloidal Systems

### 2.1 Tyndall Effect

A beam of light is passed through the above three types of sols. Observe the Tyndall effect. For comparison, test distilled water and 2% CuSO$_4$ solution in the same way, respectively.

### 2.2 Electrical Properties of Colloidas — Electrophoresis

Place colloidal Fe(OH)$_3$ in a U-tube, add electrolyte solution on the both ends until the height of electrolyte solution reaches about 2 cm with a clear boundary. Insert electrodes in the electrode solution and switched on the direct current. Regulate the voltage to 160~180 V. Observe the shift of boundary of AgI sol and estimate the charge type of the colloidal particles.

### 2.3 Purification of Colloids — Dialysis

(1) Purification of Macromolecular Solution: Add a suitable amount of starch solution into the dialysis bag. Then add two drops of saturated NaCl solution and immerse the dialysis bag in a beaker containing distilled water. Take the liquid outside the bag every 10 minutes till no chloride can be detected with silver nitrate. Then test the liquid inside and outside the bag for starch with iodine solution. Record and discuss the results.

(2) Purification of Fe(OH)$_3$ Sol: Place as-prepared Fe(OH)$_3$ sol in a dialysis bag and seal it tightly. Make sure not to contaminate the outside of the dialysis bag. Suspend the dialysis bag in a beaker with distilled water. Then change the water at an interval of 10 minutes until no Cl$^-$ (with silver nitrate) or Fe$^{3+}$ (with KCNS) can be detected in the liquid outside the bag. Keep the purified Fe(OH)$_3$ sol in a clean and dry container, after aging.

### 2.4 Coagulation of Colloids

(1) Effect of Electrolytes on Coagulation: Take three test tubes and place 2 mL of AgI sol, respectively. Add 0.01 mol·L$^{-1}$ NaCl in the first tube, 0.01 mol·L$^{-1}$ CaCl$_2$ in the second tube and 0.01 mol·L$^{-1}$ AlCl$_3$ in the third tube dropwise Count the drops and shake the tube after the addition of each drop until precipitation can be observed. Record the results and explain.

(2) Intercoagulation of Sols: Mix 2 mL of Fe(OH)$_3$ sol with 2 mL of AgI sol in a small test tube. Observe pHenomenon and explain.

(3) Effect of Heating on Coagulation: Transfer 2 mL of AgI sol into a small test tube and heat it to boil. Observe phenomenon and explain.

### 2.5 Protection of Macromolecule to Sol

Take two large test tubes. Add 2 mL of distilled water into the first tube and 2 mL of 3% gelatin solution into the second tube. Then add 4 mL of AgI sol in each test tube. Swirl the tube gently.

Add saturated NaCl solution into each test tube after 3 min. Observe phenomenon and compare the two tests. Record the result in Table 4-1.

## 【Data and Results】

Date:　　　　　　　　Temperature:　　　　　℃　　　Relative Humidity:

Table 4-1　Properties of Colloidal Systems

| Contents | | Phenomenon | Result | Discussion |
|---|---|---|---|---|
| Preparation of Fe(OH)$_3$ sol | | | | |
| Preparation of AgI sol | | | | |
| Tyndall effect | | | | |
| Electrical properties - electrophoresis | | | | |
| Purification of colloids - dialysis | macromolecular solution | | | |
| | Fe(OH)$_3$ sol | | | |
| Coagulation of colloids | Effect of electrolytes | | | |
| | Intercoagulation of sols | | | |
| | Effect of heating | | | |
| Protection of Macromolecule to Sol | | | | |

## 【Questions】

(1) Why shall FeCl$_3$ solution be added dropwise into boiling water for the preparation of Fe(OH)$_3$ sol?

(2) How gelatin stabilizes the sols?

(3) Explain the cause of Tyndall effect.

（刘国杰　申小爱）

# Chapter 2

# Experiments of Titration Analysis

### Experiment 5   Acid-Base Titration Analysis

【Objectives】

(1) To learn the principle and procedures of acid-base titration.
(2) To learn standardized titration operation and determine the end point of the titration.
(3) To learn how to prepare standard solutions and how to analyze acidic and basic substances.
(4) To learn the basic operations of back titration.
(5) To analyze purity of aspirin using back titration.

【Pre-lab Assignments】

(1) What is the difference between titration and standardization?
(2) How to prepare and standardize a standard solution?
(3) Why are HCl and NaOH solutions often used as standard solutions in acid-base titration?
(4) Which is better in the standardization of HCl solution when methyl orange is used as the indicator, $Na_2B_4O_7 \cdot 10H_2O$ or $Na_2CO_3$?
(5) What are criteria for the selecting on indicator in acid-base titration?
(6) Is the color change range of an acid-base indicator calculated from $pK_{HIn}$?
(7) Does the amount of indicator used influence the titration error?
(8) Please use a flow chart to explain the procedures of the following experiments and write down the notes that need attention in each step.

【Principles】

Acid-base titration is also known as neutralization titration. It is a titration analysis based on proton transfer. It can be used to determine the concentration of acids, bases, and substances that can react with acid or base. When an acidic sample is analyzed, a strong basic solution with known concentration is often used. The concentration of the acidic compound is calculated based on the amount of the base consumed. When a basic sample is analyzed, a strong acidic solution with known concentration is usually used. The concentration of the basic substance is calculated by the amount of the acid consumed.

The insoluble acidic substance can be neutralised by adding a certain excess of alkali. Thus the remained alkali can be titrated by another standard acidic solution. This is known as back titration. For the insoluble basic substance the method is also applicable.

Generally, titration analysis involves three steps: preparing the standard solutions, standardizing concentration of the standard solutions, and determining the content of the samples.

In acid-base analysis, a hydrochloric acid or sodium hydroxide solution is usually used as a standard solutions. Due to the instability of both hydrochloric acid and sodium hydroxide, their standard solutions cannot be prepared directly. A solution with approximate concentration required is prepared first. Its accurate concentration is then standardized with a primary standard substance or another standard solution of a known accurate concentration. Then the concentration can be calculated according to the ratio of $V(NaOH)/V(HCl)$ obtained by titration and the mass of primary standard substance or the concentration of standard solution.

Commonly used primary standard substances for the standardization of HCl solution include anhydrous sodium carbonate ($Na_2CO_3$) and borax ($Na_2B_4O_7 \cdot 10H_2O$). Sodium carbonate is commercially available in high purity form and less expensive. Since it absorbs moisture and $CO_2$ from the air, however, it must be dried at 270~300 ℃ for about one hour and kept it in a desiccator after cooling down. Borax has a large relative molar mass ($M_r = 381.4$) but contains lattice water. Thus it must be kept in an environmental chamber at a relative humidity of about 60%.

Meanwhile, commonly used primary standard substances for the standardization of NaOH solution include oxalic acid ($H_2C_2O_4 \cdot 2H_2O$) and potassium acid phthalate ($KHC_8H_4O_4$, KHP). The most commonly used primary standard substance is potassium acid phthalate because of its large relative molar mass ($M_r = 204.2$), high stability, and availablity in high purity form.

## 1. Standardization of the Concentration of HCl Solution

### 【Principles】

$Na_2CO_3$ is often used to titrate HCl solution. The reaction is as follows:

$$Na_2CO_3 + 2HCl = 2NaCl + H_2O + CO_2 \uparrow$$

Take the titration of 0.1 000 mol·L$^{-1}$ $Na_2CO_3$ with 0.1 000 mol·L$^{-1}$ HCl for example. At the stoichiometric point, pH equals to 3.9. The interval of inflection point of titration corresponds to pH 5.0~3.5. So methyl orange or methyl red fits well since their color change intervals are pH 3.2~4.4 and pH 4.8~6.0. At the end point of titration, the solution should be boiled to reduce the effect of $CO_2$.

Accordingly, the accurate concentration of HCl solution can be calculated according to the following equation.

$$\frac{1}{2} n(HCl) = n(Na_2CO_3) \tag{5-1}$$

$$c(HCl) = \frac{2 \times m(Na_2CO_3)}{M_r(Na_2CO_3) V(HCl)} \text{ (mol·L}^{-1}\text{)} \tag{5-2}$$

### 【Apparatus, Reagents and Materials】

**Apparatus** 0.1mg-Electronic balance, acidic buret or PTFE buret (25 mL), volumetric flask (100 mL,

1 000 mL), volumetric pipette (20 mL), conical flask (250 mL × 3), beaker (100 mL), graduated cylinder (10 mL), weighing bottle, buret holder, washing bottle, stirring rod

**Reagents and Materials** Anhydrous sodium carbonate (A.R), concentrated hydrochloric acid solution (12 mol·L$^{-1}$), 0.05% methyl orange indicator solution

## 【Procedures】

### 1.1 Preparation of 0.1 mol·L$^{-1}$ HCl Solutions

Calculate the volume of concentrated HCl solution needed to prepare 1 000 mL of 0.1 mol·L$^{-1}$ HCl solution. Then measure it by measuring cylinder and pour the acid into the liter volumetric flask containing about 300 mL distilled water. Rinse the measuring cylinder 2~3 times with a small amount of distilled water and add the washings to the volumetric flask. Make up to the calibration mark with distilled water, shake the flask sufficiently, transfer the solution to liter reagent bottle and insert the stopper.

### 1.2 Preparation of Na$_2$CO$_3$ Standard Solution

Weigh accurately pure Na$_2$CO$_3$ 0.4~0.6 g (0.0 001 g precisely) with analytic balance using the differential weighing method. Pour the anhydrous sodium carbonate into a 100 mL beaker. Add 20~30 mL distilled water. Stir the solution with a glass rod until all Na$_2$CO$_3$ is dissolved. Then transfer the solution to a 100 mL volumetric flask. Rinse the beaker for several times with a small amount of distilled water and transfer it to the volumetric flask. Add distilled water to the calibration mark and shake the flask sufficiently to homogenize the solution.

### 1.3 Standardization the Concentration of HCl Solution

Transfer 20.00 mL of Na$_2$CO$_3$ standard solution into a 250 mL conical flask with a pipette (see also *Chapter 2 Ordinary Instruments in Chemical Experiments*). Wash the inner walls of the conical flask with a little distilled water. Add 2 drops of methyl orange indicator and mix the solution sufficiently to homogenization. Then the solution should become yellow.

Rinse a clean buret three times with a small amount of the acid to be standardized. Fill the buret to a point 2~3 cm above the zero mark and open the stopcock momentarily in order to fill the tip of buret with liquid. Examine the jet to insure that no bubbles enclosed. If there are, more liquid must be run out until the tip of buret is completely filled. Re-fill, if necessary, to bring the level between the 0.00 mL and 0.50 mL marks. Read the position of the meniscus to 0.01 mL. Place the conical flask containing the sodium carbonate solution upon a piece of unglazed white paper or a white tile beneath the acidic buret, and drop the acid slowly from the burette. During the addition of the acid, the flask must be constantly rotated with one hand while the other hand controls the stopcock. Continue the addition until the solution in the conical flask just change from yellow to orange. Then stop the addition and boil the solution in the conical flask to 2~3 minutes. Shake it to remove $CO_2$. The color of the solution changes from orange to yellow again. After cool down, continue the titration until the solution became faint yellow. Wash the inner walls of the conical flask down with a little distilled water from a washing bottle, and continue the titration very carefully by adding the acid dropwisely until the color of the methyl orange becomes orange or a faint pink and the color does not disappear in 30 seconds. This marks the end point of the titration. To ensure the accuracy of the end point, we must add fractions of a drop of titrant using the following methods: allow a drop to begin to form on the burette tip. Then touch the burette tip to the inner surface of the flask and rinse the inner surface with a stream of distilled water from a wash bottle. Finally, continuously shake the flask to ensure the complete mixing. Record the volume of the standard hydrochloric acid solution used in the titration. Record the volume of the standard hydrochloric acid solution used in the titration in Table 5-1. Repeat the titration process till the relative deviation of the analytical results is less than 0.2%. From the weights of sodium carbonate and the volumes of hydrochloric acid employed, the concentration of the acid can be calculated for each titration.

## 【Data and Results】

Date:　　　　　　　Temperature:　　　　℃　　　Relative Humidity:

(1) Use _____ mL of concentrated HCl solution to prepare 1 000 mL of 0.1 mol·L$^{-1}$ HCl solution.

(2) Standardization the concentration of HCl solution. (Table 5-1)

Table 5-1　Standardization the concentration of HCl solution

| Experiment No | 1 | 2 | 3 |
|---|---|---|---|
| $\Delta m(Na_2CO_3)$ / g | | | |
| $m_{Determined}(Na_2CO_3)$ / g | | | |
| Indicator | | | |
| Color of solution at the end point of titration | | | |
| $V(Na_2CO_3)$ / mL | | | |
| $V_{Final}(HCl)$ / mL | | | |
| $V_{Initial}(HCl)$ / mL | | | |
| $\Delta V_{Consumed}(HCl)$ / mL | | | |
| $c(HCl)$ / mol·L$^{-1}$ | | | |
| $\bar{c}(HCl)$ / mol·L$^{-1}$ | | | |
| $\bar{d}_r$ / % | | | |

## 【Questions】

(1) Does the volumetric flask need to be dried before using it for the preparation of $Na_2CO_3$ solution. Why or why not?

(2) Is it necessary to rinse the pipettes prior to use?

(3) In the titration of HCl by $Na_2CO_3$, what is the reason for remove $CO_2$ by boiling the solution when the color first changes? What will happen if $CO_2$ hasn't been removed?

## 2. Standardization of the Concentration of NaOH Solution

### 【Principles】

Potassium acid phthalate ($KHC_8H_4O_4$, KHP) is often used for the titration of NaOH solution. The reaction is as follows:

$$\text{C}_6\text{H}_4(\text{COOH})(\text{COOK}) + \text{NaOH} = \text{C}_6\text{H}_4(\text{COONa})(\text{COOK}) + \text{H}_2\text{O}$$

At the stoichiometric point pH of the solution is 9.1. The interval between inflection points of titration in pH is 8.2~10.0. Phenolphthalein can be chosen as the indicator. According to the reaction equation above, at the stoichiometric point, the following equations can be used for the calculation of the accurate concentration of NaOH solution.

$$n(\text{NaOH}) = n(\text{KHC}_8\text{H}_4\text{O}_4) \tag{5-3}$$

$$c(\text{NaOH}) = \frac{m(\text{KHC}_6\text{H}_4\text{O}_4)}{M_r(\text{KHC}_8\text{H}_4\text{O}_4)V(\text{NaOH})} \, (\text{mol} \cdot \text{L}^{-1}) \tag{5-4}$$

## 【Apparatus, Reagents and Materials】

**Apparatus** 0.1mg-electronic balance, electronic balance, basic burette or PTFE buret (25 mL), volumetric flask (100 mL, 250 mL), volumetric pipette (20 mL), conical flask (250 mL × 3), beaker (100 mL), weighing bottle, buret holder, washing bottle

**Reagents and Materials** Solid of sodium hydroxide, potassium hydrogen phthalate (KHP, A.R), 0.1% phenolphthalein indicator

## 【Procedures】

### 2.1 Preparation of 0.1 mol·L$^{-1}$ NaOH Solution

Weigh out solid sodium hydroxide needed to prepare 250 mL of 0.1 mol·L$^{-1}$ NaOH solution in a 100 mL beaker. Add 50 mL of distilled water to dissolve it, and then transfer the basic solution into a 250 mL volumetric flask after it is cooled down to room temperature. Add distilled water to the calibration mark and mix it sufficiently to homogenization. Then transfer the solution into a 500 mL reagent bottle and insert the rubber stopper.

### 2.2 Preparation of Standard Solution of Potassium Acid Phthalate (KHC$_8$H$_4$O$_4$)

Weigh out the mass of potassium acid phthalate between 1.0~1.3 g (0.0001 g precisely) with indirect weighing method, and put it into a 100 mL beaker, dissolve it in 20~30 mL of distilled water, stir with a glass rod until all potassium acid phthalate is dissolved. Transfer the contents to a 100 mL volumetric flask. Wash out the beaker several times with a small amount of distilled water and transfer the washings to the original solution. Add distilled water to the calibration mark and shake the volumetric flask sufficiently to homogenization.

### 2.3 Standardization the Concentration of NaOH Solution with Potassium Acid Phthalate

Precisely transfer 20.00 mL of potassium acid phthalate standard solution into a 250 mL conical flask with a 20 mL pipette which is rinsed three times with a small amount of potassium acid phthalate solution before use. Wash the inner walls of the conical flask with a little distilled water. Add 2 drops of phenolphthalein indicator.

Rinse a clean base buret or PTFE buret three times with a small amount of NaOH solution, fill the buret to a point 2~3 cm above the zero mark and open the stopcock momentarily in order to fill completely the tip of buret with liquid. Examine the tip of buret to insure that no air bubbles are enclosed. If there are, more liquid must be run out until the tip is completely filled. Re-fill, if necessary, bring the level of titrant between the 0.00 and 0.50 mL marks. Read the position of the meniscus to the nearest 0.01 mL. Place the conical flask containing the potassium acid phthalate solution upon a piece of unglazed white paper or a white tile beneath the buret, and drop the NaOH solution slowly from the basic buret. During the addition of the NaOH solution, the conical flask must be constantly rotated with one hand while the other hand controls the stopcock, close to end point of titration, wash the inner surface of the flask with a little distilled water, and continue the titration very carefully until the color of the solution becomes a faint pink. The endpoint is reached when the color does not disappear at least in 30 seconds. Record the volume of the standard NaOH solution consumed in the titration in Table 5-2.

Repeat the titration process till the relative deviation of the determination is less than 0.2%.

The concentration of the standard NaOH solution can be calculated according to the mass of the potassium acid phthalate and the volumes of sodium hydroxide solution consumed in titration.

## 【Notes】

(1) NaOH solution is toxic and corrosive and can cause burns. Prevent contact with eyes, skin, and clothing. Do not ingest the solution. If you spill any solution, immediately notify your lab instructor.

(2) The standard NaOH solution should be stored in plastic bottles instead of glass reagent bottles.

## 【Data and Results】

Date:　　　　　　　　　　Temperature:　　　　　℃　　　Relative Humidity:

(1) Use _____ g of solid of NaOH to prepare 250 mL of 0.1 mol·L$^{-1}$ NaOH solution.

(2) Standardization the concentration of NaOH solution. (Table 5-2)

Table 5-2　Standardization the concentration of NaOH solution with potassium acid phthalate

| Experiment No. | 1 | 2 | 3 |
|---|---|---|---|
| $\Delta m$ (KHC$_8$H$_4$O$_4$) / g | | | |
| $m_{Determined}$(KHC$_8$H$_4$O$_4$) / g | | | |
| Indicator | | | |
| Color of the solution at the end point of titration | | | |
| $V$(KHC$_8$H$_4$O$_4$) / mL | | | |
| $V_{Final}$(NaOH) / mL | | | |
| $V_{Inatial}$(NaOH) / mL | | | |
| $\Delta V_{Consumed}$(NaOH) / mL | | | |
| $c$(NaOH) / mol·L$^{-1}$ | | | |
| $\overline{c}$(NaOH) / mol·L$^{-1}$ | | | |
| $\overline{d_r}$ / % | | | |

## 【Questions】

(1) What are the prerequisites for a primary standard substance?

(2) Why wash the inner surface of conical flask with distilled water near the end point of titration?

(别子俊)

# 3. Determination of Acetic Acid Content in Vinegar

【 Pre-lab Assignments 】

(1) Why is it needed to dilute vinegar before the determination?

(2) After the completion of titration, will a suspending drop of solution on the tip of the buret affect the titration result?

【 Principles 】

Acetic acid ($CH_3COOH$) is the major ingredient in commercial vinegar. Its content is in the range of 30~50 g·L$^{-1}$. As a weak monoacid, its acid dissociation constant $K_a = 1.75 \times 10^{-5}$. When standard sodium hydroxide is used to titrate acetic acid solution, the following reaction takes place:

$$HAc + NaOH = NaAc + H_2O$$

According to the formula above, pH at the stoichiometric point is 8.73. Phenolphthalein can be chosen as the indicator. Hence the content of acetic acid in vinegar can be calculated according to the mass of vinegar titrated, the volume and the concentration of NaOH solution consumed.

$$\rho_{(HAc)} = \frac{c_{(NaOH)} \times V_{(NaOH)} \times M_{(HAc)}}{V_{(sample)}} \quad (g \cdot L^{-1}) \tag{5-5}$$

Where $c(NaOH)$ is the molarity (mol·L$^{-1}$) of the standard NaOH solution used in the titration, $V(NaOH)$ is the volume of the standard NaOH solution consumed (mL), $M(HAc)$ is the molar mass of acetic acid (g·mol$^{-1}$), and $V(sample)$ is the volume of the vinegar used in each titration, respectively.

【 Apparatus, Reagents and Materials 】

**Apparatus**  base buret (25 mL), or PTFE buret (25 mL), volumetric flask (100 mL), pipette (20 mL, 25 mL), conical flask (250 mL × 3), buret holder, washing bottle, glass rod

**Reagents and Materials**  vinegar, standard sodium hydroxide, 0.1% phenolphthalein indicator

【 Procedures 】

Rinse a clean transfer pipette three times with a small amount of the sample vinegar. Transfer 25.00 mL of vinegar into a 100 mL volumetric flask, and then add distilled water to the mark. Mix the solution thoroughly by shaking and inverting the flask repeatedly.

Rinse a clean transfer pipette three times with a small amount of distilled water and dilute vinegar in turn. Transfer 20.00 mL of the above vinegar solution into a 250 mL conical flask. Wash the inside walls of the flask with a little distilled water. Add 2 drops of phenolphthalein indicator.

Titrate with standard sodium hydroxide solution till light pink color shows and doesn't disappear within 30 seconds. Record the volume of standard sodium hydroxide solution used in the titration in Table 5-3.

Repeat the titration 2~3 times until the relative deviation of the analytical results is less than 0.2%.

The total quantity of acetic acid (in grams) contained in this sample can be calculated from the volume and the concentration of the sodium hydroxide solution used in the titration.

You may also repeat the titration with methyl orange indicator and compare the results. What conclusion can you draw?

**【 Data and Results 】**

Date:　　　　　　　Temperature:　　　　　℃　　　Relative Humidity:

Table 5-3　Determination Content of Acetic Acid in Vinegar

| Experiment No. | 1 | 2 | 3 |
|---|---|---|---|
| $V_{Sample}$(original vinegar) | | | |
| $V_{Total}$(diluted vinegar) | | | |
| Indicator | | | |
| Color of the solution at end point of titration | | | |
| $V_{React}$(diluted vinegar) / mL | | | |
| $V_{Final}$(NaOH)/ mL | | | |
| $V_{Initial}$(NaOH) / mL | | | |
| $\Delta V_{Consume}$(NaOH) / mL | | | |
| $c$(NaOH) / mol·L$^{-1}$ | | | |
| $\rho$(HAc) / g·L$^{-1}$ | | | |
| $\overline{\rho}$(HAc) / g·L$^{-1}$ | | | |
| $\overline{d_r}$ / % | | | |

**【 Questions 】**

(1) Can phenolphthalein indicator be replaced by methyl red or methyl orange for the experiment? Why or why not?

(2) The volumetric pipette must be rinsed for three times with acetic acid before use. Is it necessary for conical flask?

## 4. Determination of Borax Content

**【 Pre-lab Assignments 】**

(1) Which type of buret shall be used for the titration of borax, acid type or basis type?

(2) Which titration method shall be used in this experiment, direct titration or indirect titration?

(3) Does temperature affect the color change sensitivity of the indicator in this experiment? If it does, which temperature shall we choose?

**【 Principles 】**

Borax, sodium borate $Na_2B_4O_7 \cdot 10H_2O$, is soluble in water and produces $Na^+$ and $B_4O_7^{2-}$ ions. According to Brönsted definition, $B_4O_7^{2-}$ is a proton acceptor. It is a base since it takes protons from water molecules. Therefore, the quantity of borax can be determined through acid-base titration.

When standard hydrochloric acid solution is used to titrate borax solution, the following reaction takes place:

$$Na_2B_4O_7 \cdot 10H_2O + 2HCl = 2NaCl + 4H_3BO_3 + 5H_2O$$

or

$$B_4O_7^{2-} + 2H_3O^+ + 3H_2O = 4H_3BO_3$$

According to the equations above, at the stoichiometric point,

$$\frac{1}{2}n(HCl) = n(Na_2B_4O_7 \cdot 10H_2O) \tag{5-6}$$

$$\omega_{(Na_2B_4O_7 \cdot 10H_2O)} = \frac{c_{(HCl)} \times V_{(HCl)} \times M_{(Na_2B_4O_7 \cdot 10H_2O)}}{m_{(sample)} \times 2 \times 1\,000} (g \cdot g^{-1}) \tag{5-7}$$

where $c(HCl)$ is the molarity ($mol \cdot L^{-1}$) of the standard HCl solution consumed in the titration, $V(HCl)$ is the volume of the standard HCl solution consumed (mL), $m$(sample) is the mass of the borax used in each titration (g), $Mr$ ($Na_2B_4O_7 \cdot 10H_2O$) is the relative molar mass of the borax.

At the endpoint, pH of the solution is 5.1 when standard hydrochloric acid solution is used to titrate borax solution. Hence methyl red can be chosen as the indicator.

## 【Apparatus, Reagents and Materials】

**Apparatus** Analytic balance, acidic or PTFE buret (25 mL), volumetric pipette (20 mL), conical flask (250 mL × 3), volumetric flask (100 mL), beaker(100 mL), buret holder, weighing bottle, washing bottle

**Reagents and Materials** Borax sample, standard HCl solution, 0.1% methyl red indicator

## 【Procedures】

Weigh about 1.0~1.4 g $Na_2B_4O_7 \cdot 10H_2O$ (0.0 001 g precisely) with electronic balance using the weight difference method. Put the sample in a 100 mL beaker. Add 20~30 mL of distilled water. Heat and stir the solution till all borax is dissolved. cool to room temperature, and then transfer the solution into a 100 mL volumetric flask. Wash out the beaker several times with a small amount of distilled water and all of them transfer to volumetric flask. Add water to the calibration mark and shake the flask sufficiently to homogenize the solution.

Rinse a clean volumetric pipette three times with a small amount of borax solution. Transfer 20.00 mL of the above borax solution by volumetric pipette and pour it into a 250 mL conical flask. Wash the inner wall of the conical flask with a little distilled water. Add 2 drops of 0.1% methyl red indicator and the solution becomes yellow.

Titrate with the standard hydrochloric acid solution until the solution in the flask turns light pink. The color shouldn't disappear within 30 seconds. It is the endpoint of the titration. Record the volume of the standard hydrochloric acid solution used in the titration in Table 5-4.

Repeat the titration process 2~3 times until the relative deviation of the analytical results is less than 0.2%.

Calculate the content of borax using the above equation.

## 【Recording and treating Data】

Date:　　　　　　　　Temperature:　　　　　℃　　　Relative Humidity:

Table 5-4　Determination of Borax Content

| Experiment No. | 1 | 2 | 3 |
|---|---|---|---|
| $m_{\text{sample}}$ ($Na_2B_4O_7 \cdot 10H_2O$) / g | | | |
| $m_{\text{Determine}}$ ($Na_2B_4O_7 \cdot 10H_2O$) / g | | | |
| Indicator | | | |
| Color of solution at end point of titration | | | |
| $V(Na_2B_4O_7)$ / mL | | | |
| $V_{\text{Final}}(HCl)$ / mL | | | |
| $V_{\text{Inatial}}(HCl)$ / mL | | | |
| $\Delta V_{\text{Consume}}(HCl)$ / mL | | | |
| $c(HCl)$ / mol·L$^{-1}$ | | | |
| $\omega(Na_2B_4O_7 \cdot 10H_2O)$ / g·g$^{-1}$ | | | |
| $\overline{\omega}(Na_2B_4O_7 \cdot 10H_2O)$ / g·g$^{-1}$ | | | |
| $\overline{d_r}$ / % | | | |

## 【Questions】

When titrating borax with standard HCl solution, which indicator do you think is better, methyl red or methyl orange? Why?

（石婷婷）

## 5. Determination of Aspirin Content by Back Titration

### 【Objectives】

(1) To master the principle and operation of learning back titration.

(2) To understand the determination method of aspirin content in tablets by acid-base titration.

### 【Pre-lab Assignments】

(1) Why shouldn't direct method be used for titration in this experiment?

(2) Why should the sample be cooled rapidly after the reaction with excess NaOH solution in water bath?

(3) Please use a flow chart to briefly describe the experimental procedures. Mark steps to be paid attention for each stage.

### 【Principles】

Aspirin used to be a widely used antipyretic analgesic. Its main ingredient is acetyl salicylic acid, an organic weak acid ($K_a = 1 \times 10^{-3}$, $M_r = 180.16$). It is slightly soluble in water, soluble in ethanol, and can be hydrolyzed to salicylic acid and acetic acid in strong alkaline solution as follows:

Since a certain amount of excipients, such as magnesium stearate, starch and other insoluble substances are typically added to the tablets, it is not able to directly determine the content of aspirin in tablets. Hence back titration should be employed.

After grinding the tablets into powdery, Add excess standard NaOH solution [$n$(NaOH)] and heat the mixture to complete the hydrolysis. As 1 mole of acetyl salicylic acid typically consume 2 moles of NaOH, with phenolphthalein as the indicator, by back titration of excess NaOH with standard HCl solution [$n$(HCl)], the content of aspirin in tablets can be calculated with the following equation.

$$n(\text{aspirin}) = \frac{1}{2}[n(\text{NaOH}) - n(\text{HCl})] \tag{5-8}$$

The end point reaches when pink of the solution just disappears.

## 【Reagents and Materials】

**Apparatus**  Electronic balance (0.1 mg), base buret (50 mL), acid buret (50 mL), amphotericity buret (50 mL), conical flask (250 mL × 3), water bath, cylinder (25 mL), washing bottle, glass petri dish, mortar and pestle

**Reagents and Materials**  Aspirin tablets, 0.1 mol·L$^{-1}$ NaOH standard solution (accurate to 0.0001 mol·L$^{-1}$), 0.1 mol·L$^{-1}$ HCl standard solution (accurate to 0.0001 mol·L$^{-1}$), phenolphthalein (2 g·L$^{-1}$ solution in ethanol), ethanol (95%), zeolite

## 【Procedures】

### 5.1  Preparation of Aspirin Sample Solution

Grind aspirin tablets into powder with a mortar. Then accurately weigh three samples of aspirin powder of 0.27~0.33 g (accurate to 0.0001 g) into three 250 mL conical flasks, respectively. For each sample, add 20.0 mL of ethanol and 3 drops of phenolphthalein. Then gently shake the flasks to dissolve the samples.

### 5.2  Reaction of Aspirin with Excess NaOH Standard Solution

Accurately add about 40 mL of NaOH standard solution with base buret into conical flasks containing sample solutions. Record the total volume of NaOH solution as $V_{\text{Total}}$ (NaOH). Add some zeolites in, cover with glass petri dish, then heat with a water bath. Pay attention to avoid boiling of the solution and by gently rotating the conical flasks from time to time. After 15 min, stop heating and swiftly rinse outside of the flasks with flowing tap water to cool the solution to room temperature.

### 5.3  Back titration with HCl Standard Solution

Titrate the remaining NaOH solution in the conical flasks with HCl standard solution. Determine the end point when the pink color of the solution just disappears. Record the volume of the consumed HCl as $V(HCl)$ in Table 5-5.

【Data and Results】

Date:    Temperature:    ℃    Relative Humidity:

Table 5-5  Determination of Aspirin using Back Titration

| Number | 1 | 2 | 3 |
|---|---|---|---|
| $c(HCl)$ / mol·L$^{-1}$ | | | |
| $c(NaOH)$ / mol·L$^{-1}$ | | | |
| $m$(sample) / g | | | |
| $V_{Initial}(NaOH)$ / mL | | | |
| $V_{Final}(NaOH)$ / mL | | | |
| $V_{Total}(NaOH)$ / mL | | | |
| $V_{Initial}(HCl)$ / mL | | | |
| $V_{Final}(HCl)$ / mL | | | |
| $V(HCl)$ / mL | | | |
| $\omega$(aspirin)/ g·g$^{-1}$ | | | |
| $\bar{\omega}$(aspirin)/ g·g$^{-1}$ | | | |
| $\bar{d}_r$ / % | | | |

【Exercises】

(1) Analyze the main factors causing errors in this experiment.

(2) Ethanol, a weak acid that can react with NaOH, is used in the solutions to help dissolve the aspirin. Design a blank experiment to determine the amount of NaOH reacted with ethanol in solution. How can the results of a blank experiment be used in the data analysis?

(3) Why use ethanol to dissolve the sample? Analysis the reasons for the turbidity of the dissolved solution.

（林　毅）

# Experiment 6　Oxidation-Reduction Titration

## 1. Potassium permanganate method

【Objectives】

(1) To master the method of preparation and standardization of potassium permanganate standard solution.

(2) To learn the principles and methods for the determination content of $Na_2C_2O_4$ and $H_2O_2$ with potassium permanganate titration.

(3) To consolidate the specification operation of electronic balance, buret and pipette.

## 【 Pre-lab Assignments 】

(1) What are the hazards of hydrogen peroxide to human health? Why can it still be used in medical care?

(2) What is the basis for the determination of hydrogen peroxide in its aqueous solution? What methods can be used for the determination of hydrogen peroxide?

(3) A sintered glass funnel is often used for filtering in the preparation of a standard $KMnO_4$ solution. Can filter paper be used for filtering $KMnO_4$ solution? Why or why not?

(4) Why should 10 mL of $KMnO_4$ standard solution be added into the conical flask prior to reaction with $Na_2C_2O_4$ solution before its titration? Can this operation be exempted?

(5) How to accurately read colored solution in the buret?

(6) How to reduce experimental errors to assure accuracy of the measurement?

(7) According to the principle of oxidation-reduction titration and potassium permanganate method and reference to this experimental procedures, please describe the experimental design ideas with a flow chart and label the reaction and operation conditions of titration that can be ensure the accuracy of the determination on the flow chart.

## 【 Principles 】

$KMnO_4$ is a vigorous oxidant and is often used as standard solution in oxidation-reduction titration methods. $KMnO_4$ can be reduced to $Mn^{2+}$ under strong acidic solution. According to Nernst equation, its half reaction and standard electrode potential $\varphi^\ominus$ are

$$MnO_4^- + 8H_3O^+ + 5e \Longrightarrow Mn^{2+} + 12H_2O \quad \varphi^\ominus = 1.507 \text{ V}$$

Commercial $KMnO_4$ reagent usually contains impurities such as manganese dioxide and others and therefore it cannot be prepared directly as standard solution. However, an approximate concentration solution can be prepared and then standardized by primary standard substance. $KMnO_4$ readily reacts with reductive substances such as organic impurities in water and ashes in air and so on, and it easily decomposed when exposed to light. So solution of permanganate can be prepared by boiling for an hour and filtering through a sintered glass filter after cooling down and aging for 2~3 days. Finally, keep it in a dark brown reagent bottle in the dark.

The primary standard substance used to standardize $KMnO_4$ solution is sodium oxalate ($Na_2C_2O_4$), the reaction between $KMnO_4$ and $Na_2C_2O_4$ under acidic solution is

$$2MnO_4^- + 5C_2O_4^{2-} + 16H_3O^+ \Longrightarrow 2Mn^{2+} + 10CO_2\uparrow + 24H_2O$$

Here, $H_2SO_4$ is often used to adjust the acidity of solution within 0.5~1.0 mol·L$^{-1}$. But if its concentration is more than 1.0 mol·L$^{-1}$, $KMnO_4$ solution would decompose as following

$$4MnO_4^- + 12H_3O^+ \Longrightarrow 4Mn^{2+} + 5O_2\uparrow + 18H_2O$$

$H_2O_2$ is a common medicament disinfector. The commercial hydrogen peroxide is about 3% (g·mL$^{-1}$) or 30% (g·mL$^{-1}$). In acidic solution, $KMnO_4$ oxidizes $H_2O_2$ to form $O_2$ and colorless $Mn^{2+}$. The reaction is as follows

$$2MnO_4^- + 5H_2O_2 + 6H_3O^+ \Longrightarrow 2Mn^{2+} + 14H_2O + 5O_2\uparrow$$

The rate of the two reactions above is slow at room temperature. So prior to the titration, about 10 mL of titrant KMnO$_4$ solution is often added into the Na$_2$C$_2$O$_4$ solution, which is then heated at a temperature below 75 ℃ in order to produce Mn$^{2+}$ that catalyze the reaction. However, if the temperature is higher than 90 ℃, the Na$_2$C$_2$O$_4$ would decompose. At the beginning of titration, the speed of reaction is slow. Thus the KMnO$_4$ solution must be titrated dropwisely to gradually produce Mn$^{2+}$ in the solution. Because H$_2$O$_2$ readily decomposes on heating, the reaction of H$_2$O$_2$ with KMnO$_4$ must occur at room temperature.

In the titration of KMnO$_4$ method, all species are colorless except KMnO$_4$. When the reaction reaches its stoichiometric point, the addition of a slightly excessive KMnO$_4$ solution makes the solution pink and indicates the end point of titration.

## 【 Apparatus, Reagents and Materials 】

**Apparatus**  0.1mg-electronic balance, acid buret (25 mL) or PTFE buret (25 mL), volumetric flask (100 mL), Erlenmeyer flask (250 mL × 3), beakers (100 mL), pipette (20 mL), graduated pipette (1 mL), graduated cylinder (10 mL), pipette bulb, weighing bottle, washing bottle, glass rod, medicine dropper, sintered glass filter funnel

**Reagents and Materials**  KMnO$_4$(s, A.R), 0.004 mol·L$^{-1}$ KMnO$_4$ standard solution, Na$_2$C$_2$O$_4$(s, A.R) (dried in oven for 1 h under 105 ℃), 3 mol·L$^{-1}$ H$_2$SO$_4$, commercial H$_2$O$_2$ solution (about 3% or 30%)

## 【 Procedures 】

### 1.1  Preparation and Standardization of KMnO$_4$ Standard Solution

Weigh out about 0.16 g solid KMnO$_4$ by platform balance and dissolve it in a beaker with distilled water. Dilute it to 250.0 mL and boil for about 1 hour. Allow it settle down for 2~3 days in dark before filtration by sintered-glass filter funnel. The KMnO$_4$ solution should be stored in a clean brown glass reagent bottle in the dark.

Weigh precisely primary standard substance Na$_2$C$_2$O$_4$ 0.13~0.14 g (precisely 0.0 001 g) by an electronic balance and put it in a small beaker. Dissolve it with about 20.0 mL of distilled water, and then transfer it to a 100 mL of volumetric flask. Wash the beaker for 2~3 times with distilled water, transfer all the rinsing solution to the volumetric flask, add distilled water to the mark, and homogenize the solution.

Transfer 20.00 mL of Na$_2$C$_2$O$_4$ solution to a 250 mL Erlenmeyer flask, add 5 mL of 6 mol·L$^{-1}$ H$_2$SO$_4$ and about 10 mL of KMnO$_4$ solution, then heat the mixed solution at about 40 ℃ until the red color disappears. Continually use KMnO$_4$ solution to titrate the mixed solution dropwisely till the solution shows a pink color that do not fade within 30 seconds. Record the reading in Table 6-1 and repeat the above performances for two times. According to the volume of KMnO$_4$ consumed and the mass of Na$_2$C$_2$O$_4$, the concentration of KMnO$_4$ can be calculated with the following equation.

$$c(\text{KMnO}_4) = \frac{2 \times m_{\text{Total}}(\text{Na}_2\text{C}_2\text{O}_4) \times \dfrac{20.00}{100.00} \times 1\,000}{5 \times V(\text{KMnO}_4) \times M_r(\text{Na}_2\text{C}_2\text{O}_4)} \quad (\text{mol·L}^{-1}) \tag{6-1}$$

$$M_r(Na_2C_2O_4) = 134.0$$

## 1.2 Determination of the Content of $H_2O_2$ in Commercial $H_2O_2$ Solution

Transfer 1.00 mL $H_2O_2$ solution to 100 mL volumetric flask by graduated pipette and dilute it to 100 mL with distilled water to prepare a sample solution. Transfer 20.00 mL the sample solution to 250 mL conical flask by volumetric pipette, add 5.0 mL 6 mol·L$^{-1}$ $H_2SO_4$ in it. Titrate the sample solution by the $KMnO_4$ standard solution until a pink color appears and persists for at least 30 seconds. Record the buret reading in Table 6-2. Repeat the titration for two times. According to the volume of $KMnO_4$ consumed and the concentration of $KMnO_4$ calculate the percentage $H_2O_2$ in the commercial $H_2O_2$ solution with the following equation (6-2)

$$\rho(H_2O_2) = \frac{5 \times c(KMnO_4) \times V(KMnO_4) \times M_r(H_2O_2)}{2 \times 1.00 \times \frac{20.00}{100.00}} \quad (g \cdot L^{-1}) \tag{6-2}$$

## 【Data and Results】

Date:　　　　　　　　Temperature:　　　　　℃　　　Relative Humidity:

**Table 6-1　Standardizing of $KMnO_4$ solution**

| Experiment No. | 1 | 2 | 3 |
|---|---|---|---|
| Indicator | | | |
| Color change at the end point of titration | | | |
| $m_{Total}(Na_2C_2O_4)$ / g | | | |
| $V(Na_2C_2O_4)$ / mL | | | |
| $V_{Final}$ $(KMnO_4)$ / mL | | | |
| $V_{Initial}$ $(KMnO_4)$ / mL | | | |
| $\Delta V(KMnO_4)$ / mL | | | |
| $c$ $(KMnO_4)$ / mol·L$^{-1}$ | | | |
| $\bar{c}(KMnO_4)$ / mol·L$^{-1}$ | | | |
| $\bar{d_r}$ / % | | | |

**Table 6-2　Determine the content of $H_2O_2$ in the commercial $H_2O_2$ solution**

| Experiment No. | 1 | 2 | 3 |
|---|---|---|---|
| Indicator | | | |
| Color change at the end point of titration | | | |
| $V_{Commercial}(H_2O_2)$ / mL | | | |
| $V_{Determined}$ $(H_2O_2 diluted)$ / mL | | | |
| $c(KMnO_4)$ / mol·L$^{-1}$ | | | |
| $V_{Final}$ $(KMnO_4)$ / mL | | | |
| $V_{Initial}(KMnO_4)$ / mL | | | |
| $\Delta V(KMnO_4)$ / mL | | | |
| $\rho(H_2O_2)$ / g·L$^{-1}$ | | | |
| $\bar{\rho}(H_2O_2)$ / g·L$^{-1}$ | | | |
| $\bar{d_r}$ / % | | | |

## 【Questions】

(1) When prepare KMnO₄ solution is prepared, what conditions shall be paid attention to?

(2) When $KMnO_4$ solution is standardized, why should the solution be heated to 75~85℃? Is it okay to heat the solution to above 90℃? Why?

(3) Can the solution be heated when the content of $H_2O_2$ in commercial $H_2O_2$ solution is determined with potassium permanganate standard solution? Why?

（席晓岚）

## 2. Iodimetry

### 【Objectives】

(1) To understand the principle and method for the determination of vitamin C by iodimetry.

(2) To learn the preparation and standardization of iodine standard solution and sodium thiosulfate standard solution.

(3) To learn to determine the endpoints with starch solution.

### 【Pre-lab Assignments】

(1) What are pharmacological functions and clinical applications of vitamin C?

(2) What methods can be used to standardize $Na_2S_2O_3$ solution?

(3) Why should an iodimetric titration be performed in weak acidic condition?

(4) Please use a flow chart to illustrate how to design procedures to determine the content of vitamin C in fruits and vegetables by iodimetry in the experiment. Remember to label the notes and give the reasons why should do that in each step.

### 【Principles】

Iodimetric titrations are based on redox reactions when $I_2$ is used as oxidizing titrant or $I^-$ is used as reducing agent in titrations. The electrode reaction for iodimetric titration is

$$I_2 + 2e \rightleftharpoons 2I^- \quad \varphi^{\ominus} = 0.5355 \text{ V}$$

A reducing agent whose standard electrode is lower than 0.5355 V can be titrated directly with a standard iodine solution. This method is called "iodimetry". The method can work under acidic, neutral and weak basic condition.

Solid iodine is volatilizable and slightly soluble in water, but it easily dissolves in an aqueous solution of potassium iodide. Usually, standard iodine solution is prepared by dissolving iodine in potassium iodide solution and storing it in a clean brown glass bottle. Iodine solution may be standardized using arsenic trioxide or sodium thiosulfate as the primary standard.

Commercial $Na_2S_2O_3 \cdot 5H_2O$ effloresces or deliquesces when exposed to air. Therefore, it cannot be used as a primary standard. For the preparation of standard $Na_2S_2O_3$ solution, freshly boiled distilled water is usually used as solvent. A small amount of sodium carbonate is added to keep pH 9~10. Then the $Na_2S_2O_3$ solution is kept in darkness for 7~10 days and standardized with $KIO_3$ before use. Briefly, iodine will be generated stoichiometrically by the $KIO_3$ reacting KI in slightly acidic solution, and then is titrated with standard $Na_2S_2O_3$ solution with starch used as the indicator. The titration reactions are as follows

$$IO_3^- + 5I^- + 6H^+ = 3I_2 + 3H_2O$$
$$I_2 + 2S_2O_3^{2-} = 2I^- + S_4O_6^{2-}$$

According to the above reactions,

$$n(Na_2S_2O_3) = 6n(KIO_3) \qquad (6\text{-}3)$$

The concentration of the sodium thiosulfate standard solution can be calculated by the following equation

$$c(Na_2S_2O_3) = \frac{6W(KIO_3) \times 1\,000}{M_r(KIO_3) \times V(Na_2S_2O_3)} \; (mol \cdot L^{-1}) \qquad (6\text{-}4)$$

In the titration, starch solution as the indicator must be added just before the endpoint (the solution appears light yellow). The endpoint of titration reaches when the blue color of the solution disappears.

Vitamin C ($C_6H_6O_6$) ($M_r = 176.12$), also called ascorbic acid, is a water-soluble vitamin that widely exists in both fruits and vegetables.

Vitamin C is a strong reducing agent. It can be oxidized to form dehydroascorbic acid. Thus its content can be directly determined by iodimetry. The reaction is as follows

$$\begin{array}{c}\text{O}\\ \text{C—C=C—C—C—CH}_2\text{OH} \\ \text{O OH OH H OH}\end{array} + I_2 \longrightarrow \begin{array}{c}\text{O} \quad\; H\\ \text{C—C—C—C—C—CH}_2\text{OH} \\ \text{O O O H OH}\end{array} + 2HI$$

where the molar ratio between the reactants is 1:1. Vitamin C can be easily oxidized by air, especially in a basic solution due to its strong reducing properties. Therefore the reaction should be carried out in a dilute acetic acid solution (pH 3~4) to prevent side reactions. The endpoint reaches when the solution appears blue by excess iodine reacts with starch. The content of vitamin C can be calculated by the following equation (6-5)

$$\text{Percentage of vitamin C \%} = \frac{c(I_2)V(I_2)M_r(C_6H_8O_6)}{m(\text{sample}) \times 1\,000} \times 100\% \; (g \cdot g^{-1}) \qquad (6\text{-}5)$$

where, the $m$(sample) is the mass of sample containing vitamin C in each titration.

## 【Apparatus, Reagents and Materials】

**Apparatus** 0.1mg-analytical balance, platform balance, acid buret or PTFE buret (50 mL), volumetric pipette (25 mL × 2), volumetric flask (250 mL), brown reagent bottle (500 mL × 2), measuring cylinder (10 mL, 50 mL, 100 mL), Erlenmeyer flask (250 mL × 3), beakers (100 mL, 250 mL), pipette bulb

**Reagents and Materials** Vitamin C tablets, fruits or vegetables (tomato, orange, strawberry, *etc.*), $KIO_3$ (A.R.), $I_2$ (A.R.), $Na_2S_2O_3 \cdot 5H_2O$ (A.R.), $Na_2CO_3$(A.R.), 0.2% starch aqueous solution (freshly prepared), 2 mol·L$^{-1}$ HAc solution, 1 mol·L$^{-1}$ $H_2SO_4$ solution, 20% KI solution

## 【Procedures】

### 2.1　Preparation and Standardization of 0.1 mol·L$^{-1}$ Sodium Thiosulfate Solution

Weigh about 12.5 g of $Na_2S_2O_3 \cdot 5H_2O$ ($M_r$ = 248.17) with a platform balance. Dissolve it with cold and newly boiled distilled water and add 0.1 g of $Na_2CO_3$. Dilute to 500.00 mL in a volumetric flask and mix it homogeneously. Then transfer it into a brown reagent bottle. Keep it in darkness for 7～10 days before standardization.

Weigh primary standard substance $KIO_3$ 0.9～1.0 g (accurate to 0.0 001g) into a small beaker. Dissolve it with 20 mL distilled water, and then transfer it into a 250 mL volumetric flask. Wash the small beaker with distilled water for 2～3 times. Transfer all the rinsing solution to the volumetric flask. Then add distilled water to the mark and homogenize the solution.

Transfer 25.00 mL of standard $KIO_3$ solution with a volumetric pipette to a 250 mL Erlenmeyer flask, add 5.0 mL of 1 mol·L$^{-1}$ $H_2SO_4$, 20.0 mL of 20% KI and 50 mL of distilled water subsequently with measuring cylinder. Titrate with $Na_2S_2O_3$ solution till the solution appears pale yellow, add 5.0 mL of 0.2% starch solution and titrate till the blue color disappears. Record the data and perform two more titrations. Calculate the concentration of the $Na_2S_2O_3$.

### 2.2　Preparation and Standardization of 0.05 mol·L$^{-1}$ Iodine Solution

Weigh about 6.6 g of solid iodine (A.R.) and 10.0 g KI into a mortar. Add litter distilled water and grind it in a hood till all the iodine has been dissolved, dilute it with distilled water to 500 mL and transfer it to a brown reagent bottle. Keep it in a cool and dark place.

Transfer 25.00 mL of $Na_2S_2O_3$ solution with pipette to a 250 mL Erlenmeyer flask and dilute it with 50 mL of water, add 5.0 mL of 0.2% starch solution in it. Then titrate it with iodine standard solution till a blue color persists for at least 30 seconds. Record the data in Table 6-4 and perform two more titrations. Then calculate the concentration of the iodine solution.

### 2.3　Preparation of 0.005 mol·L$^{-1}$ Iodine Solution

Transfer 25.00 mL of 0.05 mol·L$^{-1}$ iodine solution with pipette into a 250 mL volumetric flask and add distilled water to the mark and homogenize the solution.

### 2.4　Determination of Vitamin C Content in Tablets

Weight 2.3～2.7 g of vitamin C tablets powder (accurate to 0.0 001 g) in a 250 mL Erlenmeyer flask. Dissolve it with 100 mL of freshly boiled distilled water, add 10.0 mL of 2.0 mol·L$^{-1}$ HAc solution and 5.0 mL of 0.2% starch solution, respectively. Titrate the solution with standard iodine solution immediately till a blue color appears and persists for 30 seconds. Record the data in Table 6-5 and perform two more titrations and calculate the content of Vitamin C in tablets.

### 2.5　Determination of Vitamin C Content in Fruits or Vegetables

Weight about 50 g (accurate to 0.0 001g) squeezed fruits or vegetables in a 100 mL dry beaker. Then transfer it entirely into a 250 mL Erlenmeyer flask, add 10.0 mL of 2.0 mol·L$^{-1}$ HAc solution and 5 mL of 0.2% starch solution successively. Titrate the solution immediately with standard iodine solution till the color is changed from red to bluish violet. Record the data in Table 6-6 and perform two more titrations. Then calculate the content of vitamin C.

## 【Notes】

(1) In the standardization of sodium thiosulfate solution, the starch indicator should be added near the endpoint when the color of $I_2$ solution almost disappears. Otherwise the endpoint will be delayed because of the wrapping of $I_2$ by starch.

(2) Pretreatment of an analyte depends on the kind of fruit or vegetable. For example, for oranges or grapefruits, only their pulp should be used for squeezing juice. Shell, seed, and fiber must be removed before titration. Furthermore, the color of pulp should be taken into consideration when observing the change of color at the endpoint.

## 【Data and Results】

Date:                Temperature:             ℃        Relative Humidity:

Table 6-3  Standardization of 0.1 mol·L$^{-1}$ sodium thiosulfate solution

| Experiment No. | 1 | 2 | 3 |
|---|---|---|---|
| Indicator | | | |
| The change of color at the endpoint | | | |
| $m(KIO_3)$ / g | | | |
| $c(KIO_3)$ / mol·L$^{-1}$ | | | |
| $V(KIO_3)$ / mL | | | |
| $V_{Final}$ (Na$_2$S$_2$O$_3$)/ mL | | | |
| $V_{Initial}$ (Na$_2$S$_2$O$_3$) / mL | | | |
| $\Delta V_{Consume}$ (Na$_2$S$_2$O$_3$) / mL | | | |
| $c$(Na$_2$S$_2$O$_3$) / mol·L$^{-1}$ | | | |
| $\bar{c}$(Na$_2$S$_2$O$_3$) / mol·L$^{-1}$ | | | |
| $\bar{d}_r$ / % | | | |

Table 6-4  Standardization of 0.05 mol·L$^{-1}$ iodine solution

| Experiment No. | 1 | 2 | 3 |
|---|---|---|---|
| Indicator | | | |
| The change of color at the endpoint | | | |
| $c$(Na$_2$S$_2$O$_3$) / mol·L$^{-1}$ | | | |
| $V$(Na$_2$S$_2$O$_3$) / mL | | | |
| $V_{Final}$ (I$_2$) / mL | | | |
| $V_{Initial}$ (I$_2$) / mL | | | |
| $\Delta V_{Consume}$ (I$_2$) / mL | | | |
| $c$(I$_2$) / mol·L$^{-1}$ | | | |
| $\bar{c}$(I$_2$) / mol·L$^{-1}$ | | | |
| $\bar{d}_r$ / % | | | |

Table 6-5  Determination of Vitamin C content in tablets

| Experiment No. | 1 | 2 | 3 |
|---|---|---|---|
| Indicator | | | |
| The change of color at the endpoint | | | |
| $m(V_C)$ / g | | | |
| $V_{Final}(I_2)$ / mL | | | |
| $V_{Initial}(I_2)$ / mL | | | |
| $\Delta V_{Consume}(I_2)$ / mL | | | |
| $\omega(V_C)\%$ / mg·g$^{-1}$ | | | |
| $\overline{\omega}(V_C)\%$ / mg·g$^{-1}$ | | | |
| $\overline{d_r}$ /% | | | |

Table 6-6  Determination of vitamin C content in fruits or vegetables

| Experiment No. | 1 | 2 | 3 |
|---|---|---|---|
| Indicator | | | |
| The change of color at the endpoint | | | |
| $m(\text{sample})$ / g | | | |
| $V_{Final}(I_2)$ / mL | | | |
| $V_{Initial}(I_2)$ / mL | | | |
| $\Delta V_{Consume}(I_2)$ / mL | | | |
| $\omega(V_C)\%$ / mg·g$^{-1}$ | | | |
| $\overline{\omega}(V_C)\%$ / mg·g$^{-1}$ | | | |
| $\overline{d_r}$ / % | | | |

## 【Questions】

(1) Why add excess KI when preparing iodine solution?

(2) Why is it necessary to add newly distilled cold water in dissolving Vitamin C samples?

(3) What factors may cause errors and how to eliminate errors in iodimetry?

(丁 琼)

# Experiment 7　Complexometric Titration

## 【Objectives】

(1) To learn the applications of Eriochrome Black T and xylenol orange indicators.

(2) To learn the preparation and standardization of standard EDTA solution.

(3) To know the basic procedures of complexometric titration.

(4) To learn the applications of complexometry to determine the concentration of metal ions in samples.

## 【Principles】

Ethylendiaminetetraacetic acid, abbreviated as EDTA or $H_4Y$, is a widely used complexing agent in titrimetric analysis. EDTA strongly reacts with metal ions. Thus the concentration of EDTA will change in impure distilled water. Standard EDTA solution is usually prepared by indirect preparation and standardization with primary substance. The prepared EDTA solution should be stored in polyethene bottles in order to avoid the reaction between EDTA and traces of metal ions yielded from the surface of glass vessels.

The primary standard substances for the standardization of an EDTA solution include purified Zn, Cu, Bi, ZnO, $CaCO_3$, $MgCO_3$, $MgSO_4$, $ZnSO_4$, etc., Zn, ZnO, $ZnSO_4 \cdot 7H_2O$ and $CaCO_3$ are the most often used. In order to reduce the systematic errors, the conditions of determining a sample should be the same as the conditions of standardizing the EDTA solution. If, for example, an unknown ion such as $Bi^{3+}$, $Zn^{2+}$, $Pb^{2+}$, or $Al^{3+}$ is determined, pH of the EDTA solution should be adjusted to pH 5~6, with xylenol orange as the indicator.

To determine $Ca^{2+}$ and $Mg^{2+}$ present in a sample, $CaCO_3$ should be utilized as the primary standard. pH of the solution should be adjusted to about 10 with $NH_3$-$NH_4Cl$ buffer solution, with Eriochrome Black T as the indicator. However, no sharp endpoint can be observed with calcium ions and Eriochrome Black T (EBT). On the contrary, Mg-EBT in that pH range shows sharp wine-red in color compared with the blue color of the free indicator. So $MgCO_3$ can be used as an ideal primary standard substance.

Traces of many metal ions, e.g., $Fe^{3+}$, $Al^{3+}$, $Cu^{2+}$, $Zn^{2+}$, $Pb^{2+}$, etc., will interfere the determination of $Ca^{2+}$ and $Mg^{2+}$ or water hardness using Eriochrome Black T indicator. Their interferance can be overcome by the addition of a little triethanolamine (masking $Fe^{3+}$ and $Al^{3+}$) and $Na_2S$ (masking $Cu^{2+}$, $Zn^{2+}$, and $Pb^{2+}$).

The metallic ions, which in the complicated sample react slowly with EDTA, e.g., $Al^{3+}$, can be determined by back titration or replacement titration.

## 1. Determination of Water Hardness

### 【Pre-lab Assignments】

(1) What is water hardness? What is the total hardness of water? How to express the total hardness of water? What is the limitation the total hardness for water which cannot drink?

(2) What are the primary standard substances that can be used for the standardization of EDTA solution in the determination of water hardness?

(3) In the determination of water hardness, why adding 1~2 drops of 3 mol·$L^{-1}$ HCl to the water sample? And adding triethanolamine solution and $Na_2S$ solution, respectively?

(4) In the determination of the total hardness of water, which indicator can be used? What kind of solution is needed to control the acidity of water sample?

(5) To determine the content of $Ca^{2+}$ and $Mg^{2+}$ separately in a water sample, why adjusting pH to 12?

(6) Describe the color change of solution when Eriochrome Black T serves as the indicator in the determination of total hardness of water.

(7) Describe the color change of solution when Cal-Red serves as the indicator in the determination

of total hardness of water.

(8) Use a flow chart to illustrate how to design the experimental procedures of this experiment. Write down the notes that should be paid attention in each step.

## 【Principles】

Historically, "hardness" was defined in terms of the capacity of a water sample to precipitate soap, an undesirable quality. In natural waters, the concentration of calcium and magnesium ions generally far exceeds that of any other metal ion. Therefore, water hardness, $c$(Total hardness) (mmol·L$^{-1}$), has come to mean the total concentration of calcium and magnesium.

The determination of hardness is a useful analytical process for measuring the quality of water for household and industrial uses. The importance for the latter is due to the fact that hard water, upon heating, precipitates calcium carbonate that clogs boilers and pipes.

Water hardness is usually determined with a standard EDTA solution after the sample has been buffered to pH 10, with Eriochrome Black T serves as the indicator.

To determine the individual $Ca^{2+}$ or $Mg^{2+}$ content, NaOH solution is added to precipitate $Mg^{2+}$. Then Cal-Red indicator, an indicator that only binds with $Ca^{2+}$ to form a red complex, is added. After that, standard EDTA solution is used to titrate $Ca^{2+}$ in the water sample. Color of the solution will turn blue at the endpoint of the titration. Thus $Ca^{2+}$ content can be determined by titration with standard EDTA solution. The content of $Mg^{2+}$ can be evaluated by subtracting the amount of $Ca^{2+}$ from the total hardness of the water sample.

In order to reduce the systematic errors in this experiment, $MgCO_3$ can be used as the primary standard substance for the standardization of EDTA.

## 【Apparatus, Reagents and Materials】

**Apparatus**  0.1mg-Electronic balance, platform balance, acid or PTFE buret (25 mL), Erlenmeyer flask (250 mL × 3), volumetric pipette (20 mL, 25 mL, 100 mL), volumetric flask (25 mL, 250 mL, 1 L), beaker (150 mL, 250 mL), polyethylene reagent bottle (1 000 mL), graduated cylinder (5 mL, 10 mL × 2, 50 mL), stirring rod

**Reagents and Materials**  $Na_2H_2Y·2H_2O$ (A.R.), 3 mol·L$^{-1}$ HCl solution, $MgCO_3$ (A.R.), 0.5% Eriochrome Black T, $NH_3$-$NH_4Cl$ buffer solution (pH 10), 20% triethano-lamine solution, 2% $Na_2S$ solution, calconcarboxylic acid indicator, 10% NaOH solution, universal pH test paper

## 【Procedures】

### 1.1  Preparation of 0.01 mol·L$^{-1}$ EDTA Solution

Weigh about 3.8 g of $Na_2H_2Y·2H_2O$ with in a clean beaker. Dissolve it with distilled water. Transfer the solution into a polythene bottle and dilute to about 1.0 L. Mix the solution thoroughly and label the bottle.

### 1.2  Standardization of Standard EDTA Solution

Weigh 0.20~0.22 g (accurate to 0.0 001 g) of primary standard $MgCO_3$ that has been previously dried at 110 ℃ into a 150 mL clean beaker. Add 5 drops of water to moisten the solid and cover with a watch glass. Add 3.0 mL of 3 mol·L$^{-1}$ HCl solution dropwise from the mouth of the beaker. After the solid is dissolved, add 50.0 mL of distilled water and boil gently for a few minutes to remove $CO_2$. Cool the solution and rinse the surface of watch glass with distilled water sprayed from a washing bottle. The distilled water rinsing the watch glass is collected into the beaker. Transfer it quantitatively to a 250 mL volumetric flask and dilute to the mark, mix and homogenize the solution.

Transfer 20.00 mL of $MgCO_3$ standard solution into a 250 mL Erlenmeyer flask. Add 10 mL of $NH_3$-$NH_4Cl$ buffer solution (pH 10) and 2~3 drops of Eriochrome Black T indicator. Titrate with standard EDTA solution till the color of solution changes from wine-red to pure blue. Record the data in Table 7-1 and perform two more titrations. Calculate the concentration of EDTA solution by the equation (7-1).

$$c(\text{EDTA}) = \frac{m(MgCO_3) \times \frac{20.00}{250.00} \times 1\,000}{M_r(MgCO_3) \times V(\text{EDTA})} \quad (\text{mol} \cdot \text{L}^{-1}) \qquad (7\text{-}1)$$

### 1.3 Determination of Water Hardness

(1) Determination of total hardness in water samples: Transfer 100.00 mL of water sample into a 250 mL Erlenmeyer flask, add 1~2 drops of 3 $mol \cdot L^{-1}$ HCl solution and boil gently for a few minutes to remove $CO_2$. Cool the solution. Then add 5.0 mL of triethanolamine, 10.0 mL of $NH_3$-$NH_4Cl$ buffer solution (pH 10), 10 drops of $Na_2S$ solution, and 2~3 drops of Eriochrome Black T solution. Titrate with standard EDTA solution till the color changes from wine-red to pure blue. Record the data in Table 7-2 and perform two more titrations. Calculate the total hardness of the water sample by the equation (7-2)

$$c(\text{Total hardness}) = \frac{c(\text{EDTA}) \times V_1(\text{EDTA})}{V(\text{water sample})} \times 1\,000 \ (\text{mmol} \cdot \text{L}^{-1}) \qquad (7\text{-}2)$$

(2) Determination of $Ca^{2+}$ and $Mg^{2+}$ concentration individually: Transfer 100.00 mL of water sample into a 250 mL Erlenmeyer flask. Add 6 drops of 6 $mol \cdot L^{-1}$ HCl and boil gently for a few minutes. Cool the flask and add 5.0 mL of triethanolamine solution, 10.0 mL of 10% NaOH solution, and a small amount of Cal-Red indicator. Titrate with standard EDTA solution till the color of the solution changes from wine-red to a pure blue. Record the data in Table 7-3 and perform two more titrations. Calculate the content of the $Ca^{2+}$ and $Mg^{2+}$, respectively.

$$c(Ca^{2+}) = \frac{c(\text{EDTA}) \times V_2(\text{EDTA})}{V(\text{water sample})} \times 1\,000 \ (\text{mmol} \cdot \text{L}^{-1}) \qquad (7\text{-}3)$$

$$c(Mg^{2+}) = \frac{c(\text{EDTA}) \times [V_1(\text{EDTA}) - V_2(\text{EDTA})]}{V(\text{water sample})} \times 1\,000 \ (\text{mmol} \cdot \text{L}^{-1}) \qquad (7\text{-}4)$$

or

$$c(Mg^{2+}) = c(\text{Total hardness}) - c(Ca^{2+}) \ (\text{mmol} \cdot \text{L}^{-1}) \qquad (7\text{-}5)$$

## 【Data and Results】

Date:　　　　　　　Temperature:　　　　℃　　　Relative Humidity:

Table 7-1　Standardization of EDTA solution

| Experiment No. | 1 | 2 | 3 |
|---|---|---|---|
| $m_1$(Weighing bottle + Mass of $MgCO_3$) / g | | | |
| $m_2$(Weighing bottle + Mass of $MgCO_3$) / g | | | |
| $m_1 - m_2$(Mass of $MgCO_3$) / g | | | |
| $m_{\text{Determine}}$(Mass of $MgCO_3$) / g | | | |

续表

| Experiment No. | 1 | 2 | 3 |
|---|---|---|---|
| Indicator | | | |
| The change of color at the endpoint of titration | | | |
| $V(MgCO_3)$ / mL | | | |
| $V_{Initial}(EDTA)$ / mL | | | |
| $V_{Final}(EDTA)$ / mL | | | |
| $\Delta V_{Consume}(EDTA)$ / mL | | | |
| $c(EDTA)$ / mol·L$^{-1}$ | | | |
| $\overline{c}(EDTA)$ / mol·L$^{-1}$ | | | |
| $\overline{d_r}$ / % | | | |

Table 7-2  Determination of of water hardness

| Experiment No. | | 1 | 2 | 3 |
|---|---|---|---|---|
| Indicator | | | | |
| The change of color at the endpoint of titration | | | | |
| $V$(water sample) / mL | | 100.00 | 100.00 | 100.00 |
| Preparation of water sample | 3 mol·L$^{-1}$ HCl / drop | Boil gently to remove CO$_2$ | | |
| | Triethanolamine / mL | | | |
| | NH$_3$-NH$_4$Cl buffer solution / mL | | | |
| | Na$_2$S / drop | | | |
| Determination | $V_{Initial}(EDTA)$ / mL | | | |
| | $V_{Final}(EDTA)$ / mL | | | |
| | $\Delta V_{Consume}(EDTA)$ / mL | | | |
| | $c$(Total hardness) / mmol·L$^{-1}$ | | | |
| | $\overline{c}$(Total hardness) / mmol·L$^{-1}$ | | | |
| $\overline{d_r}$ / % | | | | |

Table 7-3  Determination of Ca$^{2+}$ and Mg$^{2+}$ concentration individually

| Experiment No. | | 1 | 2 | 3 |
|---|---|---|---|---|
| Indicator | | | | |
| The change of color at the endpoint of titration | | | | |
| $V$(water sample) / mL | | 100.00 | 100.00 | 100.00 |
| Preparation of water sample | 3 mol·L$^{-1}$ HCl / drop | Boil gently to get rid of CO$_2$ | | |
| | Triethanolamine / mL | | | |
| | 10% NaOH solution / mL | | | |
| | Na$_2$S / drop | | | |
| $V_{Initial}(EDTA)$ / mL | | | | |
| $V_{Final}(EDTA)$ / mL | | | | |

| Experiment No. | 1 | 2 | 3 |
|---|---|---|---|
| $\Delta V_{\text{Consume}}$(EDTA) / mL | | | |
| $c(Ca^{2+})$ / mmol·L$^{-1}$ | | | |
| $c(Mg^{2+})$ / mmol·L$^{-1}$ | | | |
| $\overline{c}(Ca^{2+})$ / mmol·L$^{-1}$ | | | |
| $\overline{c}(Mg^{2+})$ / mmol·L$^{-1}$ | | | |
| $\overline{d}_r$ / % | | | |

## 【Questions】

(1) Why adding buffer solution of pH 10 to the sample solution in the determination of the total hardness of water?

(2) In the determination the content of $Ca^{2+}$ and $Mg^{2+}$, are there disturbances if the analyte contains little impure ions of $Fe^{3+}$, $Al^{3+}$, $Cu^{2+}$, $Zn^{2+}$, $Pb^{2+}$, and so on? What methods can be used to eliminate the disturbance of the ions above?

(3) Describe the principles of determining the content of $Ca^{2+}$ and $Mg^{2+}$ in limestone with compleximetry.

（庄海旗）

## 2. Determination of $KAl(SO_4)_2 \cdot 12H_2O$ in Alum

### 【Objectives】

(1) To know the general process of back titration.

(2) To learn the preparation and standardization method of EDTA solution.

(3) To master the using method and condition of xylenol orange indicator.

(4) To learn the determination principle and method of $KAl(SO_4)_2 \cdot 12H_2O$ in alum.

### 【Pre-lab Assignments】

(1) How to determine $KAl(SO_4)_2 \cdot 12H_2O$ in Alum?

(2) When shall back titration be adopted in the titration?

(3) Why is the back titration adopted in the determination of $KAl(SO_4)_2 \cdot 12H_2O$ in alum?

(4) How to apply xylenol orange indicator in the complex titration?

(5) Draw the design scheme by flow chart according to the coordination titration principle and experiment procedure. Then label the reaction and operation condition of titration to ensure the accuracy of the determination on the flow chart.

### 【Principles】

Alum is commonly called as white alum, which is extracted from alumstone. In the determination of $Al^{3+}$ content in a sample by EDTA titration, a difficulty is that the complexation of $Al^{3+}$ with EDTA takes place very slowly, especially when $Al^{3+}$ is present in the form of hydroxide complex. Furthermore, the xylenol orange indicator would be blocked by $Al^{3+}$. Thus $Al^{3+}$ is often determined by back titration instead of direct titration. The procedures can be described as follows: add a measured excess of standard EDTA solution into the sample solution containing $Al^{3+}$, then boil to ensure the complete formation of Al-EDTA complex. Cool the solution to room temperature and add HAc-NaAc buffer solution to adjust the solution pH 5~6. Using xylenol orange as the indicator (the xylenol orange indicator would not be blocked by $Al^{3+}$ because of the formation of complex $AlY^-$), titrate the excess of EDTA with standard $Zn^{2+}$ solution. When the excessive EDTA has been coordinated with $Zn^{2+}$, the xylenol orange indicator will react with $Zn^{2+}$ to produce purplish red $ZnH_3In^{2-}$, so the color of solution changes from yellow (the color of xylenol orange) to orange (the combination color of yellow and purplish red). That is, the orange is just the color of the end point of titration. Then the content of $Al^{3+}$ can be determined according to the concentration and consumed volume of the two standard solutions. The reactions are as follows

$$Al^{3+} + H_2Y^{2-}(excess) + 2H_2O = AlY^- + 2H_3O^+$$
$$H_2Y^{2-}(rest) + Zn^{2+} + 2H_2O = ZnY^{2-} + 2H_3O^+$$
$$Zn^{2+} + H_3In^{4-} = ZnH_3In^{2-}$$
$$\text{(yellow)} \quad \text{(purplish red)}$$

The stability order of complex ions during the titration is: $AlY^- > ZnY^{2-} > ZnH_3In^{2-}$.

## 【Apparatus, Reagents and Materials】

**Apparatus** Electronic balance, acid or amphotericity buret (25 mL), conical flask (250 mL × 3), volumetric pipette (25 mL × 3), volumetric flask (100 mL, 250 mL), beaker (100 mL × 2), reagent bottle (250 mL), cylinder (10 mL × 3, 100 mL), glass rod

**Reagents and Materials** 0.05 mol·L⁻¹ EDTA solution, $ZnSO_4 \cdot 7H_2O$ (A.R or C.P.), 2 mol·L⁻¹ HCl, HAc-NaAc buffer solution (pH 4.5), 0.5% xylenol orange indicator solution, alum sample

## 【Procedures】

### 2.1 Preparation of 0.05 mol·L⁻¹ $Zn^{2+}$ standard solution

Weigh 3.6~3.8 g (0.0001 g precisely) primary standard $ZnSO_4 \cdot 7H_2O$ ($M_r$ = 287.56) into a 100 mL beaker, add 2.0 mL of 2 mol·L⁻¹ HCl solution with cylinder and a little distilled water, dissolve it and transfer quantitatively into a 250 mL volumetric flask, dilute it to the mark, and mix well. Calculate the concentration of $Zn^{2+}$ by the equation (7-6)

$$c(ZnSO_4 \cdot 7H_2O) = \frac{m(ZnSO_4 \cdot 7H_2O)}{M_r(ZnSO_4 \cdot 7H_2O) \times \frac{250.00}{1000}} \text{(mol·L}^{-1}\text{)} \tag{7-6}$$

### 2.2 Determination the concentration of EDTA solutionsolution

Transfer 25.00 mL EDTA solution into a 250 mL volumetric flask, then add 100.0 mL of distilled water, 5 mL of HAc-NaAc buffer solution and 1 mL of xylenol orange indicator. Titrate with $Zn^{2+}$ standard solution until the color of the solution changes from yellow to orange. Record the date in Table 7-4. Repeat the titration twice and calculate the concentration of EDTA solution by the equation (7-7).

$$c(EDTA) = \frac{c(ZnSO_4) \times V(ZnSO_4)}{V(EDTA)} \text{(mol·L}^{-1}\text{)} \tag{7-7}$$

## 2.3 Determination of KAl(SO₄)₂·12H₂O content in alum

Weigh precisely 1.3~1.4 g ground alum into a 100 mL beaker, dissolve it with distilled water and transfer quantitatively to a 100 mL volumetric flask, dilute to the mark and mix throughly. Transfer 25.00 mL of the solution and 25.00 mL of standardized EDTA solution into a conical flask, and warm the solution in a boiling water bath for 10 minutes. Cool to room temperature; add 100 mL of distilled water and 5.00 mL of HAc-NaAc buffer solution and 1 mL of xylenol orange indicator. Titrate with the $Zn^{2+}$ standard solution until the color changes from yellow to orange. Record the date in Table 7-5. Repeat the titration twice and calculate the mass fraction of $KAl(SO_4)_2 \cdot 12H_2O$ in the alum by the following equation (7-8)

$$\omega[KAl(SO_4)_2 \cdot 12H_2O] = \frac{[c(EDTA)V(EDTA) - c(ZnSO_4)V(ZnSO_4)] \times M(KAl(SO_4)_2 \cdot 12H_2O) \times 10^{-3}}{m(\text{sample}) \times \frac{25.00}{100.00}} (g \cdot g^{-1}) \quad (7-8)$$

## 【Notes】

(1) In order to accelerate the reaction between the $Al^{3+}$ and EDTA, the mixture must be boiled after the addition of excess EDTA into the $Al^{3+}$ sample solution.

(2) After cooling, adjust pH of the reaction solution to 5~6 to ensure the quantitative coordination between $Al^{3+}$ and EDTA.

(3) Xylenol orange indicator must be added after the completion of reaction. Otherwise, the excess EDTA will be titrated by the $Zn^{2+}$ standard solution.

## 【Data and Results】

Date:　　　　　　　　Temperature:　　　　　℃　　　Relative Humidity:

Table 7-4　Determination of sodium EDTA solution

| Experiment No. | 1 | 2 | 3 |
|---|---|---|---|
| $m_{\text{Weighing}}(ZnSO_4 \cdot 7H_2O)$ / g | | | |
| $m_{\text{Determine}}(ZnSO_4 \cdot 7H_2O)$ / g | | | |
| Indicator | | | |
| The change of color at the end point of titration | | | |
| $V_{\text{Initial}}(ZnSO_4)$ / mL | | | |
| $V_{\text{Final}}(ZnSO_4)$ / mL | | | |
| $\Delta V_{\text{Consume}}(ZnSO_4)$ / mL | | | |
| $c(EDTA)$ / mol·L$^{-1}$ | | | |
| $\overline{c}(EDTA)$ / mol·L$^{-1}$ | | | |
| $\overline{d_r}$ / % | | | |

Table 7-5　Determination of KAl(SO₄)₂·12H₂O in alum

| Experiment No. | 1 | 2 | 3 |
|---|---|---|---|
| $m_{\text{Weighing}}(\text{Alum})$ / g | | | |
| $m_{\text{Determine}}(\text{Alum})$ / g | | | |

续表

| Experiment No. | 1 | 2 | 3 |
|---|---|---|---|
| $V$(Alum) / mL | | | |
| Indicator | | | |
| The change of color at the end point of titration | | | |
| $V_{Initial}$(ZnSO$_4$) / mL | | | |
| $V_{Final}$(ZnSO$_4$) / mL | | | |
| $\Delta V_{Consume}$(ZnSO$_4$) / mL | | | |
| $\omega$[KAl(SO$_4$)$_2 \cdot$12H$_2$O] / g·g$^{-1}$ | | | |
| $\overline{\omega}$[KAl(SO$_4$)$_2 \cdot$12H$_2$O] / g·g$^{-1}$ | | | |
| $\overline{d_r}$ / % | | | |

## 【Questions】

(1) Can direct titration be adopted in the determination of Al$^{3+}$ with EDTA? Why or why not?

(2) Can eriochrome black T be used as the indicator in the Al$^{3+}$ determination? Why or why not?

(3) Why can xylenol orange be used as the indicator in the Al$^{3+}$ determination, although Al$^{3+}$ will block it?

## 3. Determination of Calcium Gluconate

### 【Objectives】

(1) To know the general process of titration.

(2) To learn the preparation and standardization method of EDTA solution.

(3) To master the using method and condition of eriochrome black T indicator.

(4) To learn the determination principle and method of calcium gluconate.

### 【Pre-lab Assignments】

(1) How to prepare EDTA standard solution?

(2) What's the principle of the calcium gluconate determination?

(3) Which indicator will be used in the calcium gluconate determination?

(4) How to apply eriochrome black T in the coordination titration? Why shall an assistant indicator be added in this experiment?

(5) Draw the design scheme by flow chart according to the coordination titration principle and experiment procedure. Then label the reaction and operation condition of titration to ensure the accuracy of the determination on the flow chart.

### 【Principles】

Calcium gluconate is commonly used as tablets or injections in clinic. Usually, the content of calcium gluconate is determined by coordination titration as follows: adjust the acidity of the calcium gluconate solution with $NH_3$-$NH_4Cl$ buffer solution (pH10), add eriochrome black T as the indicator, then titrate with standard EDTA solution until the color of solution changes from purplish-red to pure blue at the end point. In this experiment, however, no sharp color variation can be observed without the presence of $Mg^{2+}$. A common procedure is to add a small amount of $Mg^{2+}$ to the buffer solution. In that case, the additional $Mg^{2+}$ must be determined with EDTA solution first. After this additional procedure, the sample of calcium gluconate can be added in the solution, and then continue the titration to get the consumption of EDTA solution for the calcium gluconate sample. Thus, the percentage of calcium gluconate can be calculated by the molarity of the standard EDTA solution and the volume of EDTA solution required for titration from the first to the second end point. Hence the steps and the reactions of this experiment are as follows:

(1) Add a small amount of $MgSO_4$ solution and some drops of eriochrome black T indicator into the $NH_3$-$NH_4Cl$ buffer solution (pH10). Titrate with standard EDTA solution until a color changes from purplish-red to pure blue as the first end point.

(2) Add sample solution of calcium gluconate in the solution above (1). Then continuously titrate with the same EDTA solution until the color changes from purplish-red to pure blue again as the second end point.

The titration for the calcium gluconate needs only the volume of EDTA solution from the first end point to the second.

## 【Apparatus, Reagents and Materials】

**Apparatus**  Electronic balance, base or PTFE buret (25 mL), conical flask (250 mL × 3), volumetric pipette (10 mL, 20 mL), volumetric flask (250 mL × 3), beaker (250 mL × 2), measuring cylinder (10 mL × 2, 50 mL), washing bottle

**Reagents and Materials**  $Na_2H_2Y \cdot 2H_2O$ (s, A.R.), $MgSO_4$ (1%), $NH_3$-$NH_4Cl$ buffer solution (pH10), eriochrome black T indicator (0.5%), calcium gluconate (tablets or injections)

## 【Procedures】

### 3.1 Preparation of 0.01 mol·$L^{-1}$ standard EDTA solution

Weigh 1.0~1.1 g (accurate to 0.0 001 g) primary standard substance $Na_2H_2Y \cdot 2H_2O$ ($M_r$=372.26) into a 250 mL beaker. Dissolve it with 50.0 mL of distilled water, then transfer quantitatively to a 250 mL volumetric flask, dilute it to the mark and mix well.

$$c(\text{EDTA}) = \frac{m(\text{EDTA}) \times 1\,000}{M_r(\text{EDTA}) \times 250.00} (\text{mol} \cdot L^{-1}) \qquad (7\text{-}9)$$

### 3.2 Preparation of the of calcium gluconate sample

**For tablets**  Weigh precisely 1.0~1.1 g powder of calcium gluconate (from the tablet). Put it into a 250 mL beaker, dissolve it with 50.0 mL distilled water and transfer quantitatively into a 250 mL volumetric flask. Dilute it to the mark with distilled water, and mix well.

**For injections**  Transfer 10.00 mL of injection solution with pipette into a 250 mL volumetric flask. Dilute it to the mark and mix well.

### 3.3 Determination of calcium gluconate content

Add 10 mL of distilled water and 10.0 mL of $NH_3$-$NH_4Cl$ buffer solution (pH10) into a 250 mL conical flask. Add 1~2 drops of 1% $MgSO_4$ and 2~3 drops of eriochrome black T indicator and mix well. Titrate with standard EDTA solution until a color changes from purplish-red to pure blue. Record the first end point reading of EDTA in the buret as $V_1$. Transfer 20.00 mL of sample solution into the above conical flask. Continue the titration with the same EDTA solution until the color changes from purplish-red to pure blue again. Record the second end point reading of EDTA in the buret as $V_2$. in Table 7-6 Repeat the titration twice. Calculate the mass fraction ($\omega$) for tablets by the equation (7-10) and mass concentration ($p$) for injections of the calcium gluconate by the equation (7-11).

For tablets

$$\overline{\omega}(\text{calcium gluconate}) = \frac{c(\text{EDTA}) \times [V_2(\text{EDTA}) - V_1(\text{EDTA})] \times M_r(\text{calcium gluconate})}{m \times \dfrac{20.00}{250.00} \times 1\,000} \text{ mg} \cdot \text{g}^{-1} \quad (7\text{-}10)$$

where $m$ is mass of calcium gluconate tablet.

For injections

$$\overline{\rho}(\text{calcium gluconate}) = \frac{c(\text{EDTA}) \times [V_2(\text{EDTA}) - V_1(\text{EDTA})] \times M_r(\text{calcium gluconate})}{10.00 \times \dfrac{20.00}{250.00} \times 1\,000} \text{ g} \cdot \text{L}^{-1} \quad (7\text{-}11)$$

## 【Notes】

(1) Eriochrome black T indicator must be used in the solution of pH 10.

(2) A small amount of $MgSO_4$ is added into the solution to avoid the advance of the titration end point owing to the instability of calcium eriochrome black T complex.

(3) There are two titration end points in the titration. Hence the volume of EDTA solution consumed by the calcium gluconate is recorded from the first end point to the second.

## 【Data and Results】

Date:　　　　　　　　Temperature:　　　　　℃　　　Relative Humidity:

Table 7-6　Determination of the calcium gluconate in injections or tablets

| Experiment No. | 1 | 2 | 3 |
|---|---|---|---|
| $m(Na_2H_2Y \cdot 2H_2O)$ / g | | | |
| $c(\text{EDTA})$ / mol·L$^{-1}$ | | | |
| Indicator | | | |
| The change of solution color at the first end point of titration | | | |
| The change of solution color at the second end point of titration | | | |
| $V$(diluted injection solution) / mL | | | |
| $V_{\text{End point 1}}(\text{EDTA})$ / mL | | | |
| $V_{\text{End point 2}}(\text{EDTA})$ / mL | | | |

续表

| Experiment No. | 1 | 2 | 3 |
|---|---|---|---|
| $V_{\text{End point 2}} - V_{\text{end point 1}}$(EDTA) / mL | | | |
| $\Delta \overline{V}_{\text{Consumed}}$(EDTA for calcium gluconate) / mL | | | |
| $\omega$(calcium gluconate) / mg·g$^{-1}$ | | | |
| $\overline{\omega}$(calcium gluconate) / mg·g$^{-1}$ | | | |
| $\rho$(calcium gluconate) / g·L$^{-1}$ | | | |
| $\overline{\rho}$(calcium gluconate) / g·L$^{-1}$ | | | |
| $\overline{d}_r$ / % | | | |

## 【Questions】

(1) Why can a small amount of $Mg^{2+}$ be added into the calcium gluconate sample solution? Does it influence the result of the determination?

(2) Can $Mg^{2+}$ be added into the solution as an assistant indicator when the EDTA standard solution is being prepared?

（陈志琼）

# Chapter 3
# Spectrophotometry

## Experiment 8  Determination of $Fe^{3+}$ ($Fe^{2+}$) Content in Water Sample with Visible Spectrophotometry

### 【Objectives】

(1) To learn the principles and methods for the determination of $Fe^{3+}$ ($Fe^{2+}$) content in a water sample with visible spectrophotometry.

(2) To practice the operation for a 722-spectrophotometer.

### 【Pre-lab Assignments】

(1) What is Lambert-Beer law? How to express it with two different expressions?

(2) What are the most common quantitative analysis methods with spectrophotometry?

(3) How to obtain an absorption spectrum? Why should the maximum absorption wavelength be found?

(4) What are the basic components of a spectrophotometer? What can they be used for, respectively?

(5) What are the basic principles for the determination of $Fe^{3+}$ ($Fe^{2+}$) content in water sample with O-phenanthroline method, sulfocyanate method, and sulfosalicylic acid method, respectively? Summarize the similarities and differences among them.

(6) Please use a flow chart to illustrate how to design the experimental procedures for the determination of $Fe^{3+}$ ($Fe^{2+}$) content in water sample with spectrophotometry? Mark the steps that should be paid attention to.

### 【Principles】

Spectrophotometry is a modern instrumental analysis method. It base on the absorption spectrum of matter and the absorption law of light. According to a light source, spectrophotometry can be divided into visible spectrophotometry (380~760 nm), ultraviolet spectrophotometry (200~380 nm), and infrared spectrophotometry (780~3×10⁵ nm). Lambert-Beer law is the basic law for absorption, which describes the relationship between the absorption intensity of a substance and its concentration ($c$) and thickness of the cell ($b$). Lambert-Beer law provides the mathematical correlation between absorbance and concentration as following

$$A = \varepsilon bc \tag{8-1}$$

where $A$ is the absorbance, $\varepsilon$ is the analyte's molar absorptivity or extinction coefficient (L·mol⁻¹·cm⁻¹) that characterizes the medium. When the concentration of the light-absorbing medium is expressed in mass concentration with units of g·L$^{-1}$, $A = abc$. Here $a$ is the mass absorptivity (L·g$^{-1}$·cm$^{-1}$), and $b$ is the path length of the radiation that is identical with thickness of the absorption cell.

When the light-absorbing materials, the wavelength of incident light, the temperature and the solvent are all fixed, $\varepsilon$ is a constant. Then the absorbance $A$ is proportional to the concentration of the absorbing species ($c$) and the thickness of the cell ($b$), $A \propto c$. The Lambert-Beer law can be used as the basis for the quantitative analysis and to determine the content of the absorbing species.

However, since the Lambert-Beer law can be applied only for the monochromatic light, an optimum wavelength of incident radiation should be chosen by absorption spectrum of the light-absorbing species in experiments. A plot of absorbance *versus* wavelength is called absorption spectrum. The maximum absorption wavelength, $\lambda_{max}$, can be located in absorption spectrum and used as the incident radiation in spectrophotometry.

The spectrophotometry method is only suitable for the analysis of microcomponents. Generally, $A$ within 0.2~0.7 is the best.

The quantitative analysis methods commonly used in spectrophotometry include standard comparison method and standard curve method.

For the comparison method, standard solution of absorbent species and sample solution of absorbent species are prepared. Then absorbance is determined under the same conditions.

$$A_{Standard} = abc_{Standard} \tag{8-2}$$

$$A_{Sample} = abc_{Sample} \tag{8-3}$$

When cuvettes in the spectrophotometer have the same thickness with $b$, and absorbent species of standard solution and sample solution are the same, they have the same constant of $a$. Thus the following equations can be used for the determination of the sample concentration.

$$\frac{A(\text{Standard})}{A(\text{Sample})} = \frac{abc(\text{Standard})}{abc(\text{Sample})} \tag{8-4}$$

$$\frac{A(\text{Standard})}{A(\text{Sample})} = \frac{c(\text{Standard})}{c(\text{Sample})} \tag{8-5}$$

$$c(\text{Sample}) = \frac{A(\text{Sample})}{A(\text{Standard})} \times c(\text{Standard}) \tag{8-6}$$

In practice, in order to simplify the calculation and increase the accuracy of the determination, a standard curve (working curve) of absorbance versus concentration is plotted by determining the absorbance of a series of standard solutions with known concentration. Then the concentration of sample solution containing the same absorbent substance with standard species can be obtained by determining its absorbance under the same conditions and locating its value on the working curve. This way called the working curve method. This plot can be represented by the regression equation $y = ax + b$.

## 1. O-phenanthroline Method

In spectrophotometric determination of trace $Fe^{2+}$, o-phenanthroline is a sensitive color-developing agent, with which a complex of $Fe^{2+}$ is formed to give orange red color ($\lg K_s = 21.3$). The colored solution can be measured with a spectrophotometer, maximum absorbance is observed at 510 nm. Molar absorption coefficient is equal to $1.1 \times 10^4$ L·mol$^{-1}$·cm$^{-1}$. In the range of pH 3 to 9, the complex is very stable. Iron must be in ferrous state and hence a reducing agent is added before the color is developed. Hydroxylamine hydrochloride can be used to reduce $Fe^{3+}$ to $Fe^{2+}$. These reactions are given below:

$$2Fe^{3+} + 2NH_2OH \cdot HCl + 2H_2O = 2Fe^{2+} + N_2 \uparrow + 4H_3O^+ + 2Cl^-$$

【Apparatus, Reagents and Materials】

**Apparatus** 722-Spectrophotometer, 0.1mg-electronic balance, volumetric flask (50 mL × 7), measuring pipette (1 mL, 2 mL × 2, 5 mL, 10 mL), dropper, adjustable transfer gun(5 mL, 10 mL)

**Material and Reagents** 8 mmol·L$^{-1}$ o-Phenanthroline solution (freshly prepared), 1.5 mol·L$^{-1}$ hydroxylamine hydrochloride solution (freshly prepared), 1 mol·L$^{-1}$ sodium acetate solution, 2.000 mmol·L$^{-1}$ standard ferrous sulfate solution[*]

【Procedures】

### 1.1 Preparation of Standard Solutions and Sample Solution of $Fe^{2+}$

According to Table 8-1, transfer reagent solution into each of the seven volumetric flasks. Then dilute to the mark with distilled water and homogenize.

Table 8-1 Preparation of the working curve and determination of trace $Fe^{2+}$ in a sample solution

| Experiment No. | 1 (blank) | 2 | 3 | 4 | 5 | 6 | 7 |
|---|---|---|---|---|---|---|---|
| $Fe^{2+}$(standard) / mL | 0 | 0.40 | 0.80 | 1.20 | 1.60 | 2.00 | — |
| $Fe^{2+}$(water sample) / mL | — | — | — | — | — | — | 10.00 |
| Hydroxylamine hydrochloride / mL | 1.00 | 1.00 | 1.00 | 1.00 | 1.00 | 1.00 | 1.00 |
| o-Phenanthroline solution / mL | 2.00 | 2.00 | 2.00 | 2.00 | 2.00 | 2.00 | 2.00 |
| NaAc / mL | 5.00 | 5.00 | 5.00 | 5.00 | 5.00 | 5.00 | 5.00 |
| $V_{Total}$(diluted)/ mL | 50.00 | 50.00 | 50.00 | 50.00 | 50.00 | 50.00 | 50.00 |
| $c$($Fe^{2+}$ diluted) / μmol·L$^{-1}$ | | | | | | | |

### 1.2 Determination of Absorption Spectrum

Before starting the experiment, read the operation manual of spectrophotometer carefully.

The absorption spectrum of the complex of $Fe^{2+}$ can be obtained by measuring the absorbance of standard solution No.4, and the reading of absorbance, $A$, is recorded in Table 8-2.

---

[*] Weight out 0.7 842 g of $(NH_4)_2Fe(SO_4)_2 \cdot 6H_2O$ accurately, put it into a beaker, and add 120 mL of hydrochloric acid (6 mol·L$^{-1}$) and small volume distilled water to dissolve it. Then transfer the solution to a 1 000 mL volumetric flask and fill to mark with distilled water. Homogenize the solution by shaking the flask.

### 1.3 Plotting of a Standard Curve

Select $\lambda_{max}$ in the absorption spectrum. Use it as the wavelength of incident light for the determination, and determine absorbance of a serial standard solution. Record the reading of the absorbance, $A$, into Table 8-3. Draw a working curve. The abscissa is the concentration of $Fe^{2+}$ in standard solutions, and the ordinate is the absorbance of the standard solutions.

### 1.4 Determination of the Content of trace $Fe^{2+}$ in a Water Sample

Under the same conditions as determination of the working curve, determine the $A$ of the water sample. Record the reading of the absorbance, $A$, into Table 8-3.

【Data and Results】

Date:                Temperature:           ℃          Relative Humidity:

(1) Draw an absorption spectrum and locate $\lambda_{max}$

Table 8-2  Determination of absorption spectrum

| $\lambda$ / nm | 450 | 460 | 470 | 480 | 490 | 500 | 510 | 520 | 530 | 540 | 550 | 560 |
|---|---|---|---|---|---|---|---|---|---|---|---|---|
| $A$ | | | | | | | | | | | | |
| $\lambda_{max}$ / nm | | | | | | | | | | | | |

(2) Plot a standard curve

Table 8-3  Determination of the absorbance of $Fe^{2+}$ standard solutions and sample solution

| Experiment No. | 1 (blank) | 2 | 3 | 4 | 5 | 6 | 7 (water sample) |
|---|---|---|---|---|---|---|---|
| $c(Fe^{2+}$ diluted$)$ / $\mu mol \cdot L^{-1}$ | | | | | | | |
| $A$ | | | | | | | |

(3) Determine the content of $Fe^{2+}$ in sample solution

1) Standard curve method: The concentration of the sample solution from the working curve is _____ $\mu mol \cdot L^{-1}$, The concentration of the $Fe^{2+}$ in sample solution is _____ $\mu mol \cdot L^{-1}$.

2) Standard comparison method

$c_{Standard}$ = _____ $\mu mol \cdot L^{-1}$, $A_{Standard}$ = _____, $A_{Sample}$ = _____, $c_{Sample}$ = _____ $\mu mol \cdot L^{-1}$.

【Questions】

(1) Why is it better for controlling the absorbance of the solution in the A range of 0.2~0.7? How can it be controlled?

(2) Is Are the $Fe^{2+}$ determined by the working curve? $Fe^{2+}$ the original sample solution?

(3) According to this experimental result, identify the advantages and disadvantages of comparison method and working curve method.

## 2. Sulfocyanate Method

【Principles】

Visible spectrophotometry only applies for colored materials. But dilute solution of $Fe^{3+}$ is almost colorless. KSCN is used as a color developing reagent in the experiment. $Fe^{3+}$ can react with $SCN^-$ and produce red $[Fe(SCN)_6]^{3-}$ ($lgK_s = 6.4$):

$$Fe^{3+} + 6 SCN^- \rightleftharpoons [Fe(SCN)_6]^{3-}$$

According to the Lambert Beer law, the color of the solution is in proportion to the concentration of $Fe^{3+}$. Thus in this experiment $SCN^-$ must excess so that it can react with $Fe^{3+}$. $HNO_3$ should be added into the sample solution to prevent $Fe^{3+}$ from hydrolysis. Furthermore, $Fe^{3+}$ can be slowly reduced to $Fe^{2+}$ by $SCN^-$ $(NH_4)_2S_2O_8$ should also be added as a strong oxidant to prevent $Fe^{3+}$ from being reduced.

## 【Apparatus, Reagents and Materials】

**Apparatus**  722-spectrophotometer, volumetric flasks (50 mL × 7, 1 000 mL), measuring pipettes (1 mL, 5 mL, 10 mL), dropper

**Material and Reagents**  0.2 mol·L$^{-1}$ KSCN solution, 2 mol·L$^{-1}$ HNO$_3$ solution, concentrated H$_2$SO$_4$ solution, 0.1 000 g·L$^{-1}$ Fe$^{3+}$ standard solution*, 25 g·L$^{-1}$ (NH$_4$)$_2$S$_2$O$_8$ solution

## 【Procedures】

### 2.1 Preparation of Standard Solutions and the Sample Solution

According to Table 8-4, transfer the solutions into volumetric flasks, respectively. Add distilled water to the mark and homogenize. Then the series of standard solutions, the blank, and the sample solution are prepared.

Table 8-4  Preparation of Fe$^{3+}$ standard solutions and the sample solution

| Experiment No. | 1 (blank) | 2 | 3 | 4 | 5 | 6 | 7 |
|---|---|---|---|---|---|---|---|
| Fe$^{3+}$ (standard solution) / mL | 0 | 0.50 | 1.00 | 1.50 | 2.00 | 2.50 | — |
| Fe$^{2+}$ (water sample) / mL | — | — | — | — | — | — | 10.00 |
| 2 mol·L$^{-1}$ HNO$_3$ / mL | 1.00 | 1.00 | 1.00 | 1.00 | 1.00 | 1.00 | 1.00 |
| 0.2 mol·L$^{-1}$ KSCN / mL | 5.00 | 5.00 | 5.00 | 5.00 | 5.00 | 5.00 | 5.00 |
| 25 g·L$^{-1}$ NH$_4$Fe(SO$_4$)$_2$ / drop | 1 | 1 | 1 | 1 | 1 | 1 | 1 |
| $V_{total}$ / mL | 50.00 | 50.00 | 50.00 | 50.00 | 50.00 | 50.00 | 50.00 |
| $\rho(Fe^{3+})$ / mg·L$^{-1}$ | | | | | | | |

### 2.2 Determination of $\lambda_{max}$

The absorption spectra of the solution can be determined by determining absorbance for the No. 4 standard solution in Table 8-4 at the homogeneous light from 450 nm to 520 nm interval of 10 nm, vicinal maximum absorbance at a 5 nm interval. Use the blank as the reference. Record the absorbance into Table 8-5.

### 2.3 Plotting the Standard Curve

Select the $\lambda_{max}$ from the absorption spectrum determined above. Use blank as the reference to determine the absorbance of each standard solution and the sample solution. Record the results in

---

*Preparation of 0.1 mg / 1.00 mL Fe$^{3+}$ standard solution: Weigh 0.8 640 g NH$_4$Fe (SO$_4$)$_2$·12H$_2$O, dissolve it with a little distilled water and then add 5mL concentrated H$_2$SO$_4$ to the solution. Allow the solution to cool to room temperature, pour it into a volumetric flask (1 000 mL), and add distilled water until the meniscus descends to the ring mark. (0.1 mg / 1.00 mL Fe$^{3+}$ solution)

Table 8-6. Plot the standard curve.

### 2.4 Determination of Concentration

The concentration of $Fe^{3+}$ ($mg \cdot L^{-1}$) in the water sample obtained according the absorbance of sample by the comparison method and the working curve method.

【Data and Results】

Date:　　　　　　　　Temperature:　　　　　℃　　　Relative Humidity:

(1) Plot the absorption spectrum and determine the $\lambda_{max}$: According to Table 8-5, plot the absorption spectrum. Find out the $\lambda_{max}$ in the curve.

Table 8-5　Determination of absorption spectrum

| $\lambda$ / nm | 450 | 460 | 470 | 480 | 490 | 500 | 510 | 520 |
|---|---|---|---|---|---|---|---|---|
| $A$ | | | | | | | | |
| $\lambda_{max}$ / nm | | | | | | | | |

(2) Plotting the curve: According to Table 8-6, plot the working curve.

Table 8-6　Absorbance of $Fe^{3+}$ standard solutions and water sample

| Experiment No. | 1(blank) | 2 | 3 | 4 | 5 | 6 | 7 (sample) |
|---|---|---|---|---|---|---|---|
| $\rho(Fe^{3+})$/ $mg \cdot L^{-1}$ | | | | | | | |
| $A$ | | | | | | | |

(3) Determination of the concentration $Fe^{3+}$ in sample solution: The concentration of $Fe^{3+}$ ($mg \cdot L^{-1}$) in the water sample can be determined with the comparison method or the standard curve method.

1) Standard curve method

The concentration of the sample from the standard curve is_____$mg \cdot L^{-1}$.

The original concentration of $Fe^{3+}$ in sample is _____$mg \cdot L^{-1}$.

2) Standard comparison method

$\rho_{Standard}$ = _____$mg \cdot L^{-1}$, $A_{Standard}$ = _____, $A_{Sample}$ = _____, $\rho_{Sample}$ = _____$mg \cdot L^{-1}$.

【Questions】

Why should excess KCNS be added in the solution?

## 3. Sulfosalicylic Acid Method

【Principles】

Sulfosalicylic acid ($H_2Ssal$) is the color developing reagent in this experiment. $Fe^{3+}$ can react with $H_2Ssal$ to produce many kinds of complex ions, with different colors under different pHs. For example, violet red $[FeSsal]^+$ forms under pH 1.8~2.5, brown $[Fe(Ssal)_2]^-$ is forms under pH 4~8, and stable yellow $[Fe(Ssal)_3]^{3-}$ forms under pH 8.0~11.5. $Fe^{3+}$ is easy to hydrolyze and produce precipitation Fe(OH)$_3$ when pH > 12. Therefor this method can only be used when pH < 12 and pH remains a constant.

In the experiment, brown complex of $[Fe(Ssal)_2]^-$ forms in HAc-NaAc buffer solution (pH5). Determine the absorbance at a $\lambda$ of 466nm. This reaction is given below

$$Fe^{3+} + 2 \underset{HOOC}{\overset{HO}{\bigcirc}}-SO_3^- \xrightleftharpoons{pH=5} \left[Fe\left(\underset{-OOC}{\overset{HO}{\bigcirc}}-SO_3^-\right)_2\right]^- + 2H^+$$

In the experiment, absorption spectrum first should be plotted to determine $\lambda_{max}$ in a range of 420~490 nm. The concentration of an unknown solution can be determined by comparison method or the working curve method.

【Apparatus Reagents and Materials】

**Apparatus**  722 Spectrophotometer, volumetric flasks (25 mL × 7), pipette (2 mL × 2, 1 mL), dropper

**Reagents and Materials**  standard solution of $NH_4Fe(SO_4)_2 \cdot 12H_2O$ (0.1 g·L$^{-1}$ Fe$^{3+}$ or 0.8 634 g·L$^{-1}$ $NH_4Fe(SO_4)_2 \cdot 12H_2O$ solution), 100 g·L$^{-1}$ sulfosalicylic acid solution, HAc-NaAc buffer solution (pH5), Fe$^{3+}$ sample solution, coordinate paper

【Procedures】

### 3.1 Preparation of the Standard Solutions and the Sample Solution

Transfer each solution into 7 of 50 mL volumetric flasks respectively according to Table 8-7. Add HAc-NaAc buffer solution (pH 5) until the meniscus descends to the ring mark and homogenize.

Table 8-7  Preparation of Fe$^{3+}$ standard solutions and sample solution

| Experiment No. | 1 (blank) | 2 | 3 | 4 | 5 | 6 | 7 (water sample) |
|---|---|---|---|---|---|---|---|
| Fe$^{3+}$ standard solution / mL | 0.00 | 0.40 | 0.60 | 0.80 | 1.00 | 1.20 | — |
| Fe$^{2+}$ (water sample) / mL | — | — | — | — | — | — | 1.00 |
| 100g·L$^{-1}$ H$_2$Ssal solution / mL | 2.00 | 2.00 | 2.00 | 2.00 | 2.00 | 2.00 | 2.00 |
| $V_{Total}$ / mL | 25.00 | 25.00 | 25.00 | 25.00 | 25.00 | 25.00 | 25.00 |
| $\rho(Fe^{3+})$ / mg·L$^{-1}$ | | | | | | | |

### 3.2 Plot the Spectrum

The absorption spectrum of the complex of $[FeSsal]^+$ can obtained by determining the absorbance of standard solution No.4. Use the blank solution as a reference. Record $A$ into Table 8-8.

### 3.3 Plot the Standard Curve

Select the $\lambda_{max}$ as incident radiance from the absorption spectrum. With the blank solution as the reference, determine absorbance of each standard solution and sample solution, and record the results in Table 8-9. Plot the standard curve.

### 3.4 Determination of Content in Sample Solution

The concentration of trace Fe$^{3+}$ (mg·L$^{-1}$) in the sample solution can be calculated with the comparison method or the standard curve method.

【Data and Results】

Date:　　　　　Temperature:　　　　°C　　　Relative Humidity:

(1) Plot the absorption spectrum: According to Table 8-8, plot the absorption spectrum. Determine the maximum absorption wavelength $\lambda_{max}$.

Table 8-8  Plotting the absorption spectrum

| $\lambda$ / nm | 420 | 430 | 440 | 450 | 460 | 470 | 480 | 490 |
|---|---|---|---|---|---|---|---|---|
| $A$ | | | | | | | | |
| $\lambda_{max}$ / nm | | | | | | | | |

(2) Plot the standard curve: Plot the standard curve according to Table 8-9. Fit with linear equation.

Table 8-9  Absorbance of the standard solutions and the sample solution

| Experiment No. | 1 (blank) | 2 | 3 | 4 | 5 | 6 | 7 (water sample) |
|---|---|---|---|---|---|---|---|
| $\rho(Fe^{3+})$ (solution diluted)/ mg·L$^{-1}$ | | | | | | | |
| $A$ | | | | | | | |

(3) Determine $Fe^{3+}$ content in the sample solution

The concentration of trace $Fe^{3+}$ (mg·L$^{-1}$) in the sample solution can be calculated by the absorbance of sample solution with the comparison method or the working curve method.

1) Standard curve method

The concentration of the unknown solution from the working curve is_____mg·L$^{-1}$. The concentration of the $Fe^{3+}$ in the sample solution is _____mg·L$^{-1}$.

2) Standard comparison method

$\rho_{Standard}$ = _____mg·L$^{-1}$, $A_{Standard}$ = _____, $A_{Sample}$ = _____, $\rho_{Sample}$ = _____mg·L$^{-1}$.

## 【Questions】

(1) Why should buffer solution be added in the preparation of the solution?

(2) How to prepare the blank solution in the experiment?

（王金铃）

## Experiment 9　Determination of Aspirin Content in Tablets by Visible Spectrophotometry

## 【Objectives】

(1) To learn the method for the determination of aspirin content in tablets by visible spectrophotometry.

(2) To master the operation of 722(721)-spectrophotometer.

## 【Pre-lab Assignments】

(1) Please give the structure formula of aspirin and its properties by reviewing literatures search?

(2) Can we use distilled water instead of alcohol as solvent in this experiment?

(3) Why is it necessary to adhere to the order by which the reagents are added in this experiment?

(4) What are the factors that affect the accuracy of the experiment?

(5) Please describe the experimental design ideas with a flow chart. Mark the cautions and give the

reasons steps that should be paid attention to.

(6) Review the ways that may be used to determine the content of aspirin in tablets according to literatures.

## 【 Principles 】

Aspirin, $CH_3COOC_6H_4COOH$, is a white and crystalline compound that can be derived from salicylic acid. Aspirin is powerful in the relief of pain, reducing fever and inflammation. It is commonly used in medicine in the form of tablet.

The major composition of aspirin tablet is acetylsalicylic acid. Its ethoxycarbonyl structure can react with hydroxylamine to form hydroxamic acid in alkaline condition. The maximum absorbance wavelength for the complex is $\lambda_{max} = 520$ nm. Lambert-Beer law can be applied to determine the content of aspirin within a certain range of concentration of aspirin by the standard curve method.

## 【 Apparatus Reagents and Materials 】

**Apparatus**  722-Spectrophotometer, volumetric flask (25 mL × 6), measuring pipette (1 mL × 5, 2 mL)

**Reagents and Materials**  2 mol·L$^{-1}$ NaOH, 4 mol·L$^{-1}$ HCl, 10% FeCl$_3$, 0.500 g·L$^{-1}$ standard acetylsalicylic acid solution in ethanol, 7% hydroxylamine hydrochloride in ethanol solution, and aspirin sample solution (20 aspirin tablets in 1 L of ethanol solution)

## 【 Procedures 】

### 1. Preparation of Standard Solutions and a Sample Solution

Prepare the blank solution, a series of standard solutions according to Table 9-1, and the sample solution.

Table 9-1  Preparation of the standard solution and the sample solution

| Experiment No. | 1 (Blank) | 2 | 3 | 4 | 5 | 6 | 7 (Sample) |
|---|---|---|---|---|---|---|---|
| 0.500 g·L$^{-1}$ Standard aspirin / mL | 0.00 | 0.50 | 1.00 | 1.50 | 2.00 | 2.50 | — |
| Aspirin sample / mL | — | — | — | — | — | — | 1.00 |
| 7% Hydroxylamine / mL | 1.00 | 1.00 | 1.00 | 1.00 | 1.00 | 1.00 | 1.00 |
| 2 mol·L$^{-1}$ NaOH / mL | 1.00 | 1.00 | 1.00 | 1.00 | 1.00 | 1.00 | 1.00 |
| 4 mol·L$^{-1}$ HCl / mL | 1.00 | 1.00 | 1.00 | 1.00 | 1.00 | 1.00 | 1.00 |
| 10%FeCl$_3$ / mL | 1.00 | 1.00 | 1.00 | 1.00 | 1.00 | 1.00 | 1.00 |
| $V_{Total}$ / mL | 25.00 | 25.00 | 25.00 | 25.00 | 25.00 | 25.00 | 25.00 |
| $c$(solution diluted) / mg·L$^{-1}$ | | | | | | | |

### 2. Determination of Absorbance (A)

Select a wavelength of 520 nm as the incident radiation. Use the blank solution as reference solution to determine absorbance of each solution. Record the results in Table 9-2.

## 【Data and Results】

(1) Absorbance (*A*) of the standard solutions and the sample solution

Date:　　　　　　　　　Temperature:　　　　　℃　　Relative Humidity:

(2) Plot the working curve of acetylsalicylic acid

According to Table 9-2, plot a standard curve. Linear fitting equation and the correlation coefficient should be given.

Table 9-2　Determination of absorbance of standard solutions and sample solution

| Experiment No. | 1(blank) | 2 | 3 | 4 | 5 | 6 | 7(sample) |
|---|---|---|---|---|---|---|---|
| $c$(diluted solution) / mg·L$^{-1}$ | | | | | | | |
| $A$ | | | | | | | |

(3) Determine the content of aspirin in tablets

The concentration of the aspirin sample solution from the standard working curve: $\rho_1$ =_____mg·L$^{-1}$.

The original concentration of aspirin in the initial sample solution: $\rho_2 = \rho_1 \times 25$ =_____mg·L$^{-1}$.

The content of aspirin in commercial aspirin tablets:_____mg·Tablet$^{-1}$.

## 【Questions】

(1) Why does the acidity of the test solution change during the determination procedure?

(2) How to eliminate the influence of the purple color formed by the salicylic acid and ferric chloride?

（尹计秋）

# Experiment 10　Determination of the Formula and the Stability Constant of Sulfosalicylate Iron (Ⅲ) by Spectrophotometry

## 【Objectives】

(1) To understand the basic principles of spectrophotometry.

(2) To learn the operation of 722 or 721 Spectrophotometer.

(3) To lean to determine the formula and the stability constant for the complex by concentration gradient method.

## 【Pre-lab Assignments】

(1) How to select color developing reagent in spectrophotometry?

(2) How to control the composition of colored complexes in this experiment?

(3) Why choose perchloric acid to control acidity of solution in this experiment?

(4) How to select the measurement wavelength? How to select a blank solution? What is the reference solution in this experiment?

(5) What methods can be used to determine the formula and the stability constant of the complex? Which method is selected in this experiment?

(6) Why does the complex have the maximum concentration when the composition of the complex is measured by equimolar series method and the ratio of the concentration of components of the solution and ratio of central atom and ligand in complex are same?

(7) Illustrate the experimental procedures by a flow chart. Describe how to determine the wavelength of maximum absorption, the composition and the stability constant of the complex, and then label conditions of the determination. Give the reasons why you consider to do that on the flow chart.

【Principles】

A complex can be yielded by a metal ion reacting with a ligand as follows.

$$M^{n+} + nL^- \rightleftharpoons ML_n$$

$$K_s = \frac{[ML_n]}{[M][L]^n} \tag{10-1}$$

where $n$ is the coordination number of the complex and $K_s$ is the stability constant of the complex.

If M and L are colorless and $ML_n$ is colored, according to Lambert-Beer law, $A = \kappa bc$, the absorbance of the solution is proportional to the concentration of the complex. The experiment is carried out with an equimolar series method.

The equimolar series method is also called the continuous variation method or the concentration ratio variation method. In this method, a serial of solutions are prepared by mixing the solution and ligands solution with different volume ratios (molarity ratios), and then determine their absorbance under the wavelength of maximum absorption. When the concentration of the complex reaches the maximum, the coordination number $n$ can be obtained according to the following equation

$$n = \frac{c(L)}{c(M)} = \frac{1-f}{f} \tag{10-2}$$

where $c(M)$ and the $c(L)$ are the concentrations of the metals ion and the ligands, respectively. $f$ is fraction of the metals ion in the total concentration.

$$c(M) + c(L) = c = \text{constant} \tag{10-3}$$

$$f = \frac{c(M)}{c} \tag{10-4}$$

As can be seen in Figure 10-1, if $f=0$ or $f=1$, no metal ion or ligand exists since no complex forms. When the molar ratio of metal ions to ligands is the same as the composition of the complexes, the concentration of the complexes, the absorbance and the corresponding $f$ all reach their maximal value. For example, in a 1:1 type complex, the value of $f$ is 0.5 at the maximum absorbance. For the 1:2 type complex, $f = 0.34$.

For complex ML, its maximum absorbance $A$ is somewhat lower than the absorbance $A'$ at the junction B of elongation lines as the complex slightly dissociate in an aqueous solution. The smaller difference between the $A$ and $A'$, the more stable the complex. According to the description above, the stability constant of complex can be calculated as following

$$K_s = \frac{[ML]}{[M][L]} \tag{10-5}$$

Because the absorbance of the complex solution is proportional to the concentration of the complex, we have

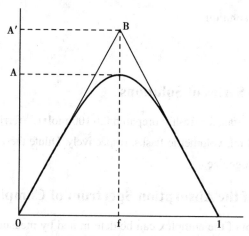

Figure10-1　Concentration gradient plot

$$\frac{A}{A'} = \frac{[ML]}{c'} \tag{10-6}$$

where $c'$ is the concentration of the complex when it does not dissociate completely.

$$c' = c(M) = c(L) \tag{10-7}$$

and

$$[M] = [L] = c' - [ML] = c' - c'\frac{A}{A'} = c'\left[1 - \frac{A}{A'}\right] \tag{10-8}$$

taking equations (10-5) and (10-6) into (10-4), we have

$$K_s = \frac{\dfrac{A}{A'}}{\left[1 - \dfrac{A}{A'}\right]^2 c'} \tag{10-9}$$

The composition of $Fe^{3+}$ complex with sulfosalicylic acid varies with pH. Under pH 2~3, a fuchsia complex with one ligand is formed; under pH 4~9, complex containing two ligands forms. Under pH 9~11.5, the yellow complex containing three ligands forms; When pH > 12:00, Fe(OH)$_3$ will precipitate by the destruction of the colored complex. Since sulfosalicylic acid is colorless and $Fe^{3+}$ solution is diluted, it can also be considered as colorless. Therefore, only the iron sulfosalicylic acid ion is colored. The absorbance of the solution is proportional only to the concentration of the complex. The composition of the complex ion can be obtained by determining the absorbance of the solution. The composition and stability constant of the complex of $Fe^{3+}$ and sulfosalicylic acid at pH 2~3 can be determined. pH of the solution can be controlled by adding a certain amount of $HClO_4$ solution, the main advantage of which is that $HClO_4$ can prevent $Fe^{3+}$ from hydrolysis and from being reduced.

## 【Apparatus, Reagents and Materials】

**Apparatus**　722 or 721-Spectrophotometer, volumetric flask (50 mL × 7), measuring pipette (10 mL × 2), pipette bulb

**Reagents and Materials**　0.0100 mol·L$^{-1}$ H$_2$Ssal solution, 0.0100 mol·L$^{-1}$ Fe(NH$_4$)SO$_4$

solution, 0.1 mol·L$^{-1}$ HClO$_4$ solution

## 【Procedures】

### 1. Preparation of a Series of Solutions

According to Table 10-1, transfer freshly prepared 0.0 100 mol·L$^{-1}$ ferric sulfate and 0.0 100 mol·L$^{-1}$ sulfosalicylic acid to seven 50 mL volumetric flasks, respectively. Dilute these solutions with 0.100 mol·L$^{-1}$ HClO$_4$ to the marks and homogenize.

### 2. Determination of the Absorption Spectrum of Complex

The absorption spectrum of the complex can be determined by measuring the No. 4 solution over a range of 400-700 nm with an interval of 10 nm. Take distilled water as the blank solution. In the vicinity of maximum absorbance, the wavelength interval is 5 nm. Record absorbance into Table 10-2. Plot of the absorbance versus wavelength and locate the peaks to determine the wavelength ($\lambda_{max}$).

### 3. Determination of Absorbance of the Series Solutions

Determine the absorbances of the seven solutions at the maximum absorption wavelength.

## 【Data and Results】

(1) Plot the absorption spectrum to find out measurement wavelength

Date:　　　　　　　　Temperature:　　　　　℃　　　Relative Humidity:

Table 10-1　Determination the absorption spectrum of complex

| $\lambda$ / nm | |
|---|---|
| $A$ | |
| $\lambda_{max}$ / nm | |

Plot the absorption spectrum of the complex

Measurement wavelength: $\lambda_{max}$ = _____ nm

(2) The composition of the complex

According to Table 10-2, plot of concentration gradient of the complex in which the abscissa is $f$ and the ordinate is the absorbance, determine the composition of the complex and give the formular of the complex under 2~3 of pH aqueous solution.

(3) Calculation of the stability constant: Get maximum absorbance $A$ from $A$ - $f$ graphs, and then extend the straight lines on both sides of the curve at one point. Calculate $c'$ and the stability constant of sulfosalicylate iron (Ⅲ) by substituting the formula (10.7).

Table 10-2　Preparation and determination of sulfosalicylate iron(Ⅲ) complex solution

| Experiment No. | Blank | 1 | 2 | 3 | 4 | 5 | 6 | 7 |
|---|---|---|---|---|---|---|---|---|
| 0.0 100 mol·L$^{-1}$ sulfosalicylic / mL | 10 | 1.00 | 2.00 | 3.00 | 5.00 | 7.00 | 8.00 | 9.00 |
| 0.0 100 mol·L$^{-1}$ ferric sulphata / mL | | 9.00 | 8.00 | 7.00 | 5.00 | 3.00 | 2.00 | 1.00 |

续表

| Experiment No. | Blank | 1 | 2 | 3 | 4 | 5 | 6 | 7 |
|---|---|---|---|---|---|---|---|---|
| 0.1 mol·L$^{-1}$ HClO$_4$ / mL | | | | | 40 | | | |
| $A$ | | | | | | | | |
| $f = \dfrac{c(M)}{c}$ | | | | | | | | |

## 【Questions】

(1) What are the conditions in the determination of the stability constant of the complex by the continuous variations method?

(2) What is the effect of acidity on the composition of sulfosalicylate iron (Ⅲ)?

（王美玲）

# Experiment 11　Determination of Vitamin B$_2$ by Fluorescence Analysis

## 【Objectives】

(1) To master the principle and method of fluorometry.

(2) To understand the basic structure of a fluorescence photometer.

(3) To learn the operation of a 930-fluorometer.

## 【Pre-lab Assignments】

(1) What is excitation spectrum? What is emission spectrum?

(2) In the measurement process, why is the excitation wavelength shorter than the emission wavelength? Is it possible for the wavelength of the excitation and the emission to be the same in fluorescence analysis?

(3) What's the prerequisite for a molecule producing fluorescent?

(4) Can spectrophotometry be used for measuring riboflavin's concentration in a multivitamin?

(5) Based on the principle of fluorescence analysis and the experimental procedures described in the following sections, illustrate the design ideas of the experiment by a flow chart.

## 【Principles】

When a substance is excited by a beam of light, usually ultraviolet light excites the electrons in the molecules of the substances causing them to emit certain frequency light that is called fluorescence. The method for the qualitative or quantitative analysis based on using fluorescence is called fluorometry.

Different fluorescent substances have different characteristic excitation wavelength and emission wavelength.

If the concentration of fluorescent substance is very low, the intensity of fluorescence ($F$) and the solution mass concentration ($\rho$) correlates with each other

$$F = 2.3\Phi I_0 ab\rho \qquad (11\text{-}1)$$

Where $\Phi$ is the fluorescence efficiency, $I_0$ is the intensity of incidence, $a$ is the mass absorptivity, and $b$ is the thickness of sample cell.

For a fluorescent substance, when $I_0$ and $b$ are fixed, the intensity of fluorescence is proportional to the mass concentration of the substance. Therefore,

$$F = K'\rho \qquad (11\text{-}2)$$

Vitamin $B_2$, $C_{17}H_{20}N_4O_6$ ($M_r = 376.37$) also known as riboflavin, is an orange-yellow crystal powder. Its structure formula is as following:

$$\begin{array}{c}
H_2C-(CHOH)_3-CH_2OH \\
\text{[structure of riboflavin]}
\end{array}$$

Vitamin $B_2$ in 0.1 mol·$L^{-1}$ HAc solution can be excited by ultraviolet and quantitatively determined by its emitting of yellow-green florescence. The excitation light wavelength may be 360 nm, 400 nm, or 420 nm, with an emission wavelength of 530 nm.

## 【Apparatus, Reagents and Materials】

**Apparatus**  930 Fluorometer, 0.1 mg-analytic balance, volumetric flask (25 mL × 6, 50 mL, 1 000 mL), pipette (1 mL, 5 mL), beaker (100 mL), stirring rod

**Reagents and Materials**  Vitamin $B_2$ (biochemistry reagent), vitamin $B_2$ tablet, 0.1 mol·$L^{-1}$ HAc solution

## 【Procedures】

### 1. Preparation of Standard Vitamin $B_2$ Solution

Weigh vitamin $B_2$ of about 10.0 mg into a small beaker, dissolve with a little 0.1 mol·$L^{-1}$ HAc solution. Transfer the solution into a 1 000 mL volumetric flask. Then use 0.1 mol·$L^{-1}$ HAc solution to dilute to the mark and homogenize. Store the vitamin $B_2$ standard solution at a low temperature in darkness.

Take 6 volumetric flasks of 25 mL, add 0.00 mL, 0.50 mL, 1.00 mL, 1.50 mL, 2.00 mL and 2.50 mL vitamin $B_2$ standard solution, respectively. Then use 0.1 mol·$L^{-1}$ of HAc solution to dilute them to the mark, and homogenize to get a series of standard solutions of vitamin $B_2$.

### 2. Plotting the Standard Curve

Turn on the power button to pre-heat the instrument for 10 minutes before the determination. Select 360 nm or 400 nm as the excitation wavelength and 530 nm as the emission wavelength. Use 0.1 mol·$L^{-1}$ of HAc solution as the reference to calibrate the instrument. Then use standard solution with the highest concentration to adjust the scale to maximum reading. Fix the conditions, following the order from diluted solution to concentrated solution to determine the intensity of fluorescence for the series of standard solutions. Record the data in Table 11-1. According to the data, plot the standard curve.

### 3. Determination of Vitamin B$_2$ in tablet Sample

Take one vitamin B$_2$ tablet and weigh it (accurate to 0.1mg). The powder was prepared by mortar, dissolve it in 0.1 mol·L$^{-1}$ HAc solution, and then store in brown reagent bottle with a constant volume of 1 000 mL.

Transfer 2.50 mL of the solution into a 50 mL volumetric bottle and diluted with 0.1 mol·L$^{-1}$ HAc solution. Fluorescence intensity of the sample can be determined under the same conditions as for the standard solution. Find out the corresponding concentration $\rho_{determined}$ from the standard curve. Then the concentration of the sample solution and that in the tablets can be calculated. Record the data in Table 11-1.

## 【Data and Results】

(1) The Standard Curve of Vitamin B$_2$

Date:　　　　　　　　Temperature:　　　　　　℃　　　　Relative Humidity:

Table 11-1　Standard curve and measuring sample of vitamin B$_2$

| Volumetric flask No. | 0 | 1 | 2 | 3 | 4 | 5 | Sample |
|---|---|---|---|---|---|---|---|
| $m$(vitamin B$_2$) / mg | | | 10.0 | | | | — |
| $\rho_{Stock}$(vitamin B$_2$) / μg·mL$^{-1}$ | | | 10.0 | | | | — |
| $V_{Standard}$(stock vitamin B$_2$) / mL | 0 | 0.50 | 1.00 | 1.50 | 2.00 | 2.50 | — |
| $V_{Total}$(diluted by 0.1 mol·L$^{-1}$ of HAc) / mL | | | 25.00 | | | | 50.00 |
| $\rho_{Standard}$ (vitamin B$_2$) / μg·mL$^{-1}$ | 0 | 0.20 | 0.40 | 0.60 | 0.80 | 1.00 | — |
| $F$ | | | | | | | |
| $m$(sample of vitamin B$_2$) / mg | | | | | | | — |

According to the results in Table 11-1, plot a working curve.

The regression equation of the standard curve is _____.

The correlation coefficient is _____.

The concentration of the vitamin B$_2$ sample solution from the working curve is_____μg·mL$^{-1}$.

(2) The Concentration of Vitamin B$_2$ in tablets

$$\omega(\text{Vitamin B}_2)\% = \frac{\rho_{determined}\,(\mu g \cdot mL^{-1}) \times 20(\text{dilution multiple}) \times 1\,000 mL \times 10^{-3}}{m(\text{sample of vitamin B}_2)}(mg \cdot mg^{-1}) \quad (11\text{-}3)$$

## 【Questions】

(1) Why is 0.1 mol·L$^{-1}$ of HAc solution used for the preparation of vitamin B$_2$ solution in the experiment?

(2) If a radiation of 420 nm or 440 nm is selected as the excitation wavelength, does it influence the results of determination?

(周昊霏)

# Experiment 12 Identification and Content Determination of Vitamin $B_{12}$ by Ultraviolet Spectrophotometry

## 【Objectives】

(1) To learn the application of ultraviolet (UV) spectrophotometer.

(2) To master the method for qualitative analysis of vitamin $B_{12}$ by UV spectrophotometry.

## 【Pre-lab Assignments】

(1) What's the structure of vitamin $B_{12}$? Briefly introduce the main functions of vitamin $B_{12}$ in the body?

(2) Briefly describe the basic principles of UV spectrophotometry.

(3) Consult the relevant literature, and analyze what factors affect the accuracy of the determination of vitamin $B_{12}$ content by UV spectrophotometry.

(4) Refer to Chinese Pharmacopoeia (2015 edition), point out the identification and detection methods of vitamin $B_{12}$, and the requirements of UV spectrophotometer.

(5) Refer to the experimental steps; please use a flow chart to describe the design ideas, the precautions and the basis in the experimental steps.

## 【Principles】

UV spectrophotometry is a method for qualitative or quantitative analysis of the substance by determining the absorbance of the measured substance in the UV light area (200~400 nm) with specific absorption wavelength or certain wavelength range. The relationship between the absorbance and substance concentration is in accordance with Lambert-Beer law.

Vitamin $B_{12}$ is a cobalt coordination polymer, with porphyrin as the ligand (molecular formula: $C_{63}H_{88}CoN_{14}O_{14}P$, $M_r = 135\ 5.38$). It is a deep-red, hydroscopic crystal. The labeling quantities of the vitamin $B_{12}$ injection solution are 500 μg•mL$^{-1}$, 250 μg•mL$^{-1}$, 100 μg•mL$^{-1}$, and 50 μg•mL$^{-1}$, respectively. The water solution of vitamin $B_{12}$ absorb at a wavelengths of 278 nm, 361 nm, and 550 nm. According to *Chinese Pharmacopoeia* (2015 edition), the ratio of the absorbance at the three wavelengths is the basis for qualitative identification of vitamin $B_{12}$. The ranges of the ratios are:

$$A_{361nm}/A_{278nm} = 1.70 \sim 1.88 \qquad A_{361nm}/A_{550nm} = 3.15 \sim 3.45$$

Because of the strongest absorption and less interference at 361 nm, the absorbance of the vitamin $B_{12}$ solution at 361 nm can be used to calculate the concentration of vitamin $B_{12}$ with the absorption coefficient of vitamin $B_{12}$ ($E_{1cm}^{1\%} = 207$).

$$A_{361nm} = E_{1cm}^{1\%} \cdot b \cdot c \tag{12-1}$$

where, at a specific wavelength, $E_{1cm}^{1\%}$ is the absorbance, 1% (g/100 mL) concentration of the solution and 1 cm of liquid layer thickness). Here $A$ is the absorbance, $b$ is the path length, and $c$ is the analyte concentration.

To determine the content of vitamin $B_{12}$ injection, the unit of concentration should be expressed as μg·mL$^{-1}$. In the calculation, it is necessary to change the unit of the absorption coefficient $E_{1cm}^{1\%}$ 361 nm = 270 (1 g/100 mL) into μg·mL$^{-1}$.

$$E_{1cm}^{1\mu g \cdot L^{-1}} 361nm = \frac{207 \times 100}{10^6} = 207 \times 10^{-4} \tag{12-2}$$

## 【Apparatus, Material and Reagents】

**Apparatus**  754 UV-visible spectrophotometer, volumetric flask (10 mL), and pipette (1 mL)

**Material and Reagents**  Vitamin B$_{12}$ injection (500 μg·mL$^{-1}$)

## 【Procedures】

### 1. Preparation sample solution

Transfer 0.50 mL of vitamin B$_{12}$ injection solutions by pipette from 3 ampoules, respectively into 3 of 10 mL volumetric flasks and add distilled water to the mark and shake up the solution to homogenize, respectively.

### 2. Identification and Content Determination of Vitamin B$_{12}$

Transfer a certain volume of the above solution into a 1 cm quartz cell carefully. Distill water as the blank solution. Take the quartz cells in the UV-visible spectrophotometer. Next, find out the absorption peak at 278 nm, 361 nm, and 550 nm with the UV-visible spectrophotometer. Measure the absorbance of sample solution at three wavelengths, then repeat this operation three times. Take the average as the results.

## 【Notes】

(1) The test solution must be settled to clarify.

(2) Generally, the readings of test solution absorbance should between 0.3 and 0.7 for smaller errors.

(3) The absorption cell should be rinsed for 3~4 times by the test solutions, respectively. Wipe the outside of absorption cell surface with a lens cleaning paper.

(4) Don't open the cover of sample room during you measuring.

## 【Recording and Treating Data】

(1) Qualitative identifying

Calculate the ratios of absorbance at three wavelengths ($A_{361\ nm} / A_{278\ nm}$ = 1.70~1.88; $A_{361\ nm}/A_{550\ nm}$ = 3.15~3.45) and compare with values from *Chinese Pharmacopoeia*.

(2) Determining the content of vitamin B$_{12}$

Calculate the content of vitamin B$_{12}$ injection solution (μg·mL$^{-1}$) according to the absorbance at 361 nm, the absorption coefficient of vitamin B$_{12}$ (μg·mL$^{-1}$), and the diluted times of injection sample.

$$\rho_{sample}(B_{12}) = \frac{A}{E_{1cm}^{1\mu g \cdot L^{-1}} 361nm} = \frac{A}{207 \times 10^{-4}} = A \times 48.31 (\mu g \cdot L^{-1}) \tag{12-3}$$

Injection solution of vitamin B$_{12}$

$\rho_{\text{injection}} = A \times 48.31 \times 20$ (diluted times) ($\mu g \cdot mL^{-1}$).

The injection can be considered qualified, when measured value is certificated between 90% and 110% of the given value according to *Chinese Pharmacopoeia*.

## 【Questions】

(1) Why it is necessary to find out the absorption peak from the three wavelengths on the UV-visible spectrophotometer before determination?

(2) If 2 mL of injection solution is diluted to 15 times with water, the absorbance ($A$) is 0.698 at a wavelength of 361nm, calculate the content of vitamin $B_{12}$ per mL.

(3) What are the prerequisites for direct determination of sample content by absorption coefficient method?

（武世奎）

# Chapter 4

# Chemical Principles Experiment

## Experiment 13  Colligative Properties of Dilute Solution and Their Application

【Objectives】

(1) To master the principles and methods and to determine the molecular mass of solute by freezing point depression.

(2) Learn to use freezing point osmometer and to determine the molecular mass of solute from osmotic concentration of solution.

(3) Learn to use common microscope and to observe the shape of red blood cell in solutions with different osmolarity.

(4) Learn to use 0.10℃ indexing thermometer and to review the operations of electronic balance as well as pipette.

【Pre-lab Assignments】

(1) What are the general methods to measure the molecular mass of solute? Which one is the common method in tests? Why?

(2) What is the condition if $c_{os} \approx b_B$ in equation (13-5)?

(3) Why the cooling curves of pure solvent are different from the dilute solution? What does the cooling curve look like in the ideal condition?

(4) Why does the more NaCl solid must be added in the ice-water bath in freezing point depression experiment?

(5) Why is there the supercooling phenomenon in determining the osmotic concentration test? How to resolve them?

(6) To illustrate the steps for three experiments about the colligative properties of dilute solution by flow chart respectively.

(7) We can't stir glucose solution with glass rod in the test tube when we prepare it. Why?

【Principles】

Some physical properties of the solvent such as the freezing point and osmotic pressure etc. will be altered quantitatively when a non-volatile and non-electrolyte solute is dissolved in a solvent. The magnitude of the alteration is directly proportional to amount of solute dissolved but not the kind of the solute. These properties are collectively called colligative properties. The colligative properties of the solution provide useful ways to determine the molar mass of solute experimentally.

The freezing point ($T_f$) of a solution is lower than that ($T_f^0$) of the pure solvent. For a non-volatile and non-electralyte dilute solution, the relationship between the freezing point depression and molality of solution is expressed as follow

$$\Delta T_f = T_f^0 - T_f = K_f b_B \qquad (13\text{-}1)$$

$$b_B = \dfrac{m_B/M(B)}{m_A} \times 1\,000 \; (\text{mol} \cdot \text{kg}^{-1}) \qquad (13\text{-}2)$$

$b_B$ is the molality of the solution in mole per kilogram(mol·kg$^{-1}$). $M(B)$ (g·mol$^{-1}$) is the molar mass of solute in gram per mole(g·mol$^{-1}$). $m_B$ is the mass of solute dissolved in solution in gram(g). $m_A$ is the mass of solvent in gram(g). $K_f$ is the molar freezing point depression constant (K·kg·mol$^{-1}$) of solution and $\Delta T_f$ is the freezing point depression in kelvin (K) of this dilute solution.

So that

$$M(B) = \dfrac{K_f m_B}{m_A \Delta T_f} \times 1\,000 \qquad (13\text{-}3)$$

The molar mass, $m_B$ can be calculated by equation (13-3) because every terms in this equation can be determined experimentally. The relative error, $d_r$, is

$$E_r(\%) = \dfrac{M_B - M_{B(T)}}{M_{B(T)}} \times 100\% \qquad (13\text{-}4)$$

where, $M_B$ is the molar mass of solute determined in test, $M_{B(T)}$ is the true molar mass of solute.

We can calculate the relative molar mass of solute from the equations (13-3) by experiment.

You can know how to determine freezing point ($T_f$) of dilute solution and freezing point ($T_f^0$) of distilled water by Figure 13-1. The cooling curve (b) of dilute solution is different from the cooling curve (a) of pure solvent.

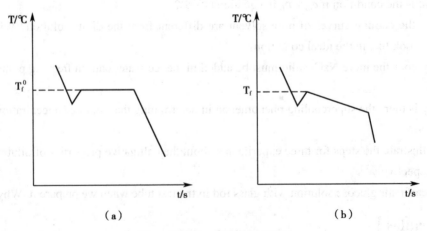

Figure 13-1  Cooling curve
(a) Pure solvent; (b) Solution

Freezing point osmometer is used to determine osmolarity of the solution. The principle to determine the osmolarity of solution is by the freezing point depression. The relationship among the freezing point depression, osmolarity of a solution and molality of solution is expressed as follow

$$c_{os} \approx b_B = \frac{\Delta T_f}{K_f} \text{ (mol·L}^{-1}\text{)} \tag{13-5}$$

$$b_B = \frac{m_B / M_B}{m_A} \times 1\,000 \text{ (mol·kg}^{-1}\text{)} \tag{13-6}$$

$c_{os}$ is osmolarity of the solution in mole per liter( mol·L$^{-1}$ ) or in mole per kilogram ( mol·kg$^{-1}$ ) if the solution is very dilute(density of dilute solution is 1g·cm$^{-3}$ nearly).

$$M_B = \frac{m_B}{m_A c_{os}} \times 1\,000 \tag{13-7}$$

When a certain mass of non-electrolyte substance is dissolved in a certain mass of solvent, Osmolarity of the solution can be determined by osmometer. The molar mass of the solute in dilute solution can be calculated by the equation (13-7). The osmotic pressure of the solution at a given temperature for dilute solution is known by following equation

$$\Pi = c_{os} R T \text{ (kPa)} \tag{13-8}$$

$\Pi$ is osmotic pressure of the solution in kilopascal (kPa). $R$ is the ideal gas constant (8.314 J·mol$^{-1}$·k$^{-1}$). $T$ is the temperature of the solution in Kelvin (K).

The osmotic pressures of different solutions can be compared directly by their osmolarity if the temperature is unchanged. The solution with higher osmolarity is called hypertonic solution and hypotonic solution with lower osmolarity in the same way. If a semi-permeable membrane is used to separate two solutions with different osmolarity, the osmosis would occur. The solvent will pass though the semi-permeable from hypotonic solution to hypertonic solution.

The osmolarity of normal plasma (280~320 mmol·L$^{-1}$) is used as a reference standard to compare the osmolarity with other solutions clinically. If the osmolarity of a solution is more than 320mmol·L$^{-1}$, it is called hypertonic solution. On the contrary, be called hypotonic solution if it is less than 280mmol·L$^{-1}$. Red blood cell would be not normal in shape when it is put into the non-isotonic solutions. The measurement of osmolarity has abroad range of applications in medicinal and other fields.

## 1. Determination of Molar Mass of Glucose by Freezing Point Depression

### 【Apparatus, Reagents and Materials】

**Apparatus**  Electronic balance, thermometer(0.10℃), dry test tube(40 mm × 150 mm), large beaker (500 mL), pipet (50 mL), rubber suction bulb, glass stirrer, magnifying glass, graduated cylinder (500 mL)

**Reagents and Materials**  Ice, salt, glucose (A. R.), distilled water

### 【Procedures】

#### 1.1 Determination of the freezing point for glucose solution

Weigh out 4.2~5.1 g (0.0 001 g) of glucose by an electronic balance and put it into a dry test tube. Add 50.00 mL of distilled water (pure solvent) into the test tube with a pipet and make glucose to dissolve completely. You can't stir glucose solution with glass rod in the test tube.

Put the test tube with solution into a large beaker which contains about half filled of mixture of ice, salt and tap water as shown in Figure 13-2. Keep the temperature of ice-salt-water system below $-5\,°C$. Stir the glucose solution continuously until some pieces of ice appear.

Continue to stir and observe the thermometer carefully as the temperature is rising. Record the temperature when the temperature rises no longer. Take out the test tube and make ice in it melt by rinsing the outside of it with running tap water. Repeat above measurement step three times until the difference between any two data among three data does not exceed $0.05\,°C$, take the average value as the freezing point of solution ($T_f$). Record the date in Table 13-1.

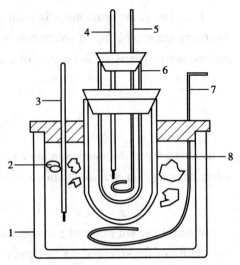

Figure 13-2 Apparatus of determining the freezing point
1. Beaker; 2. Ice; 3,4. Thermometer; 5,7. stirrer; 6,8. test tube

### 1.2 Determination of the freezing point of distilled water

Clean the test tube carefully, and then add about 50 mL of distilled water into it after cleaning test tube by distilled water three times. Repeat measurement step three times as above steps until the difference between any two data among three data does not exceed $0.05\,°C$. Take the average value as the freezing point of distilled water ($T_f^0$). Record the date in Table 13-1.

### 1.3 Calculation

Calculate freezing point depression, $\Delta T_f$, of solution, the molar mass of glucose, $M_B$, and the relative error ($E_r\%$) by equation (13-1), (13-3) and (13-4) respectively. True $M_B$ is 180 g·mol$^{-1}$.

**【Notes】**

You should use the thermometer very carefully because the temperature is low and glass instruments are easy to be broken.

**【Data and Results】**

Date:　　　　　　　　Temperature:　　　　°C　　　Relative Humidity:

Table 13-1 Determination of Molar Mass of Glucose by Freezing Point Depression

| Experiment No. | 1 | 2 | 3 |
|---|---|---|---|
| $m_A$ / g | | | |
| $m_B$ / g | | | |
| $T_f^0$ / °C | | | |
| $T_f$ / °C | | | |
| $\Delta T_f$ / °C | | | |
| $M_B$ / g·mol$^{-1}$ | | | |
| $E_r$ / % | | | |

## 【Questions】

(1) Freezing point of biological samples solution such as blood serum and urine can be determined by this test? Which kinds of substances can be determined in their molar mass by the determining Freezing Point Depression of solution?

(2) Whether the molar mass of glucose can be determined by Boiling Point Elevation of solution or not?

## 2. Determination of the Molar Mass of Glucose by Determining Osmolarity

### 【Apparatus, Reagents and Materials】

**Apparatus**  Electronic balance, freezing point osmometer, small dry test tube (2 mL), dry beaker (100 mL), pipette (50 mL), volumetric pipette (1 mL), rubber suction bulb, glass stirrer

**Reagent and Materials**  Glucose (A.R.), standard NaCl solutions ( $c_{os}=0.300$ mol·kg$^{-1}$ and $c_{os}=0.800$ mol·kg$^{-1}$ )

### 【Procedures】

#### 2.1  Preparation of glucose solution

Weigh out 9.1~9.5 g ( 0.0001 g ) of glucose by an electronic balance and put it into a dry beaker, Add 50.00 mL of distilled water into it with pipette, Make all glucose dissolves completely.

#### 2.2  Calibrate the freezing point osmometer

Read the technical manual of freezing point osmometer before experiment. Start the freezing point osmometer and keep cool cell in a proper temperature range (usually between $-8°C$ to $-10°C$). Take 0.50 mL of standard NaCl solution ($c_{os}=0.300$ mol·kg$^{-1}$) into a small dry test tube with volumetric pipette and measure the osmolarity on the freezing point osmometer. Adjust the scale and make osmolarity value just at $c_{os}=0.300$ mol·kg$^{-1}$. And then transfer 0.50 mL of another standard NaCl solution ( $c_{os}=0.800$ mol·kg$^{-1}$ ) into another small dry test tube with volumetric pipette. Measure the osmolarity of it on the freezing point osmometer again, Adjust the scale and make osmolarity value just at $c_{os}=0.800$ mol·kg$^{-1}$. Repeat the measurement step several times until the measured values are equal to the standard value exactly.

#### 2.3  Osmolarity determination of glucose solution

Transfer 0.50 mL of glucose solution into a small dry test tube with volumetric pipette. Measure the osmolarity of it on the freezing point osmometer. Record the data. Measure it twice again and take the average value as the last osmolarity of this glucose solution.

#### 2.4  Calculation of the molar mass of glucose

Calculate the molar mass of glucose by equation (13-6) and relative error by equation (13-4) respectively (True molar mass of glucose is 180 g·mol$^{-1}$). Fill the data and results in Table 13-2.

### 【Data and Results】

Date:                    Temperature:              °C       Relative Humidity:

Table 13-2  Determination of Molar Mass of Glucose by determining osmolarity

| Experiment No. | 1 | 2 | 3 |
|---|---|---|---|
| $m_A$ / g | | | |
| $m_B$ / g | | | |
| $c_{os}$ / mol·L$^{-1}$ | | | |
| $M_B$ / g·mol$^{-1}$ | | | |
| $E_r$ / % | | | |

【Problems】

(1) How to know the osmotic pressure of a solution at room temperature after getting the osmolarity of it by Freezing point osmometer?

(2) What do you need to pay attention to when you determine the osmolarity of a solution by freezing point osmometer?

(傅 迎)

## 3. Application of Freezing Point Osmometer

【Apparatus, Reagents and Materials】

**Apparatus**  Microscope, glass slides, freezing point osmometer, small dry test tube (2 mL), dry beaker (100 mL), transferring pipette (50 mL), volumetric pipette (1 mL), rubber suction bulb, stirring rod, medicine eyedropper

**Reagents and Materials**  standard NaCl solution ($c_{os}$ = 0.300 mol·L$^{-1}$, $c_{os}$ = 0.800 mol·L$^{-1}$), 0.35 mol·L$^{-1}$ NaCl, 0.05 mol·L$^{-1}$ NaCl, 9 g·L$^{-1}$ NaCl, 54 g·L$^{-1}$ glucose solutions, polyvinyl alcohol eyedrops, animal blood, anti-coagulator

【Procedures】

### 3.1  Calibration of the freezing point osmometer

Calibrate the freezing point osmometer according to the steps in 13.2.2.

### 3.2  Determination of osmotic concentrations

Transfer 0.50 mL of 0.35 mol·L$^{-1}$ NaCl solution with volumetric pipette into a small dry test tube. Measure the osmotic concentration on the freezing point osmometer and record the data. Repeat twice and take the average value as the osmotic concentration of this solution. Similarly, measure 0.05 mol·L$^{-1}$ NaCl, 9 g·L$^{-1}$ NaCl, 54 g·L$^{-1}$ glucose and polyvinyl alcohol eyedrops one by one. Record their osmotic concentrations, in Table 13-3 respectively.

### 3.3  Observation of the red blood cell shape in the different solutions under microscope

(1) Preparation of red blood cell suspension: Transfer 1 mL fresh animal blood (such as mice or rabbit) and add 20 mL isotonic solution containing anti-coagulator (0.05 mol·L$^{-1}$ sodium citrate or 10 drops of Na$_2$EDTA solution) into a beaker, stir it slightly until it is homogenous.

(2) Add 1.0 mL 0.35 mol·L$^{-1}$ NaCl, 0.05 mol·L$^{-1}$ NaCl, 9 g·L$^{-1}$ NaCl, 54 g·L$^{-1}$ Glucose and

Polyvinyl alcohol eyedrops into five test tubes respectively. Add 0.10 mL of red blood cell suspension solution into each of five test tubes respectively. Stir the mixture in test tubes slightly until they are uniformity. Wait for 15~20 min.

(3) Observation of red blood cell shape under microscope: Read the technical manual of the microscope carefully before experiment and be sure that you can observe the shape of red blood cell under the microscope. Take out one drop of mixture from one of five test tubes and drop it on glass slides. Remove superfluous liquid with another glass slides. Put the glass slides with red blood cell under a microscope and observe the shape of red blood cells. Record the results in Table 13-4. Repeat the above operation for other four samples in the rest four test tubes one by one.

【Recording and treating Data】

Date:　　　　　　　　　Temperature:　　　　　℃　　　Relative Humidity:

Table 13-3　determination of osmotic concentration of solutions

| Experiment No. | 1 | 2 | 3 |
| --- | --- | --- | --- |
| 0.35 mol·L$^{-1}$ NaCl | | | |
| 0.05 mol·L$^{-1}$ NaCl | | | |
| 9 g·L$^{-1}$ NaCl | | | |
| 54 g·L$^{-1}$ Glucose | | | |
| Polyvinyl alcohol eyedrops | | | |

Table 13-4　The shape of red blood cell in different osmotic concentration solutions

| Solutions | 0.35 mol·L$^{-1}$ NaCl | 0.05 mol·L$^{-1}$ NaCl | 9 g·L$^{-1}$ NaCl | 54 g·L$^{-1}$ Glucose | Polyvinyl alcohol eyedrops |
| --- | --- | --- | --- | --- | --- |
| Shape of red blood cell | | | | | |

【Questions】

(1) What kind of shapes will red blood cell appear in hypertonic solution, hypotonic solution and isotonic solution, respectively? Why?

(2) Why eyes would feel pain when we swim in fresh water but better in sea water?

（于　昆）

## Experiment 14　Determination of the Relative Atomic Mass of Magnesium by Replacement Reaction

【Objectives】

(1) To master the operation method of electronic balance.

(2) To learn to collect gas and determine volume of gas.

(3) To understand the principle and methods in determination of the relative atomic mass of magnesium.

## 【Pre-lab Assignments】

(1) What is the principle of determining the relative atomic mass of magnesium by displacement method?

(2) What are the factors that may lead to errors in this experiment?

(3) Are there any other methods that may be used to determine atomic weight or molecular weight? What are they? What are their principles? How to determine the molecular weight of macromolecules such as polysaccharide and protein? At least write down one method and indicate its designing route in the pre-lab assignment.

(4) Please use a flow chart to illustrate the experimental procedures of determining the relative atomic mass of magnesium by replacement reaction. Indicate the steps that should be paid attention.

## 【Principles】

Magnesium is an active metal that can replace hydrogen of dilute sulphuric acid to form $H_2$ and magnesium sulphate. The reaction is as follows.

$$H_2SO_4 + Mg = H_2 \uparrow + MgSO_4$$

According to the stoichiometric relationship, when the magnesium ribbon is used up, the amount of magnesium equals to that of $H_2$. By collecting and determining volume of gas produced in the reaction, the amount of $H_2$ can be calculated according to the ideal gas equation.

$$n(Mg) = n(H_2) = \frac{p(H_2) \times V(H_2)}{R \times T} \tag{14-1}$$

where, $p(H_2)$ / kPa is the partial pressure of hydrogen, $V(H_2)$ is the volume of gas collected, $R$ is the gas constant (8.314 J·mol$^{-1}$·K$^{-1}$), and $T$ is the absolute temperature.

$$p_{Total} = p(H_2) + p_{Saturaed}(H_2O) \tag{14-2}$$

$$p(H_2) = p_{Total} - p_{Saturaed}(H_2O) \text{ (kPa)}$$

If magnesium ribbon is weighed before reaction and the mass is $m(Mg)$ in grams, the relative atomic mass of magnesium can be determined by the equation (14-3)

$$A_r(Mg) = \frac{m(Mg)}{n(Mg)} = \frac{m(Mg) \times R \times T}{[p_{Total} - p_{Saturated}(H_2O)] \times V(H_2)} \tag{14-3}$$

## 【Apparatus, Reagents and Materials】

**Apparatus** 0.1mg-Electronic balance, thermometer, barometer, base buret(50 mL), test tube, long-stem funnel (×2), stand and clamp; pipe with plug, rubber tube, sand paper, graduated cylinder

**Reagents and Materials** 2.0 mol·L$^{-1}$ $H_2SO_4$, magnesium ribbon (25~35 mg), glycerol

## 【Procedures】

(1) Before experiments, polish magnesium ribbons with sand paper until there is no black substance on the surface. Weigh out three pieces of magnesium ribbons with an electronic balance. Each piece is about 0.0300 g and no more than 0.0350 g (accurate to 0.0001 g). Record the mass of each magnesium ribbon.

(2) Set up the apparatus as shown in Figure 14-1, pour some water into the eudiometer connected with the funnel until the level of liquid below the "0" mark slightly. Move the funnel up and down repeatedly to remove the gas bulb out of the liquid completely.

(3) Check this system is sealed with the following procedure: Connect the eudiometer and the big test tube with rubber pipe, stop them with a stopper. Move the funnel down to proper position, the liquid level will move down at the beginning. 3 min later, it will maintain at a certain position that means this system is sealed. If it is not like this, you must check all joints and repeat this test mentioned above until this system is sealed.

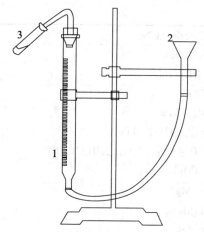

Figure 14-1  Determination of the relative atomic mass of magnesium
1. Eudiometer; 2. Funnel; 3. Test tube

(4) Carefully inject 5 mL of 2 mol·L$^{-1}$ H$_2$SO$_4$ solution into the bottom of the test tube with a funnel (do not make the acid stick to the wall of the test tube). Tilt the tube slightly and wet the strip with a little glycerin or water. Stick it to the top of the tube wall to make sure it doesn't touch sulfuric acid. Adjust the level of the pipe and make liquid surface as close as possible to the 0 scale. Install the test tube and the plug. Repeat the tightness test to ensure that the device does not leak.

(5) Move the funnel to the right side of the eudiometer and keep the liquid surfaces of the funnel and eudiometer on the same level, record the initial eudiometer reading ($V_{\text{Initial}}$).

(6) Tilt the iron stand and make the magnesium ribbon connect with solution of diluted H$_2$SO$_4$, the gases (hydrogen and water vapor) produced in reaction will flow into the eudiometer and push water into funnel. When the liquid level drops, move the funnel down correspondingly, keep both liquid levels at the same basically level.

(7) When the magnesium ribbon is used up, cool the system to room temperature, make both liquid levels on the same level, and record the position of liquid level in the eudiometer. After 1~2 minutes, record the final reading ($V_{\text{Final}}$) in Table 14-1 if the two readings are equal, meaning the temperature of the reaction system is consistent with the room temperature. Otherwise, record the final reading $V_{\text{Final}}$ until the two readings are equal.

(8) Repeat the experiment for two times using different magnesium ribbon. Calculate the relative atomic mass of magnesium and the relative error. After experiments, remove the test tube, pour out the acid out and rinse the tube with water.

【Data and Results】

Date:　　　　　　　　　Temperature:　　　　℃　　　Relative Humidity:

Table 14-1  Determination of the relative atomic mass of magnesium

| Experiment No. | 1 | 2 | 3 |
|---|---|---|---|
| $M$(Mg) / g | | | |
| $V_{\text{Initial}}$ / mL | | | |

| Experiment No. | 1 | 2 | 3 |
|---|---|---|---|
| $V_{Final}$ / mL | | | |
| $V(H_2)$ / mL | | | |
| $T$ / K | | | |
| $p_{Total}$ / kPa | | | |
| $p_{Saturated}(H_2O)$ / kPa | | | |
| $P(H_2) = p_{Total} - p_{Saturated}(H_2O)$ / kPa | | | |
| $A_r(Mg)_{Determination}$ | | | |
| $\overline{A_r(Mg)}$ | | | |
| $A_r(Mg)_{Theory}$ | | | |
| $\overline{E_r}$ / % | | | |

(1atm = 760.15 mmHg = 101.325 kPa, $p(H_2O)$ can be obtained from the appendix)

$$\overline{E_r} = \frac{A_r(Mg)_{Determination} - A_r(Mg)_{Theory}}{A_r(Mg)_{Theory}} \times 100\% \tag{14-4}$$

## 【Questions】

(1) If the oxidized film on the magnesium ribbon surface is not removed, how will it affect the results of this experiment?

(2) If the gas bubble in water is not expelled completely, how does it affect the results of this experiment?

(3) What is the principle to test if the system is sealed or not?

(4) If the reaction's system is not cooled completely to the room temperature after reaction, how will it affect the experimental results?

(5) Why must the liquid level in the eudiometer be the same as that in the funnel when recording it?

(宋 慧)

## Experiment 15  Determination of the Rate of Chemical Reaction and Activation Energy

### 【Objectives】

(1) To understand the effects of concentration, temperature, and catalyst on the rate of chemical reaction.

(2) To determinate the rate, the reaction order, the rate constant, and the activation energy of $(NH_4)_2S_2O_8$ oxidizing KI.

(3) To master graphing method to deal with experimental data.

### 【Pre-lab Assignments】

(1) What are the factors that affect the rate of chemical reaction? How do they affect?

(2) The reaction rate will be enhanced with a increase in temperature. Why?

(3) Why does the reaction rate change when adding catalyst?

(4) Using a flow chart to illustrate how to design the experimental procedures in verifying the effects of concentration, temperature and catalyst on the chemical reaction rate. How to determine the reaction order, the rate constant and the activation energy, respectively.

## 【Principles】

In a mixed solution of $(NH_4)_2S_2O_8$ and $KI$, the following redox reaction takes place

$$(NH_4)_2S_2O_8 + 3KI = (NH_4)_2SO_4 + K_2SO_4 + KI_3$$

or

$$S_2O_8^{2-} + 3I^- = 2SO_4^{2-} + I_3^- \tag{1}$$

The rate of this reaction is

$$v = k\, c(S_2O_8^{2-})^m\, c(I^-)^n \tag{15-1}$$

here, $v$ is the rate of the reaction. $c(S_2O_8^{2-})$ and $c(I^-)$ indicate the original concentration, respectively. $v$ is the original rate of the reaction and $k$ is rate constant. The sum of $m$ and $n$ is the reaction order.

In order to determine $v$, we should determine the change of the concentration of $S_2O_8^{2-}$ at a interval time of $\Delta t$. The average rate of this reaction is

$$\bar{v} = -\frac{\Delta c(S_2O_8^{2-})}{\Delta t} \tag{15-2}$$

Approximatively, the average rate can be placed by instantaneous rate

$$\bar{v} = -\frac{\Delta c(S_2O_8^{2-})}{\Delta t} \approx k\, c(S_2O_8^{2-})^m\, c(I^-)^n \tag{15-3}$$

Starch solution is used as an indicator. A certain volume of starch solution and $Na_2S_2O_3$ solution with accurately known concentration are added into the KI solution before mixing the KI solution with $(NH_4)_2S_2O_8$ solution. Then $I_3^-$ will be released from reaction (1) immediately and react with $Na_2S_2O_3$ to form the colorless $S_4O_6^{2-}$ and $I^-$

$$I_3^- + 2S_2O_3^{2-} = S_4O_6^{2-} + 3I^- \tag{2}$$

The rate of reaction (2) is much faster than that of reaction (1). As $Na_2S_2O_3$ is consumed completely, $I_3^-$ will be released from the reaction (1) immediately and react with the starch to turn the color of the solution into blue.

Comparing reaction(1) with (2), 1 mol of $S_2O_8^{2-}$ must consume 2mol of $S_2O_3^{2-}$, so

$$\Delta c(S_2O_8^{2-}) = -\frac{\Delta c(S_2O_3^{2-})}{2} \tag{15-4}$$

$\Delta c(S_2O_3^{2-})$ is the concentration of $Na_2S_2O_3$. In the experiment, we record the time interval $\Delta t$ from the beginning of the reaction to the moment that the color of the solution turns into blue.

From the above, the rate of reaction (1) can be written as follows.

$$\bar{v} = -\frac{\Delta c(S_2O_8^{2-})}{\Delta t} = \frac{c(S_2O_3^{2-})}{2\Delta t}\ (\text{mol}\cdot\text{L}^{-1}\cdot\text{s}^{-1}) \tag{15-5}$$

or

$$\bar{v} = -\frac{\Delta c(S_2O_8^{2-})}{\Delta t} = \frac{c(S_2O_3^{2-})}{2\Delta t} = k\, c(S_2O_8^{2-})^m\, c(I^-)^n \tag{15-6}$$

take logarithm on both side of formula (15.3)

$$\lg v = \lg k + m \lg c(S_2O_8^{2-}) + n \lg c(I^-) \tag{15-7}$$

When $c(I^-)$ remains constant and $c(S_2O_8^{2-})$ varies, different rates of the reaction can be obtained under different condition. In the plot of $\lg v$ versus $\lg c(S_2O_8^{2-})$, the slope of the line ($m$) is the reaction order of $S_2O_8^{2-}$. Similarly when $c(S_2O_8^{2-})$ remains constant, the plot of $\lg v$ versus $\lg c(I^-)$ can also be fixed to a straight line with its slope $n$ being the reaction order of $I^-$. Then the total reaction order ($m+n$) can be calculated. The constant $k$ at a given temperature can be calculate based on the reaction rate equation $v = k\,c(S_2O_8^{2-})^m\,c(I^-)^n$.

From Arrhenius equation, the effect of temperature on reaction rate constant can be determined.

$$\lg k = -\frac{E_a}{2.303RT} + \lg A \tag{15-8}$$

here, $A$ is the unique constant, $R$ is the gas constant, $T$ is the absolute temperature, and $E_a$ is the activation energy of the reaction.

Thus we can calculate $k$ under different temperature. Based on the equation $\lg k = -\dfrac{E_a}{2.303RT} + \lg A$, plot $\lg k$ versus $\dfrac{1}{T}$ yields a straight line. The activation energy of the reaction can be determined from the slope ($-\dfrac{E_a}{2.303R}$) of the line.

## 【Apparatus, Reagents and Materials】

**Apparatus**  Erlenmeyer flask (100 mL), measuring pipette (10 mL), test tube, cylinder (10 mL), thermometer, stopwatch, thermostat water bath, beaker

**Reagents and Materials**  0.2 mol·L$^{-1}$ (NH$_4$)$_2$S$_2$O$_8$ solution, 0.2 mol·L$^{-1}$ KI solution, 0.01 mol·L$^{-1}$ Na$_2$S$_2$O$_3$ solution, 0.2 mol·L$^{-1}$ KNO$_3$ solution, 0.02 mol·L$^{-1}$ Cu(NO$_3$)$_2$ solution, 0.2 mol·L$^{-1}$ (NH$_4$)$_2$SO$_4$ solution, 0.2% starch solution

## 【Procedures】

### 1. Effects of Concentration on the Reaction Rate

At room temperature, KI, (NH$_4$)$_2$SO$_4$, Na$_2$S$_2$O$_3$, KNO$_3$ and starch solution are added to 100 mL Erlenmeyer flasks by measuring pipette according to Table 15-1, respectively. The solution are then stirred to homogenize. Then, certain volume of (NH$_4$)$_2$S$_2$O$_8$ solution is measured out by a 10 mL cylinder (or pipette) and added into the corresponding Erlenmeyer flask. At the same time, the stopwatch is started and the Erlenmeyer flask is stired continuously. Finally, the stopwatch is stopped immediately when the color of the solution turns into blue. Record the $\Delta t$ and room temperature.

To keep the ionic strength and total volume constant, insufficient amounts of KI and (NH$_4$)$_2$S$_2$O$_8$ are supplemented by controlling the amount of KNO$_3$ and (NH$_4$)$_2$SO$_4$, respectively.

## Chapter 4  Chemical Principles Experiment

Table 15-1  The Effect of concentration on the rate of the reaction

| Experiment No. | | 1 | 2 | 3 | 4 | 5 | 6 | 7 | 8 | 9 |
|---|---|---|---|---|---|---|---|---|---|---|
| Volume / mL | $(NH_4)_2S_2O_8$ solution | 2.0 | 4.0 | 6.0 | 8.0 | 10.0 | 10.0 | 10.0 | 10.0 | 10.0 |
| | $Na_2S_2O_3$ solution | 4.00 | 4.00 | 4.00 | 4.00 | 4.00 | 4.00 | 4.00 | 4.00 | 4.00 |
| | Starch solution | 2.00 | 2.00 | 2.00 | 2.00 | 2.00 | 2.00 | 2.00 | 2.00 | 2.00 |
| | KI solution | 10.00 | 10.00 | 10.00 | 10.00 | 10.00 | 8.00 | 6.00 | 4.00 | 2.00 |
| | $KNO_3$ solution | 0 | 0 | 0 | 0 | 0 | 2.00 | 4.00 | 6.00 | 8.00 |
| | $(NH_4)_2SO_4$ solution | 8.00 | 6.00 | 4.00 | 2.00 | 0 | 0 | 0 | 0 | 0 |
| $c(NH_4)_2S_2O_{8\ \text{Initial}}$ / mol·L$^{-1}$ | | | | | | | | | | |
| $c(KI)_{\text{Initial}}$ / mol·L$^{-1}$ | | | | | | | | | | |
| $c(Na_2S_2O_3)_{\text{Initial}}$ / mol·L$^{-1}$ | | | | | | | | | | |
| Reation temperature / K | | | | | | | | | | |
| $\Delta t$ / s | | | | | | | | | | |
| $v$ / mol·L$^{-1}$·s$^{-1}$ | | | | | | | | | | |

According to the data in Table 15-1, the influence of $S_2O_8^{2-}$ concentration and $I^-$ concentration on the chemical reaction rate can be evaluated.

### 2. Effects of Temperature on the Rate of Reaction

Table 15-2  The effect of temperature on the rate of the reaction

| Experiment No. | | 10 | 11 | 12 | 13 | 14 |
|---|---|---|---|---|---|---|
| Volume / mL | $(NH_4)_2S_2O_8$ solution | 5.0 | 5.0 | 5.0 | 5.0 | 5.0 |
| | $Na_2S_2O_3$ solution | 4.00 | 4.00 | 4.00 | 4.00 | 4.00 |
| | Starch solution | 2.00 | 2.00 | 2.00 | 2.00 | 2.00 |
| | KI solution | 10.00 | 10.00 | 10.00 | 10.00 | 10.00 |
| | $(NH_4)_2SO_4$ solution | 5.00 | 5.00 | 5.00 | 5.00 | 5.00 |
| $c((NH_4)_2S_2O_8)_{\text{Initial}}$ / mol·L$^{-1}$ | | | | | | |
| $c(KI)_{\text{Initial}}$ / mol·L$^{-1}$ | | | | | | |
| $c(Na_2S_2O_3)_{\text{Initial}}$ / mol·L$^{-1}$ | | | | | | |
| Reation temperature / K | | | | | | |
| $\Delta t$ / s | | | | | | |
| $v$ / mol·L$^{-1}$·s$^{-1}$ | | | | | | |

According to the experiment No.10 in Table 15-2, the experiment is completed at room temperature.

According to the experiment No.11 in Table 15-2, KI, $(NH_4)_2SO_4$, $Na_2S_2O_3$, and starch solution are added to a 100 mL Erlenmeyer flask and mixed together. A 5.0 mL of 0.2 mol·L$^{-1}$ $(NH_4)_2S_2O_8$ is measured out by a 10 mL cylinder (or pipette) and added into a test tube. Then the Erlenmeyer flask and the test tube are heated in the constant temperature water bath kettle (room temperature + 5℃). When the temperature is 5℃ higher than the room temperature, the $(NH_4)_2S_2O_8$ solution is put into the Erlenmeyer flask, the solution is stirring and the stopwatch is started at the same time. Stopping the stopwatch immediately when the color of the solution turns into blue, the $\Delta t$ and the reaction temperature are recorded.

Repeat other experiment in the same way when the temperature is 10℃, 15℃, 20℃ higher than the room temperature and record the $\Delta t$ and the reaction temperature for experiment No.12~14.

According to the data of experiment No.10~14 in the Table 15-2, the results can reflect the influence of temperature on the chemical reaction rate.

### 3. Effect of Catalyst on the Rate of Reaction

$Cu^{2+}$ can catalyze reaction (1) mentioned above. The rate of reaction (1) increases rapidly with the addition of trace $Cu^{2+}$. The experiment can be conducted according to experiment No. 10 in Table 15-2. Here, add 2 drops 0.02 mol·$L^{-1}$ $Cu(NO_3)_2$ as the catalyst before adding $(NH_4)_2S_2O_8$ to the Erlenmeyer flask.

## 【Data and Results】

(1) Calculation of the reaction order and the rate constant

According to experimental data of Table 15-1, calculate the rate of reaction, $m$, $n$, and $k$, respectively, putting $m$ and $n$ into the equation. Record the data in Table 15-3.

Date:　　　　　　　　Temperature:　　　　　　℃　　　　Relative Humidity:

Table 15-3　Reaction order and reaction rate constant

| Experiment No. | 1 | 2 | 3 | 4 | 5 | 6 | 7 | 8 | 9 |
|---|---|---|---|---|---|---|---|---|---|
| $\lg v$ | | | | | | | | | |
| $\lg c(S_2O_8^{2-})$ / mol·$L^{-1}$ | | | | | | | | | |
| $\lg c(I^-)$ / mol·$L^{-1}$ | | | | | | | | | |
| $m$ | | | | | | | | | |
| $n$ | | | | | | | | | |
| $k$ | | | | | | | | | |
| $\bar{k}$ | | | | | | | | | |

(2) Calculation of the activation energy

Table 15-4　Activation energy of the reaction

| Experiment No. | 10 | 11 | 12 | 13 | 14 |
|---|---|---|---|---|---|
| $k$ | | | | | |
| $\lg k$ | | | | | |
| $\dfrac{1}{T}$ / $K^{-1}$ | | | | | |
| $E_a$ / kJ·$mol^{-1}$ | | | | | |

Plot $\lg k$ versus $\dfrac{1}{T}$ according to Table 15-4, and fix with a line. The activation energy of the reaction can be calculated from the slope of the line.

Based on the above experiments, discuss the influence of concentration, temperature and catalyst on the reaction rate respectively.

## 【Notes】

It is necessary to use the fresh $(NH_4)_2S_2O_8$ solution because $(NH_4)_2S_2O_8$ is easy to decompose in solution.

## 【Questions】

(1) In experiments 1~4 and 6~9 (Table 15-1), why are $(NH_4)_2SO_4$ and $KNO_3$ added?

(2) Certain amount of $Na_2S_2O_3$ and starch are added in the experiment. What is the purpose? How does amount of the $Na_2S_2O_3$ influence the experiment?

(3) Does the reaction stop when the color of solution turned blue? Why or why not?

（李振泉）

# Experiment 16　The Maximum Bubble Pressure Method to Determine the Surface Tension of Ethanol Solution

## 【Objectives】

(1) To grasp the principles and technique for determining the surface tension of solution by maximum bubble pressure method.

(2) To understand relationships between concentration, surface tension, and surface excess.

(3) To learn to calculate the maximum surface excess and the cross section area of surface-active substance.

## 【Pre-lab Assignments】

(1) What is the definition of surface tension?

(2) What factors can affect the surface tension of a liquid?

(3) Beside the maximum bubble pressure method, list at least another two methods for determination of the surface tension of a liquid, and make a comparison of all these methods.

(4) What is surface-active substance? What indexes are usually used for evaluate the performance of the surface-active substance?

(5) List at least three kinds of applications of surface-active substance in daily life.

(6) Please use a flow chart to explain the procedures of the following experiments, and list all the notes and reasons.

## 【Principles】

As the liquid molecules in the interface suffering unbalance attractive forces, the surface of the liquid tend to contract spontaneously. Surface tension can be regarded as the contract force acting on per unit length of the interface of two phases, pointing to the direction of reducing the surface area along the tangent line of the interface.

Factors that can affect the value of the surface tension include temperature, pressure, the composition of the solution, the concentration of the solute and so on. For example, at given temperature and pressure, the surface tension of water varies with the species and the concentration of the solute. Compounds that can lower the surface tension of the solvent are called surface-active substances. When the concentration of the surface-active substance is increased, the surface tension of the solution would decrease, and the concentration of the solute at the surface of the solution is usually higher than that in the bulk solution. The difference between the amount of the molecules at the surface and that in the bulk solution is usually called surface excess. And the relationship of the surface tension, concentration and the surface excess matches the Gibbs adsorption isotherm

$$\Gamma = -\frac{c}{RT} \times \frac{d\sigma}{dc} \ (\text{mol} \cdot \text{m}^{-2}) \qquad (16\text{-}1)$$

where, $\Gamma$ is the surface excess (mol·m$^{-2}$), $\sigma$ is the surface tension (N·m$^{-1}$), $T$ is the absolute temperature(K), $c$ is the concentration(mol·m$^{-3}$), $R$ is the gas constant(8.314 J·mol$^{-1}$·K$^{-1}$), $\frac{d\sigma}{dc}$ is the gradient of $\sigma = f(c)$ function under constant temperature. The $\sigma = f(c)$ curve under isothermal condition is shown as Figure 16-1. By drawing a tangent line across the K point at any concentration, $\frac{d\sigma}{dc}$ can be calculated from the slope of the tangent line. Then the surface excess can be further calculated using equation (16-1) and the $\Gamma = f(c)$ curve is shown as Figure 16-2.

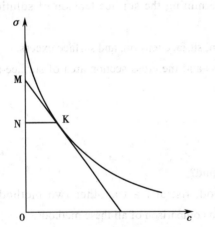

Figure 16-1 The relationship between the surface tension($\sigma$) and concentration($c$)

Figure 16-2 The relationship between surface excess ($\Gamma$) and concentration($c$)

Surface-active substance molecules bear both hydrophilic and hydrophobic groups. When dissolved in the water, the hydrophilic groups tend to point inward the bulk solution, and the hydrophobic groups tend to point to the air. As a result, the molecules tend to absorbed at the surface of the solution and decrease the surface tension. When there is a small amount of the surface-active substance in the solution, the molecules randomly distribute at the surface, as shown in Figure 16-3(a). When the concentration is increased, the surface-active substance occupied area is also increased. When the concentration is increased to an extent, all area is occupied by the surface-active substance, forming a monolayer of the molecules, as shown in Figure 16-3(b). And then the extent of the absorption of the surface-active substance on the surface reaches a saturated state, while the amount of the absorbed molecules per unit area is called maximum surface excess $\Gamma_\infty$. Further increasing the concentration of the solute only lead to the generation of the micelle, the amount of the molecules at the surface is not increased any more, as shown in Figure 16-3(c).

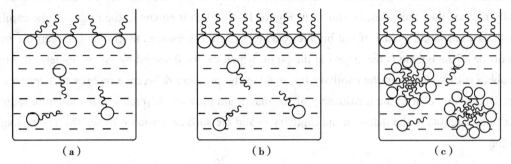

Figure 16-3  The distribution state of the surface-active substance molecules with varied concentration in solution

(a) $c < \Gamma_\infty$; (b) $c = \Gamma_\infty$; (c) $c > \Gamma_\infty$

The relationship of surface excess, concentration and maximum surface excess matches the Langmuir adsorption isotherm

$$\Gamma = \Gamma_\infty \times \frac{Kc}{1+Kc} \quad (\text{mol} \cdot \text{m}^{-2}) \tag{16-2}$$

where, $K$ is a constant, $\Gamma_\infty$ is maximum surface excess. This equation can also be written as

$$\frac{c}{\Gamma} = \frac{c}{\Gamma_\infty} + \frac{1}{K\Gamma_\infty} \tag{16-3}$$

Taking $\frac{c}{\Gamma}$ as the ordinate, and $c$ as the abscissa, we can get a straight line, the slope of which is the reciprocal of maximum surface excess.

When the extent of the absorption of the surface-active substance on the surface reaches a saturated state, the monolayer of the molecules is generated at the surface. As a result, the cross section area of single surface-active substance molecule can be calculated from the maximum surface excess

$$S = \frac{1}{\Gamma_\infty N_A} \quad (\text{m}^2) \tag{16-4}$$

where, $N_A$ is the Avogadro constant (constant to $6.02 \times 10^{23}$), $S$ is the cross section area of single surface-active substance molecule.

As the molecules are all small, taking $\text{m}^2$ as the unit of the cross section area of single surface-active substance molecule is not appropriate. Therefore, the following equation is more frequently used to calculate the cross section area of single surface-active substance molecule

$$S = \frac{10^{18}}{\Gamma_\infty N_A} \quad (\text{nm}^2) \tag{16-5}$$

Here the surface tension of ethanol solution under different concentration is determined by maximum bubble pressure method, and the surface excess, maximum surface excess and the cross section area of ethanol can be calculated.

The apparatus of maximum bubble method to determine the surface tension is shown in Figure 16-4.

When performing the experiment, we should make the tip of the capillary **b** be tangent to the liquid surface in the measuring tube **a**. Then open the stopcock of the dropping funnel **c** and let the water drip, lowering the system pressure. The atmosphere pressure would push the liquid in the capillary to the tip, and the bubble is generated. During the growth of the bubble, curvature radius of the bubble initially reduces then increases, as shown in Figure 16-5. When it equals to the radius of the capillary, the smallest curvature radius of the bubble reaches. And the pressure difference between the inside and outside of the bubble is the largest at the point, which can be detected by the manometer **d**. Finally the bubble is extruded from the capillary tip, making the pressure difference recover to a relative low value. During this course, the relationship of the maximum pressure difference $\Delta p$, curvature radius $r'_2$ of the bubble (equal to the radius of the capillary $r$) and the surface tension $\sigma$ match the Young-Laplace equation

$$\Delta p = p - p_0 = \frac{2\sigma}{r} \tag{16-6}$$

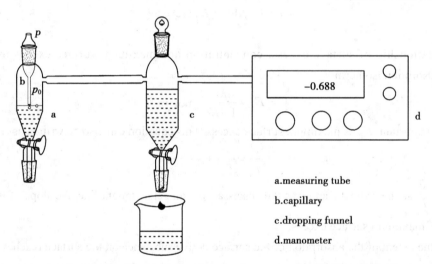

a. measuring tube
b. capillary
c. dropping funnel
d. manometer

Figure 16-4  The apparatus for determination of the surface tension through bubble method

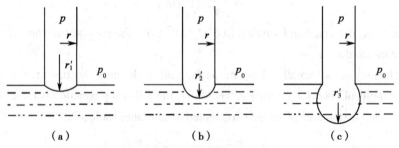

Figure 16-5  The curvature radius of the growing bubble
(a) $r'_1 > r$; (b) $r'_2 = r$; (c) $r'_3 > r$

As the radius of the bubble is hard to be determined, the standard solution, a known surface tension solution, is commonly used for determination of the surface tension of the unknown solution. Using the same capillary to determine the maximum pressure difference of each solution and according to Young-Laplace equation (equation 16-6), formula (16-7) is finally obtained and the surface tension of the unknown solution can be calculated.

$$\frac{\sigma}{\sigma'} = \frac{\Delta p}{\Delta p'} \tag{16-7}$$

where, $\sigma$ and $\sigma'$ is the surface tension of the standard solution and the sample solution (N·m$^{-1}$), $\Delta p$ and $\Delta p'$ is the maximum pressure difference of the standard solution and the sample solution (kPa).

## 【Apparatus, Reagents and Materials】

**Apparatus**   Surface tension measuring apparatus, beaker (500 mL × 2), volumetric flasks (50 mL × 8), measuring pipettes (5 mL, 10 mL), pipettes (20 mL), medicine dropper

**Reagents and Materials**   anhydrous ethanol (A. R.), pure water

## 【Procedures】

### 1. Preparation of ethanol solution

Prepare a series of ethanol solution according to Table 16-1 for following experiments.

### 2. Determination of the surface tension

(1) Add water to dropping funnel **c** (with side arm) to two thirds of its volume, rinse the measuring tube **a** and the capillary **b** three to four times with small portion of the liquid to be measured. Then add the sample solution into measuring tube **a** and plug the capillary **b** into it. Adjust the volume of the solution in measuring tube **a**, making the capillary tip be tangent to the water surface.

(2) Open the top cock of the dropping funnel **c**, power on the manometer **d** and correct it to zero. Then close the top cock and check that if the system is sealed. Open the stopcock of dropping funnel **c** and let the water drop, increasing the volume of the air in the cavity to make its pressure be lower than the atmospheric pressure. Close the stopcock of dropping funnel **c** again and wait for one to two minutes, observe if the value shown on the screen of the manometer is unchanged. If the value is not varied, it is said the system is sealed.

(3) Open the stopcock of dropping funnel **c** again and make bubble generate in the rate of eight to ten seconds for one bubble. Record the maximum pressure difference for three times, the value need to be closed to each other.

(4) Measure the surface tension of the ethanol solution from low to high concentration. For each measurement, the measuring tube **a** and the capillary **b** should be rinsed with the liquid to be measured. Record the data in Table 16-1.

## 【Notes】

(1) The temperature should be steady during the experiment.

(2) While rinsing measuring tube **a**, the water should not get into the side arm of measuring tube, or the value of the pressure in the system is not correct.

(3) All the data should be obtained by the same capillary.

**【Data and Results】**

Date:                Temperature:           ℃         Relative Humidity:

Table 16-1  Determination of surface tension of the ethanol solution

| Number | 1 | 2 | 3 | 4 | 5 | 6 | 7 | 8 |
|---|---|---|---|---|---|---|---|---|
| $V_{ethanol}$ / mL | 0.00 | 0.50 | 1.00 | 2.00 | 4.00 | 8.00 | 16.00 | 20.00 |
| $V_{total}$ / mL | 50.00 | 50.00 | 50.000 | 50.00 | 50.00 | 50.00 | 50.00 | 50.00 |
| $\Delta p_1$ / kPa | | | | | | | | |
| $\Delta p_2$ / kPa | | | | | | | | |
| $\Delta p_3$ / kPa | | | | | | | | |
| $\overline{\Delta p}$ / kPa | | | | | | | | |
| $c_{ethanol}$ / mol·m$^{-3}$ | | | | | | | | |
| $\sigma$ / N·m$^{-1}$ | | | | | | | | |
| $b = c \times \dfrac{d\sigma}{dc}$ | — | | | | | | | |
| $\Gamma$ / mol·m$^{-2}$ | — | | | | | | | |
| $c / \Gamma$ | — | | | | | | | |
| $\Gamma_\infty$ / mol·m$^{-2}$ | — | | | | | | | |
| $S$ / nm$^2$ | — | | | | | | | |

(1) Taking the surface tension $\sigma$ of the ethanol solution as the ordinate, and the concentration $c$ as the abscissa, the $\sigma = f(c)$ curve is obtained. Summarize the relationship between $\sigma$ and $c$.

(2) Make the tangent line and the parallel line of the horizontal x-axis at point K simultaneously, assumed that the two lines crosslink the y-coordinate at point M and point N (as shown in Figure 16-1), measure the distance $b$ between M and N from the plotting paper ($b = c \times \dfrac{d\sigma}{dc}$), put it into the Gibbs adsorption isotherm (16-1) and calculate the surface excess $\Gamma$ under different concentration.

(3) Taking the surface excess $\Gamma$ as the ordinate, and the concentration $c$ as the abscissa, the $\Gamma = f(c)$ curve is obtained. Summarize the relationship between the surface excess and the concentration.

(4) Taking $\dfrac{c}{\Gamma}$ as the ordinate, and the concentration $c$ as the abscissa, the $\dfrac{c}{\Gamma} = f(c)$ curve is obtained. Calculate the maximum surface excess and the cross section area of single ethanol molecule.

**【Questions】**

(1) How to make sure that the system is sealed? If gas in the system is leaked, what will happen to the experiment?

(2) Why should the capillary tip need to be tangent to the water surface?

(3) Why the ethanol solution should be determined from low to high concentration in determination of the surface tension?

(4) Whether this experiment need to be carried under isothermal conditions? Why?

（廖传安）

# Chapter 5

# Preparation of Compounds

## Experiment 17  Refining Sodium Chloride

### 【Objectives】

(1) To practice basic operations such as dissolving, filtration, evaporation, concentration, crystallization and drying.

(2) To master the principle and method of refining salt and test its purity.

(3) To understand the application of precipitation-dissolution balance in purification. of inorganic matter.

### 【Pre-lab Assignments】

(1) What are the clinic applications of NaCl?

(2) What is the purification method of solid substances?

(3) What is recrystallization? Can NaCl be purified by recrystallization?

(4) What are the methods for solid-liquid separation? How to choose in this experiment? What problems should be paid attention to for vacuum filtration?

(5) In this experiment, why is $SO_4^{2-}$ removed first, then $Ca^{2+}$, $Mg^{2+}$ and $Fe^{3+}$. Can the order of precipitants be changed or not?

(6) Why not evaporate to dry in the solution the concentrating step?

(7) Please use a flow chart to illustrate how to design the experimental steps to refine NaCl. Mark the steps that should be paid attention to in the flow chart.

### 【Principles】

NaCl and other salines are all derived from crude salt that usually contains ions such as $K^+$, $Ca^{2+}$, $Mg^{2+}$, $Fe^{3+}$, $SO_4^{2-}$, *etc.*, and some other insoluble impurities including silt, sand, and faggot, *etc.*, are also included. The insoluble material can be removed by filtrating the solution of crude salt, while soluble ones can be eliminated by adding proper reagents ($BaCl_2$, $Na_2CO_3$, HCl, *etc.*) to the solution so that precipitates can form. The general method is adding $BaCl_2$ solution into the NaCl solution to remove $SO_4^{2-}$.

$$Ba^{2+} + SO_4^{2-} \Longrightarrow BaSO_4 \downarrow$$

Filter the precipitate of BaSO$_4$, then add Na$_2$CO$_3$ solution to remove Ca$^{2+}$, Mg$^{2+}$, Fe$^{3+}$ and the excessive Ba$^{2+}$, respectively.

$$Mg^{2+} + 2OH^- = Mg(OH)_2 \downarrow$$
$$2Mg^{2+} + 2OH^- + CO_3^{2-} = Mg_2(OH)_2CO_3 \downarrow$$
$$Ca^{2+} + CO_3^{2-} = CaCO_3 \downarrow$$
$$Fe^{3+} + 3OH^- = Fe(OH)_3 \downarrow$$
$$Ba^{2+} + CO_3^{2-} = BaCO_3 \downarrow$$

Filter the precipitates, the excessive NaOH and Na$_2$CO$_3$ can be neutralized by hydrochloric acid.

$$H_3O^+ + OH^- = 2H_2O$$
$$2H_3O^+ + CO_3^{2-} = 3H_2O + CO_2 \uparrow$$

Finally, purified salt can be obtained through several steps including evaporation crystallization, filtration and drying.

## 【Apparatus, Reagents and Materials】

**Apparatus**　Platform balance, beaker(50 mL, 100 mL), cylinder(2 mL, 50 mL), test tube(10 mL × 6), Buchner funnel, suction instrument, evaporating dish, burner, tripod, spatula, stirring rod, dropper

**Reagent and Materials**　Crude salt, 6 mol·L$^{-1}$ HCl solution, 6 mol·L$^{-1}$ HAc solution, 6 mol·L$^{-1}$ NaOH solution, 1 mol·L$^{-1}$ BaCl$_2$ solution, saturated Na$_2$CO$_3$ solution, saturated (NH$_4$)$_2$C$_2$O$_4$ solution, magnesium agent, filter paper, universal pH paper, 95% ethanol

## 【Procedures】

### 1. Refining sodium chloride

#### 1.1 Weighing and dissolving

Weigh out about 5.0 g of crude salt. Put it into a 100 mL beaker, and dissolve it in 25 mL distilled water. Place the beaker over a burner and heat the solution to boiling to dissolve the salt.

#### 1.2 Removement of SO$_4^{2-}$

Add 1 mol·L$^{-1}$ BaCl$_2$ solution dropwise (about 2 mL) with stirring and continue heating for 5 minutes until BaSO$_4$ precipitate completely formed (caution: Be careful not to get the BaCl$_2$ solution in your mouth because it is poisonous). Remove the heater and cool the solution to room temperature. Then filter the solution and collect the filtrate in another beaker.

#### 1.3 Removement of Ca$^{2+}$, Mg$^{2+}$, Fe$^{3+}$ and the excessive Ba$^{2+}$

Add 2 mL of saturated Na$_2$CO$_3$ solution in the filtrate. Boil gently for 5 minutes. After cooling, take a small amount of supernatant and put it in the test tube. Drop a few drops of Na$_2$CO$_3$ solution into the test tube to check if there is any precipitation. If there is no precipitation, filtrate and collect the filtrate in the evaporating dish.

#### 1.4 Removement of the excessive CO$_3^{2-}$

Add 6 mol·L$^{-1}$ HCl dropwise in the filtrate and test the acidity with universal pH paper, until pH is 2~3.

#### 1.5 Evaporating and drying

Decant the filtrate into a evaporating dish, heat the solution to the boiling point. Carefully not to dry up. Allow to settle down and cool, transfer the crystal to the Buchner funnel with a filter paper and use suction instrument to remove the excess liquid containing $K^+$, then let it remain on the filter paper until to be the dried crystal as much as possible. Washing the crystal 2~3 times with 95% ethanol. Transfer the crystal to the evaporating dish and heat it with gently flame to dry up. Weigh the purified crystal and calculate the yield.

### 2. Test the purity of the purified salt

Take 2.0 g crude salt and purified salt in two beakers. Dissolve them with 20 mL of distilled water, respectively, test and compare the purities of them as follows. Record the result of test the purity for the purified salt in Table 17-1.

#### 2.1 Test of $SO_4^{2-}$

Add two drops of 6 mol·$L^{-1}$ HCl solution and 3~5 drops of 1 mol·$L^{-1}$ $BaCl_2$ solution to test tubes respectively. It implies that $SO_4^{2-}$ exist if white precipitate forms. Record your observations and compare the results in the two tubes.

#### 2.2 Test of $Ca^{2+}$

To another set of test tubes containing the same solutions, add two drops of 6 mol·$L^{-1}$ HAc and 3~5 drops of saturated $(NH_4)_2C_2O_4$ solution. It is indicated that $Ca^{2+}$ exist if the white precipitate forms. Record your observations and compare the results in the two tubes.

#### 2.3 Test of $Mg^{2+}$

To a third set of solutions in test tubes, add 3~5 drops of 6 mol·$L^{-1}$ NaOH solution and 1 drop of magnesium agent. It is shown that $Mg^{2+}$ exists if blue precipitate forms. Record your observations and compare the results in the two tubes.

## 【Notes】

(1) Additional water should be added to avoid the loss of NaCl during removing impurity by precipitation.

(2) boiling the solution after each adding of precipitating agents, and then wait the solution to cool down.

## 【Data and Results】

Date:　　　　　　　　　Temperature:　　　　　℃　　　Relative Humidity:

(1) Refining the sodium chloride

Crude salt $m(NaCl)=$ _____ g, purified salt $m(NaCl)=$ _____ g

Yield of product (%) = $\dfrac{\text{weight of purified salt}}{\text{weight of crude salt}} \times 100\% =$ _____

(2) Test the purity of purified salt

Table 17-1　The result of test the purity for the purified salt

| Item | Way of performance | Crude salt solution | Purified salt solution |
|---|---|---|---|
| Test $SO_4^{2-}$ | Add $BaCl_2$ solution | Phenomenon<br>Conclusion | Phenomenon<br>Conclusion |

| Item | Way of performance | Crude salt solution | Purified salt solution |
|---|---|---|---|
| Test $Ca^{2+}$ | Add saturated $(NH_4)_2C_2O_4$ solution | Phenomenon<br>Conclusion | Phenomenon<br>Conclusion |
| Test $Mg^{2+}$ | Add magnesium agent | Phenomenon<br>Conclusion | Phenomenon<br>Conclusion |

【Questions】

(1) Can we use other acids to remove $CO_3^{2-}$ in the experiment?

(2) Why is the hydrochloric solution added in during testing $SO_4^{2-}$?

（乔秀文）

## Experiment 18　Preparation of Ammonium Iron (Ⅱ) Sulfate

【Objectives】

(1) To understand the principle of preparing double salt.

(2) To practice the basic operations of water bath of heating, filtration, evaporation, concentrating, crystallization, and drying.

(3) To learn the method to inspect the quality of products with visual colorimetry.

【Pre-lab Assignments】

(1) What is the principle of preparation of double salt?

(2) How to testify that the preparation product is ammonium iron (Ⅱ) sulfate?

(3) What is the principle of visual colorimetry used to judge the grade of ammonium iron (Ⅱ) sulfate? How to definite the grade of ammonium iron (Ⅱ) sulfate?

(4) Draw a flow chart to illustrate the preparation principle and experiment procedure. Then label the reaction and operation condition on the flow chart to ensure the success of the preparation.

【Principles】

The Mohr's salt, ammonium iron (Ⅱ) sulfate[$(NH_4)SO_4 \cdot FeSO_4 \cdot 6H_2O$], a bluish hyaline crystal, is a water soluble salt (insoluble in alcohol). Because it is more stable than the other ferrous salt in the air, it can often be used as a primary standard substance in a quantitative inorganic analysis.

Like most other double salts, the solubility of $(NH_4)_2SO_4 \cdot FeSO_4$ in water is less than that of $(NH_4)_2SO_4$ or $FeSO_4$. Thus the crystal of $(NH_4)_2SO_4 \cdot FeSO_4 \cdot 6H_2O$ can be obtained from mixing the two highly concentrated solution of $FeSO_4$ and $(NH_4)_2SO_4$.

In this experiment, $FeSO_4$ solution is firstly obtained by dissolving iron metal in diluted $H_2SO_4$

solution. Then add solid $(NH_4)_2SO_4$ and heat to dissolve. After concentrating and cooling, the slightly soluble crystal of $(NH_4)_2SO_4 \cdot FeSO_4$ can be obtained.

$$Fe + H_2SO_4 \Longrightarrow FeSO_4 + H_2 \uparrow$$

$$FeSO_4 + (NH_4)_2SO_4 + 6H_2O \Longrightarrow (NH_4)_2SO_4 \cdot FeSO_4 \cdot 6H_2O$$

As $Fe^{2+}$ is easy to be oxidized to $Fe^{3+}$, therefor, it needs to be added strong acid to prevent $Fe^{2+}$ from being oxidized to $Fe^{3+}$ during the experiment.

Since the $Fe^{3+}$ ion reacts with excess $SCN^-$ to give an intensely red-colored compound of $[Fe(SCN)_6]^{3-}$, it is therefore possible to determine the content of $Fe^{3+}$, the impurity, present in the product by visual colorimetry as follow.

With the addition of excess KSCN in the same way, a series of $Fe^{3+}$ standard solutions and a sample product solution are prepared into the same colorimetric tubes. Comparing the red color of the standard solution with that of $[Fe(SCN)_6]^{3-}$ to find the standard solution with the same depth, the content of $Fe^{3+}$ in the standard solution is similar to that of impurity $Fe^{3+}$ in the product. Thus the grade of the product can be determined accordingly. The limit of $Fe^{3+}$ for 1 g ammonium ferrous sulfate for grade I, II, and III are 0.05 mg, 0.10 mg and 0.20 mg, respectively.

## 【Apparatus, Material and Reagents】

**Apparatus** Scale platform, conical flask(150 mL × 3), beakers and cylinder(25 mL × 3, 10 mL × 3), suction instrument, evaporating dish, filter papers, colorimetric tubes

**Material and Reagents** Iron filings, $(NH_4)_2SO_4$ (s), 10% $Na_2CO_3$ solution, 3 mol·L$^{-1}$ $H_2SO_4$ solution, 3 mol·L$^{-1}$ HCl solution, 25% alcohol solution, 25% KSCN solution, $Fe^{3+}$ standard solution: Grade I (containing 0.05 mg of $Fe^{3+}$), Grade II (containing 0.10 mg of $Fe^{3+}$), Grade III (containing 0.20 mg of $Fe^{3+}$)

## 【Procedures】

### 1. Cleanse the Iron Filings

Weigh 4.0 g of iron filings into a conical flask. Add 20.0 mL of 10% $Na_2CO_3$ solution and heat it in a water bath for 10 minutes. Discard the solution and rinse the iron filings with distilled water.

### 2. Preparation of FeSO$_4$

Add 25.0 mL of 3 mol·L$^{-1}$ $H_2SO_4$ into the previous conical flask containing the iron filings. Heat it in a water bath for 30 minutes. During the reaction, the conical flask should be swirled to accelerate the reaction and add some water to make up for the loss as the evaporation occasionally. As the bubbles made from the reaction have almost decreased to the end, filter the solution through the suction funnel while the solution is hot. Transfer the solution to an evaporating dish, and rinse the residues in conical flask with 1 mL of 3 mol·L$^{-1}$ $H_2SO_4$ and some of distilled water in order. Add the rinse together to above evaporating dish (Note: since $Fe^{2+}$ is stable in the strongly acidic solution, the addition of $H_2SO_4$ is to avoid the $Fe^{2+}$ invert to $Fe^{3+}$). Collect the residue of iron filings and dry it with filter paper. Weigh the residue and calculate the consumption of the iron in the reaction.

### 3. Preparation of $(NH_4)_2SO_4 \cdot FeSO_4 \cdot 6H_2O$

Weigh 9.5 g of $(NH_4)_2SO_4$(C.P.) and add it into evaporating dish containing the $FeSO_4$ solution. Place the evaporating dish above the water bath and heat it while stirring until all of the $(NH_4)_2SO_4$ has dissolved. Continue the heating and evaporating until the crystal films appear on the surface of the solution. Remove the evaporating dish from the heat and allow to settle down and cool to the room temperature, the $(NH_4)_2SO_4 \cdot FeSO_4 \cdot 6H_2O$ can be obtained. Filter it through the suction funnel and rinse with a small amount of alcohol twice with suction on. Transfer the solid to a dry piece of filter paper. Press it dry with another sheet of filter paper. Weigh it and calculate the yield of the product.

### 4. Test the Product

Protocols of preparing $Fe^{3+}$ standard solution of different grades are as follows.

The preparation of $Fe^{3+}$ standard solution: Dissolve 0.4317 g of $NH_4Fe(SO_4)_2 \cdot 12H_2O$ in 250 mL of water.

Grade I (containing 0.05 mg of $Fe^{3+}$): Add 0.25 mL of $Fe^{3+}$ standard solution, 2.0 mL of 2 mol·$L^{-1}$ HCl solution and 1.0 mL of 1 mol·$L^{-1}$ KSCN solution in a colorimetric tube. Then dilute to 25.00 mL.

Grade II (containing 0.10 mg of $Fe^{3+}$): Add 0.50 mL of $Fe^{3+}$ standard solution, 2.0 mL of 2 mol·$L^{-1}$ HCl solution and 1.0 mL of 1mol·$L^{-1}$ KSCN solution in a colorimetric tube. Then dilute to 25.00 mL.

Grade III (containing 0.20 mg of $Fe^{3+}$): Add 1.00 mL of $Fe^{3+}$ standard solution, 2.0 mL of 2 mol·$L^{-1}$ HCl solution and 1.0 mL of 1 mol·$L^{-1}$ KSCN solution in a colorimetric tube. Then dilute to 25.00 mL.

Weigh out 1.0 g of solid product into a 25.00 mL colorimetric tube, dissolve it with 15.0 mL of distilled water, add 2.0 mL of 3 mol·$L^{-1}$ HCl and 1 mL of 25% KSCN, and finally dilute with distilled water to the mark of colorimetric tube, . Homogenize and compare with the $Fe^{3+}$ standard solutions to give the grade of the product.

### 【Notes】

(1) It is necessary to cleanse the iron filings before use.

(2) The reaction between iron filings and sulfuric acid must be in the ventilating cabinet.

(3) To prevent the transformation of $Fe^{2+}$ to $Fe^{3+}$ in filtrate, 1.0 mL 3 mol·$L^{-1}$ $H_2SO_4$ and oxygen-free distilled water are used to wash the conical beaker and ion residue.

### 【Data and Results】

Date:           Temperature:           ℃     Relative Humidity:

(1) Preparation of ammomium iron(II) sulfate

Weight of iron filings $m\ (Fe)_1$ = _____ g;

The residue of iron filings $m(Fe)_2$ = _____ g;

$m[(NH_4)_2SO_4 \cdot FeSO_4 \cdot 6H_2O]$ = _____ g.

Yield of $(NH_4)_2SO_4 \cdot FeSO_4 \cdot 6H_2O(\%) = \dfrac{w(\text{product}) \text{ (g)}}{w(\text{theoretical}) \text{ (g)}} \times 100\% = $ _____

(2) Test the purity of the product

| $Fe^{3+}$ standard solution | Grade I ($Fe^{3+}$ 0.05mg) | Grade II ($Fe^{3+}$ 0.10mg) | Grade III ($Fe^{3+}$ 0.20mg) |
|---|---|---|---|
| Grade of the product | | | |

【Questions】

(1) What is double salt? What is the difference between simple and double salt?

(2) Why does the color of solution change to yellow during the evaporation and concentration? How to deal with it?

(3) How do we calculate the productive rate of ammonium ion (II) sulfate? By the amount of ion or ammonium sulfate?

(母昭德)

## Experiment 19　Preparation of Potassium Nitrate by Conversion Method

【Objectives】

(1) To learn the method for preparing the salt by double decomposition reaction and separating the substance by the effect of temperature on the solubility.

(2) To master the method and the principle for the recrystallization.

(3) To strengthen and improve the basic operation of dissolving, evaporating, crystallizing and filtrating.

【Pre-lab Assignments】

(1) What are the common preparation methods for inorganic substances?

(2) What are the preparation methods for $KNO_3$? Please point out the advantages and disadvantages of each method.

(3) What is recrystallization? What problems should attention be pain to for recrystallization?

(4) How to determine the amount of raw material added in this experiment? What is the theoretical yield of $KNO_3$?

(5) Use a flow chart to illustrate the experimental procedures. Indicate attention in each step.

【Principles】

The method based on the conversion of potassium chloride with sodium nitrate is carried out in industrial measurements. The reaction is as follows:

$$NaNO_3 + KCl \rightleftharpoons NaCl + KNO_3$$

The solubility of four salts at different temperatures is shown in Table 19-1. This reaction is a reversible reaction. Effect of solubility at different temperature stimulates the forward reaction and shift of the equilibrium condition to the right. In addition, when vaporizing a large amount of NaCl will crystallize, which will improve the yield and purity of $KNO_3$.

The solubility of $KNO_3$ is much more temperature dependent than that of NaCl (Table 19-1). These solubilities indicate the probability of a fractional crystallization process in which $KNO_3$ is crystallized from a solution containing both NaCl and $KNO_3$ by evaporating, concentrating and cooling the solution. At about 120℃, $KNO_3$ is unsaturated at a high temperature due to its high solubility. The mother liquor is evaporated and filtered while hot in order to remove NaCl. Meanwhile, $KNO_3$ will crystallize when the filtrate is cooled since it is least soluble at the room temperature. Then the crude product can be purified by recrystallization.

Table 19-1 Solubility of potassium nitrate, sodium chloride, potassium chloride and sodium nitrate at different temperature (g/100 g $H_2O$)

| T/℃ | 0 | 10 | 20 | 30 | 40 | 60 | 80 | 100 |
|---|---|---|---|---|---|---|---|---|
| $KNO_3$ | 13.3 | 20.9 | 31.6 | 45.8 | 63.9 | 110.0 | 169.0 | 246.0 |
| KCl | 27.6 | 31.0 | 34.0 | 37.0 | 40.0 | 45.5 | 51.1 | 56.7 |
| $NaNO_3$ | 73.0 | 80.0 | 88.0 | 96.0 | 104.0 | 124.0 | 148.0 | 180.0 |
| NaCl | 35.7 | 35.8 | 36.0 | 36.3 | 36.6 | 37.3 | 38.4 | 39.8 |

## 【Apparatus, Reagents, Materials】

**Apparatus**  Boiling tube (300 mL), test tube (10 mL × 2), cylinder (50 mL), beaker (500 mL, 100 mL), thermometer (200℃), ring stand, plat form balance, glass rod, glass sand funnel, filter flask, vacuum pump, radiant-cooker, evaporating dish, watch glass

**Reagents and Materials**  Sodium nitrate (Chemically Pure, CP), Potassium chloride (Chemically Pure, CP), $AgNO_3$ solution (0.1 mol·$L^{-1}$), $HNO_3$ solution (5 mol·$L^{-1}$)

## 【Procedures】

### 1. Preparation of potassium nitrate

Firstly, 22.0 g of sodium nitrate is mixed with 15.0 g of potassium chloride and 35.0 mL of deionized water in a 300 mL boiling tube. Boil the mixture in glycerin at 140~160℃. Mark at the liquid level outside the oil bath beaker and inside the test tube. Stir until the solid can be thoroughly dissolved. Boil and stir until nearly two thirds of the solution left, then amounts of crystal is formed. While still hot, the solution is rapidly filtered through a glass sand funnel under diminished pressure. Cool down the filtrate to room temperature while shaking. The $KNO_3$ crystals start to form. Vacuum filter the concentrated solution containing the crystal by the glass sand funnel. Weigh the crude crystal. Calculate the theoretical yield and the crude yield of product.

## 2. Recrystallization of potassium nitrate

Reserve a small amount of the crude product for purity, take the remainder and the deionized water into the evaporating dish. The ratio of crude product and deionized water is 2:1 (the mass ratio). Boil and stir it until dissolved, then remove from heat. Settle down and cool to the room temperature, then filter it through the glass sand funnel under diminished pressure. Transfer the solid to a watch glass and dry in a oven at 120℃. Weight the purified crystal and calculate the theoretical yield of the product. Record the data in Table 19-2.

## 3. Test of the Purity

Take 0.03 g of crude product and purified product into two test tubes, respectively. Dissolve them individually with 3.0 mL deionized water. Add one drop of 5 mol·$L^{-1}$ $HNO_3$ solution and two drops of 0.1 mol·$L^{-1}$ $AgNO_3$ solution to individual test tubes. Record your observations and compare the results in the two tubes in Table 19-2.

### 【Notes】

(1) When heating the reactants, water cannot hang outside the hard test tube because the temperature of the oil bath exceeds 100℃. Otherwise, it is easy to cause explosive boiling.

(2) Do not rapidly frozen the filtrate after heating in case the crystals are super-small.

(3) If the solution is boiling and the crystal is still not completely dissolved during the recrystallization, added a small amount of deionized water and stirred until the solid are thoroughly dissolved.

### 【Data and Results】

Date:　　　　　　　　Temperature:　　　　℃　　　Relative Humidity:

Table 19-2　Result of the yield and purity of the product

| Substance | Weight / g | Yield / % | Phenomenon of the test for the purity |
|---|---|---|---|
| $NaNO_3$ | | — | — |
| KCl | | — | — |
| $KNO_3$(Theoretical yield) | | — | — |
| $KNO_3$(Crude product) | | | |
| $KNO_3$(Purified product) | | | |

### 【Questions】

(1) In the preparation of potassium nitrate, why is the crystal obtained after the second decompression and filtration the crude product?

(2) According to the solubility, how much NaCl and $KNO_3$ crystals in this experiment can be obtained?

(3) Why can NaCl crystals be removed by immediate filtration in the experiment at a low

temperature?

(4) How to determine the proportion of crude products and water when the crude products are recrystallized?

(5) When $KNO_3$ is mixed with KCl and $NaNO_3$, how should it be purified?

（黄　静）

# Chapter 6

# Comprehensive Experiments

## Experiment 20  Determination of HAc Content and Dissociation Equilibrium Constant of Content of HAc

### 【Objectives】

(1) To learn and understand the method of determining dissociation degree and dissociation equilibrium constant.

(2) To learn and apply the techniques of using a pH meter.

(3) To learn the method of measuring the content of acetic acid in vinegar with acid-base titration.

### 【Pre-lab Assignments】

(1) What is the significance of equilibrium dissociation constant of acid?

(2) By what methods can you determine the dissociation equilibrium constant of acetic acid?

(3) When preparing HAc solution of different concentrations, is it necessary to dry the volumetric flask in advance? Is it necessary to dry the beaker in which pH value of HAc are measured?

(4) When measuring pH of HAc solutions of different concentrations, why should test the solutions from low to high concentration?

(5) Please use a flow chart to illustrate how to design experimental procedures to determine the dissociation equilibrium constant of HAc and the content of HAc in vinegar. Please mark out the precautions of experimental procedures in the flow chart and give explanations.

### 【Principles】

Acetic acid is a weak monoprotic acid. Its dissociation equilibrium in water is as follows:

$$HAc + H_2O \rightleftharpoons H_3O^+ + Ac^-$$

$$K_a = \frac{[H_3O^+][Ac^-]}{[HAc]} \approx \frac{[H_3O^+]^2}{c} \tag{20-1}$$

$$\alpha = \frac{[H_3O^+]}{c} \tag{20-2}$$

where, $K_a$ is the dissociation equilibrium constant, and $\alpha$ is the degree of dissociation. $[H_3O^+]$, $[Ac^-]$ and $[HAc]$ are the concentrations of $H_3O^+$, $Ac^-$ and HAc in equation 20-1 and 20-2 respectively. $c$ is the initial concentration of HAc. $c$ can be determined by titration with NaOH standard solution, and $[H_3O^+]$ can be obtained by determining pH of HAc solution. Thus $K_a$ and $\alpha$ can be calculated by the equations listed above.

Acetic acid (HAc) is the main composition in vinegar. Its content is in the range of 3.5~5.0 g / 100 mL. Its $K_a$ is $1.76 \times 10^{-5}$ at 25℃. So it can be titrated by NaOH standard solution directly. In this titration, phenophthalein is used as the indicator.

## 【Apparatus, Reagents and Materials】

**Apparatus**  pHS-3C type pH meter, base buret (25 mL) or PTFE buret (25 mL), conical flask (250 mL × 3), volumetric flask (50 mL × 4, 100 mL), transfer pipette (20 mL, 25 mL), measuring pipette (1 mL, 5 mL), beaker (100 mL × 4), thermometer

**Reagents and Materials**  0.1 000 mol·L$^{-1}$ NaOH standard solution, 0.1 mol·L$^{-1}$ HAc solution, commercial vinegar, phenolphthalein indicator, standard buffer solution KH$_2$PO$_4$-Na$_2$HPO$_4$ pH 6.86, potassium acid o-phthalate pH 4.00

## 【Procedures】

### 1. Determination of the dissociation degree and dissociation equilibrium constant of HAc solution

#### 1.1 Determination of the concentration of HAc solution

Transfer 20.00 mL 0.1 mol·L$^{-1}$ HAc solution to a conical flask with transfer pipette. Add 2 drops of phenolphthalein to the solution. Titrate the solution with 0.1 mol·L$^{-1}$ NaOH standard solution. The end point reaches when the pink appears and keeps for at least 30 seconds. Repeat this experiment twice. Record the data in Table 20-1 and calculate the concentration of HAc solution.

#### 1.2 Preparation of HAc solution with different concentrations

According to Table 20-2, transfer different volumes of HAc solution of known concentration to four 50 mL volumetric flasks, respectively. Add distilled water until the meniscus touch the mark in each volumetric flask carefully and mix the solution homogenously. Calculate the exact concentrations of HAc solutions in four flasks, respectively.

#### 1.3 Measuring pH of the HAc solutions

Transfer 30 mL the HAc solutions prepared above into four dry 50 mL beakers respectively. Determine their pH successively from dilute solution to concentrated solution with a pH meter. Record pH and the temperature. Calculate $\alpha$ and $K_a$, respectively. Record the data in Table 20-2.

### 2. Determination of HAc content in vinegar

#### 2.1 Preparation of a Sample Solution

According to the Table 20-3, transfer 15.00 mL of vinegar to 100 mL volumetric flask. Dilute it to 100.00 mL with distilled water, and then mix throughly.

#### 2.2 Titration of the Sample Solution with NaOH Standard Solution

Transfer 20.00 mL of the sample solution to a conical flask and add 2 drops phenolphthalein indicator in it. Titrate it with NaOH standard solution until the pink appear and keeps for at least 30 seconds. Record the volumes of NaOH standard solution in Table 20-3. Repeat the titration for another two times.

According to the following equation, calculate the content of HAc in vinegar.

$$\rho_{HAc} = \frac{c(NaOH) \times V(NaOH) \times M_r(HAc)}{20.00 \times \frac{10.00}{100.00} \times 1\,000} \times 100 (g \cdot mL^{-1}) \tag{20-3}$$

$$M_r(HAc) = 60.05$$

## 【Data and Results】

Date:　　　　　　　Temperature:　　　　　℃　　　Relative Humidity:

Table 20-1　Determination of the concentration of HAc solution

| Experiment No. | 1 | 2 | 3 |
|---|---|---|---|
| Indicator | | | |
| Change at the end point color | | | |
| $V(HAc)$ / mL | | | |
| $V_{Final}(NaOH)$ / mL | | | |
| $V_{Initial}(NaOH)$ / mL | | | |
| $\Delta V(NaOH)$ / mL | | | |
| $\overline{V}(NaOH)$ / mL | | | |
| $c(NaOH)$ / mol·L$^{-1}$ | | | |
| $c(HAc)$ / mol·L$^{-1}$ | | | |
| $\overline{d_r}$ / % | | | |

Table 20-2　Determination of the dissociation degree and dissociation constant of HAc solution

| Experiment No. | 1 | 2 | 3 | 4 |
|---|---|---|---|---|
| $c_{Original}(HAc)$ / mol·L$^{-1}$ | | | | |
| $V_{Original}(HAc)$ / mL | 2.50 | 5.00 | 25.00 | 50.00 |
| $V_{Total}(HAc\ diluted)$ / mL | 50.00 | 50.00 | 50.00 | 50.00 |
| $c(HAc\ diluted)$ / mol·L$^{-1}$ | | | | |
| pH | | | | |
| $[H_3O^+]$ / mol·L$^{-1}$ | | | | |
| $\alpha$ | | | | |
| $K_a$ | | | | |
| $\overline{K_a}$ | | | | |
| $\overline{d_r}$ / % | | | | |

Table 20-3  Determination of the content of acetic acid in vinegar

| Experiment No. | 1 | 2 | 3 |
|---|---|---|---|
| Indicator | | | |
| Change at the end point color | | | |
| $V_{Original}$ (HAc) / mL | | | |
| $V_{Total}$ (HAc diluted) / mL | | | |
| $V_{Determined}$ (HAc diluted) / mL | | | |
| $V_{Final}$ (NaOH) / mL | | | |
| $V_{Initial}$ (NaOH) / mL | | | |
| $\Delta V$ (NaOH) / mL | | | |
| $\bar{V}$ (NaOH) / mL | | | |
| $c$(NaOH) / mol·L$^{-1}$ | | | |
| $\rho$(HAc) / g·mL$^{-1}$ | | | |
| $\bar{d}_r$ /% | | | |

【Questions】

(1) What are the differences and the similarities between dissociation degree and dissociation constant to express the nature of electrolytes. Are the dissociation degrees or dissociation constants of HAc solution with different concentrations at a certain temperature the same?

(2) Does the dissociation degree or dissociation constant of HAc solution vary with temperature?

(3) Can methyl red be used as the indicator in the titration of HAc with NaOH solution? If it is chosen as the indicator, is the resulted higher or lower compared with the real concentration?

（李　蓉）

## Experiment 21　Determination of the Content of Calcium, Magnesium, and Iron in Tea

【Objectives】

(1) To master the method and the principle for the content determination of calcium and magnesium by chelometric titration.

(2) To master the method and the principle for the content determination of trace iron by visible spectrophotometer.

(3) To learn the ashing method for natural product treatment.

【Pre-lab Assignments】

(1) Please give the structure of EDTA, and then describe its major physical and chemical properties accordingly.

(2) Summarize the biological significance of calcium, magnesium and iron by reviewing the

literature.

(3) How to determine the content of calcium and magnesium respectively in this experiment?

(4) What is the specificity of o-phenanthroline?

(5) With reference to the principles of chelometric titration, spectrophotometry and the experimental procedure, please illustrate ideas for the experimental design with a flow chart. Make sure to label the reaction conditions, operation conditions, and the accuracy of all experiments on the flow chart.

## 【Principles】

Tea is an organic substance. It mainly contains elements such as C, H, O and N. Furthermore, many essential minerals for human body, such as iron, calcium, and magnesium are also contained. Place tea in an open evaporation vessel or a crucible to heat, it will be burned to ashes by oxidizing degradation (dry ashing). Using acid to dissolve the ashes, the resulted solution can be used for the analytical determination of iron, calcium, and magnesium.

Under pH 10.0 and with Eriochrome Black T as indicator, $Ca^{2+}$ and $Mg^{2+}$ can be determined by EDTA standard solution with chelometric titration. Interference by the presence of $Fe^{3+}$ and $Al^{3+}$ can be eliminated with triethylamine.

Typically, the content of iron in tea is very low. It can be determined by spectrophotometry. Under the condition of pH 2~9, $Fe^{2+}$ can combine with o-dinitrogenophenanthrene to form orange-red complex compound with a $lgK_{stable}$ of 21.3 (20℃) and a molar absorption coefficient ($\varepsilon_{510}$) of $1.10 \times 10^4$. The reaction is as follows

$$3\,(\text{phen}) + Fe^{2+} = [Fe(\text{phen})_3]^{2+}$$

Hydroxylamine hydrochloric acid can be used to reduce $Fe^{3+}$ to $Fe^{2+}$. The reaction is as follows

$$2Fe^{3+} + 2NH_2OH \cdot HCl = 2Fe^{2+} + 4H^+ + N_2\uparrow + 2Cl^- + 2H_2O$$

During the color developing, if the acidity is too high (pH < 2), the speed of the reaction slows down. On the contrary, if the acidity is too low, $Fe^{2+}$ is easy to disassociate from the complex.

## 【Apparatus, Reagents and Materials】

**Apparatus**  722-, UV800-, or TU1810- visible spectrophotometer, 0.1mg-electronic balance, volumetric flask (50 mL × 8, 250 mL × 2), volumetric pipette (25 mL, 10 mL), graduated cylinder (10 mL, 25 mL), buret, Erlenmeyer flask, beaker, evaporating dish or crucible, funnel, qualitative filter paper, stirring rod

**Reagent and Materials**  6 mol·$L^{-1}$ $NH_3 \cdot H_2O$, 6 mol·$L^{-1}$ HCl, 2 mol·$L^{-1}$ HAc, 6 mol·$L^{-1}$ NaOH, 0.01 mol·$L^{-1}$ EDTA, 0.010 g·$L^{-1}$ $NH_4Fe(SO_4)_2 \cdot 12H_2O$ ferrous standard solution, 25% (g·$g^{-1}$) triethylamine solution, pH 10 $NH_3$-$NH_4Cl$ buffer, pH 4.6 HAc-NaAc buffer, 0.1% (g·$g^{-1}$)

o-dinitrogenophenanthrene solution, 1% (g·g$^{-1}$) hydroxylamine hydrochloric acid solution, 1% (g·g$^{-1}$) Eriochrome Black T, tea, deionized water

## 【Procedures】

### 1. Tea Ashing and Reagent Preparation

Weigh 6~8 g (accurate to 0.0001 g) tea sample and put it into an evaporation vessel or a crucible, heat the tea till it turns into ashes. Allow to cool and dissolve with 10.0 mL of 6 mol·L$^{-1}$ HCl.

Transfer the solution to a small beaker, wash the evaporation vessel with 20.0 mL water for 3 times and transfer the resulted solution to the same small beaker, use 6 mol·L$^{-1}$ NH$_3$·H$_2$O to adjust pH to 6~7 to produce precipitation, heat in boiling water for 30 minutes, and then filter the solution, wash the beaker and the filter paper with deionized water thoroughly. The filtrate is directly transfer into a 250 mL volumetric flask and diluted to the mark. Label the solution as calcium and magnesium sample solution (1$^{\#}$).

Take a 250 mL volumetric flask and put it under a glass funnel. Use 10 mL of 6 mol·L$^{-1}$ HCl to re-dissolve the precipitation on the filter paper. Wash the filter paper with water twice. Label it as iron sample solution (2$^{\#}$).

### 2. Determination of the Total Content of Calcium and Magnesium in Tea

Transfer 25.00 mL of solution from 1$^{\#}$ volumetric flask into a 250 mL Erlenmeyer flask. Add 5.0 mL of triethylamine solution and 15.0 mL of NH$_3$-NH$_4$Cl buffer solution then shake to make the solution even. Add 2 drops of Eriochrome Black T indicator and use 0.01 mol·L$^{-1}$ EDTA standard solution for the titration. Record the data in Table 21-1. Calculate the total content of calcium and magnesium accordingly (designated by CaO mass fraction).

### 3. Determination of the Iron Content in Tea

#### 3.1 Plotting the Standard Curve

Transfer 0.00 mL, 1.00 mL, 2.00 mL, 3.00 mL, 4.00 mL, 5.00 mL, and 6.00 mL of ferrous standard solution into seven 50 mL volumetric flasks, respectively. Subsequently, add 5.00 mL of hydroxylamide hydrochloric acid solution, 5.00 mL of HAc-NaAc buffer, and 5.00 mL of o-dinitrogenophenanthrene solution to each flask. Dilute the solutions in the flasks with distilled water to the mark and settle down for 10 minutes after shaking up each flask. Using 1 cm colorimeter vessels and at a wavelength of 510 nm, measure the absorbance of the series of standard solutions respectively. Record the data in Table 21-2. Set the concentration of ferrous standard solution as the abscissa and the corresponding absorbance ($A$) as the ordinate to plot the standard curve.

#### 3.2 Determination of the Iron Content in Tea

Take 2.50 mL 2$^{\#}$ solution into a 250 mL volumetric flask. Perform the same steps above to determine the absorbance. Record the data in Table 21-2. Check out the Fe$^{2+}$ content in 50 mL volumetric flask from the standard curve (g·L$^{-1}$) and designated by the mass fraction of Fe$_2$O$_3$.

## 3.3 Calculation of the Content of Calcium, Magnesium, and Iron in Tea

### 【Data and Results】

(1) Determination of the total content of calcium and magnesium in tea

Date:_____ Temperature:_____ ℃ Relative Humidity:_____

$m$ (tea sample) = _____ g

Table 21-1  Determination of the content of calcium and magnesium in tea

| Experiment No. | 1 | 2 | 3 |
|---|---|---|---|
| Indicator | | | |
| The change of color at the end point of titration | | | |
| $V$(1#water sample) / mL | | | |
| Triethanolamine / mL | | | |
| $NH_3$-$NH_4Cl$ / mL | | | |
| $V_{Final}$ (EDTA) / mL | | | |
| $V_{Initial}$ (EDTA) / mL | | | |
| $\Delta V_{Consume}$ (EDTA) / mL | | | |
| $c(Ca^{2+}$ and $Mg^{2+})$ / g·L$^{-1}$ | | | |
| $\omega$(Ca and Ma) / g·g$^{-1}$ | | | |
| $\overline{d_r}$ / % | | | |

$$\omega(CaO)\% = \frac{c(EDTA) \times V(EDTA) \times M_r(CaO)}{m(sample) \times \dfrac{25.00}{250.00} \times 1\,000} (g \cdot g^{-1}) \qquad (21\text{-}1)$$

$$M_r(CaO) = 56.0774$$

(2) Determination of iron content in tea

Table 21-2  Determination of the iron content in tea

| Experiment No. | 1 | 2 | 3 | 4 | 5 | 6 | 7 | 8 |
|---|---|---|---|---|---|---|---|---|
| $V$ (Fe$^{2+}$ standard) / mL | 0 | 1.00 | 2.00 | 3.00 | 4.00 | 5.00 | 6.00 | — |
| $V$ (tea sample) / mL | | | | | | | | 2.5 |
| hydroxylamine hydrochloride / mL | 5.00 | 5.00 | 5.00 | 5.00 | 5.00 | 5.00 | | 5.00 |
| HAc-NaAc / mL | 5.00 | 5.00 | 5.00 | 5.00 | 5.00 | 5.00 | | 5.00 |
| o-Phenanthroline solution / mL | 5.00 | 5.00 | 5.00 | 5.00 | 5.00 | 5.00 | | 5.00 |
| $V_{Total}$ / mL | 50.00 | 50.00 | 50.00 | 50.00 | 50.00 | 50.00 | | 50.00 |
| $c$ (Fe$^{2+}$) / g·L$^{-1}$ | | | | | | | | |
| $\lambda$ / nm | | | | | | | | |
| $A$ | | | | | | | | |
| $c$(iron) / μmol·L$^{-1}$ | | | | | | | | |
| $\omega$(Fe$_2$O$_3$) / g·g$^{-1}$ | | | | | | | | |
| $\overline{d_r}$ / % | | | | | | | | |

Calculate the content of iron in the tea according to the absorbance of analytical solution, standard curve and the formulas as follows

$$\omega(Fe)\% = \frac{c(Fe) \times 50}{m(sample) \times \frac{2.50}{250.00} \times 1\,000} \quad (g \cdot g^{-1}) \tag{21-2}$$

$$\omega(Fe_2O_3)\% = \omega(Fe) \times \frac{M_r(Fe_2O_3)}{M_r(Fe)} \quad (g \cdot g^{-1}) \tag{21-3}$$

$M_r(Fe_2O_3) = 159.688$, $M_r(Fe_2O_3) = 55.845$

## 【Questions】

(1) What's the principle of spectrophotometry? Is the content of iron being determined in this experiment only including the content of $Fe^{2+}$ in the tea? Why?

(2) Why can $Fe^{3+}$ ions be separated completely from $Ca^{2+}$ and $Mg^{2+}$ ions at the conditions of pH 6~7?

(3) From which process do the errors come in the experiments? Compare the relative average deviation in the titration with that of spectrophotometry.

（胡　新）

# Chapter 7
# Self-Designed Experiment

## Experiment 22  Design and Research

【Subjects】

(1) Determination of dissociation equilibrium $K_b$ constant of $NH_3$.
(2) Determination of the content of $NaHCO_3$ in the sample of saleratus.
(3) Determination of the content of $Al^{3+}$ in the sample of alum.
(4) Determination of the content of $CuSO_4$ in a mixture.
(5) Determination of the content of $Ca^{2+}$ in eggshell.

【Objectives】

(1) Search for information to design and come up with an experimental program. Thereby cultivate the comprehensive experimental skills independently.

(2) To be familiar with the application of analytical methods for quantitative determination of the content of certain substances.

(3) To learn to process the experimental data.

(4) Master the general experimental operation and the use of appropriate equipments.

【Apparatus, Reagents and Materials】

**Apparatus**  Electronic balance (accurate to 0.0 001 g), platform balance, visible spectrophotometer, pH meter, electric cooker (hot plate), water bath heating pot, gauze with asbestos, rubber suction bulb, alcohol burner, transfer pipette (20 mL × 2, 25 mL × 2), measuring pipette (1 mL, 2 mL, 5 mL, 10 mL), graduated cylinder (5 mL, 10 mL, 25 mL), acid buret (25 mL), base buret (25 mL) or PTFE buret (25 mL), colorimetric tube (25 mL × 6), beaker (50 mL × 6, 100 mL, 250 mL), Erlenmeyer flask(250 mL × 3), volumetric flask (50 mL × 3, 100 mL × 3), stirring medicine dropper, buret stand, trivet, and weighing bottle, funnel, filter paper

**Material and Reagents**

(1) Samples: $NH_3 \cdot H_2O$ solution (1.0 mol·L$^{-1}$), alum, saleratus, bluestone, eggshell samples.

(2) Primary standard substance: anhydrous $Na_2CO_3$ (AR), KHP (KHC$_8$H$_4$O$_4$, potassium hydrogen phthalate) (AR), $CuSO_4 \cdot 5H_2O$ (AR), $MgSO_4 \cdot 7H_2O$ (AR), $ZnSO_4 \cdot 7H_2O$ (AR).

(3) Reagents: 6 mol·L$^{-1}$ H$_3$SO$_4$ solution, 6 mol·L$^{-1}$ HCl solution, 0.1 mol·L$^{-1}$ HCl solution, 6 mol·L$^{-1}$ HNO$_3$ solution, distilled water, buffer solution of pH 10 NH$_3$-NH$_4$Cl, buffer solution of 1.0 mol·L$^{-1}$ NH$_3$-NH$_4$Cl, 0.1 mol·L$^{-1}$ NaOH solution, buffer solution of 1.0 mol·L$^{-1}$ HAc-NaAc, 6.0 mol·L$^{-1}$ NH$_3$·H$_2$O solution, 0.01 mol·L$^{-1}$ EDTA solution.

(4) Indicator: Phenolphthalein, methyl orange, chrome black T (s), xylenol orange, cal-red indicator (s), universal pH indicator paper and so on.

## 【Requirements】

### 1. Design projects

(1) Projects and samples: Students may work in groups (4~5 persons) or act individually. According to the apparatus and reagents provided in the experiments, prepare yourself 1~2 weeks ahead of time by searching for information, please submit the best experimental projects of the 5 programs based on Table 20-1 above. Your designs must include the followings: Purpose and requirements, the principle, reagents and apparatuses, experimental procedures, phenomena, recording data and processing result, discussion and conclusion, reference (the experimental phenomena, data records, processing result and discussion are reserved).

Your designs should display the amount of the samples in Table 22-1. According to the demand for the accuracy of instruments and the error analysis, calculate the amount required for primarily standard substance. Solid samples were dubbed 100 mL solution. In every titration, 20 mL sample solution will be consumed (10 mL copper sulfate content in the mixture and 10 mL calcium content in the eggshell).

Table 22-1  Experimental projects

| Projects | Quantity of samples and requirements/g |
| --- | --- |
| Determination of Dissociation Equilibrium Constant $K_b$ of Ammonia | |
| Determination of the Content of NaHCO$_3$ in a Sample of Saleratus | 1.0~1.2 |
| Determination of the Content of Al$^{3+}$ Ions in a Sample of Alum | 0.3~0.4 |
| Determination of the Content of CuSO$_4$ in Mixture | 0.7~0.8 |
| Determination of the Content of Ca$^{2+}$ in Eggshell | 0.15~0.2 |

(2) You can calculate the required amount of the primary standard substance and other reagents based on the amount range of the test sample in Table 22-1, and improve the accuracy of the test result as much as possible according to the accuracy of the instruments and the requirements of analysis error.

### 2. Project completion and assessment of experimental skills

(1) The assessment is conducted in groups or individuals, anyone of the five items in Table 22-1 is randomly arranged by the teacher.

(2) Students within a group will perform the experiments in a "relay" mode, with standard operations and within a pre-designated time frame. Finally, Students will complete the entire experiment and submit an experiment report. The certain procedure determined by a teacher.

(3) Defense: Members of the group reports the experimental results. According to the experimental

designs and results, students will answer questions asked by the teachers and students in other groups.

(4) The final result is determined by the scores of experimental procedures, five experimental programs, experimental report (processing data, discussion and conclusions) and defense, which is shown in Table 22-2.

(5) For individual assessment, complete (4)-(5) independently.

Table 22-2  Score sheet of the experimental test

Date: _____  Week: _____  lesson: _____

| Subject | | | | Teacher | | |
|---|---|---|---|---|---|---|
| Grade | | | | Major | | |
| Project | | | | Laboratory | | |
| Test number | 1 | 2 | 3 | 4 | 5 | 6 |
| Name | | | | | | |
| Group | | | | | | |
| Student ID | | | | | | |
| Operating items | | | | | | |
| Time limited | | | | | | |
| Operation score (50 points) | | | | | | |
| Design document (25 points) | | | | | | |
| Experiment Report (20 points) | | | | | | |
| Thesis defense (5 points) | | | | | | |
| Total score (100 points) | | | | | | |

# 【Hints】

(1) At the appropriate absorption wavelength, the relationship between the concentration of $[Cu(NH_3)_4]^{2+}$ and absorbance value is shown in Table 22-3.

Table 22-3  Relationship between the concentration of $[Cu(NH_3)_4]^{2+}$ and absorbance value

| Concentration of $[Cu(NH_3)_4]^{2+}$, $c$ / mol·L$^{-1}$ | 0.002~0.01 |
|---|---|
| Absorbance ($A$) | 0.1~0.5 |

(2) While preparing the solution of $[Cu(NH_3)_4]^{2+}$, Ratio of concentrations of $Cu^{2+}$ and $NH_3$ is
$$c_{(Cu^{2+})} : c_{(NH_3)} = 1 : 200 \sim 1 : 40$$

(3) To dissolve the 0.15~0.2 g of eggshell, you can use 6 mL of 6 mol·L$^{-1}$ $HNO_3$ and heat the solution slightly.

(4) Weighting range of primarily standard substance is X± 10%X, X is calculation value.

(5) The volume of titrated solution in a conical flask is generally about 20 mL, and the consumed volume of the titrant is generally about 20 mL when using a 25 mL buret. The concentrations of titrant solution will be 0.01~0.1 mol·L$^{-1}$.

(6) When designing the experimental program, it is recommended that the solids were sampled into 100.00 mL solution, and only 20.00 mL of sample solution was taken each time for determination (the

content of copper sulfate in the mixture is 10.00 mL, calcium content in the eggshell is 10.00 mL).

## 【Questions】

(1) Why do we have to determine pH of solutions with various concentrations from low to high when using pH meter?

(2) Why should the solution be shaken violently near the end point when the content of $NaHCO_3$ in baking soda is determined by titration?

（李福森）

# 附 录

## 附录 I
## 国际相对原子量表

| | | | | | | | |
|---|---|---|---|---|---|---|---|
| 1 | 氢 | H | 1.007 94 | 27 | 钴 | Co | 58.933 195 |
| 2 | 氦 | He | 4.002 602 | 28 | 镍 | Ni | 58.693 4 |
| 3 | 锂 | Li | 6.941 | 29 | 铜 | Cu | 63.546 |
| 4 | 铍 | Be | 9.012 182 | 30 | 锌 | Zn | 65.38 |
| 5 | 硼 | B | 10.811 | 31 | 镓 | Ga | 69.723 |
| 6 | 碳 | C | 12.010 7 | 32 | 锗 | Ge | 72.64 |
| 7 | 氮 | N | 14.006 7 | 33 | 砷 | As | 74.921 60 |
| 8 | 氧 | O | 15.999 4 | 34 | 硒 | Se | 78.96 |
| 9 | 氟 | F | 18.998 403 2 | 35 | 溴 | Br | 79.904 |
| 10 | 氖 | Ne | 20.179 7 | 36 | 氪 | Kr | 83.798 |
| 11 | 钠 | Na | 22.989 769 28 | 37 | 铷 | Rb | 85.467 8 |
| 12 | 镁 | Mg | 24.305 0 | 38 | 锶 | Sr | 87.62 |
| 13 | 铝 | Al | 26.981 538 6 | 39 | 钇 | Y | 88.905 85 |
| 14 | 硅 | Si | 28.085 5 | 40 | 锆 | Zr | 91.224 |
| 15 | 磷 | P | 30.973 762 | 41 | 铌 | Nb | 92.906 38 |
| 16 | 硫 | S | 32.065 | 42 | 钼 | Mo | 95.96 |
| 17 | 氯 | Cl | 35.453 | 43 | 锝 | Tc | (98) |
| 18 | 氩 | Ar | 39.948 | 44 | 钌 | Ru | 101.07 |
| 19 | 钾 | K | 39.098 3 | 45 | 铑 | Rh | 102.905 50 |
| 20 | 钙 | Ca | 40.078 | 46 | 钯 | Pd | 106.42 |
| 21 | 钪 | Sc | 44.955 912 | 47 | 银 | Ag | 107.868 2 |
| 22 | 钛 | Ti | 47.867 | 48 | 镉 | Cd | 112.411 |
| 23 | 钒 | V | 50.941 5 | 49 | 铟 | In | 114.818 |
| 24 | 铬 | Cr | 51.996 1 | 50 | 锡 | Sn | 118.710 |
| 25 | 锰 | Mn | 54.938 045 | 51 | 锑 | Sb | 121.760 |
| 26 | 铁 | Fe | 55.845 | 52 | 碲 | Te | 127.60 |

续表

| 53 | 碘 | I | 126.904 47 | 86 | 氡 | Rn | (222) |
|---|---|---|---|---|---|---|---|
| 54 | 氙 | Xe | 131.293 | 87 | 钫 | Fr | (223) |
| 55 | 铯 | Cs | 132.905 451 9 | 88 | 镭 | Ra | (226) |
| 56 | 钡 | Ba | 137.327 | 89 | 锕 | Ac | (227) |
| 57 | 镧 | La | 138.905 47 | 90 | 钍 | Th | 232.038 06 |
| 58 | 铈 | Ce | 140.116 | 91 | 镤 | Pa | 231.035 88 |
| 59 | 镨 | Pr | 140.907 65 | 92 | 铀 | U | 238.028 91 |
| 60 | 钕 | Nd | 144.242 | 93 | 镎 | Np | (237) |
| 61 | 钷 | Pm | (145) | 94 | 钚 | Pu | (244) |
| 62 | 钐 | Sm | 150.36 | 95 | 镅 | Am | (243) |
| 63 | 铕 | Eu | 151.964 | 96 | 锔 | Cm | (247) |
| 64 | 钆 | Gd | 157.25 | 97 | 锫 | Bk | (247) |
| 65 | 铽 | Tb | 158.925 35 | 98 | 锎 | Cf | (251) |
| 66 | 镝 | Dy | 162.50 | 99 | 锿 | Es | (252) |
| 67 | 钬 | Ho | 164.930 32 | 100 | 镄 | Fm | (257) |
| 68 | 铒 | Er | 167.259 | 101 | 钔 | Md | (258) |
| 69 | 铥 | Tm | 168.934 21 | 102 | 锘 | No | (259) |
| 70 | 镱 | Yb | 173.054 | 103 | 铹 | Lr | (262) |
| 71 | 镥 | Lu | 174.967 | 104 | 𬬻 | Rf | (265) |
| 72 | 铪 | Hf | 178.49 | 105 | 𬭊 | Db | (268) |
| 73 | 钽 | Ta | 180.947 88 | 106 | 𬭳 | Sg | (271) |
| 74 | 钨 | W | 183.84 | 107 | 𬭛 | Bh | (270) |
| 75 | 铼 | Re | 186.207 | 108 | 𬭶 | Hs | (277) |
| 76 | 锇 | Os | 190.23 | 109 | 鿏 | Mt | (276) |
| 77 | 铱 | Ir | 192.217 | 110 | 𫟼 | Da | (281) |
| 78 | 铂 | Pt | 195.084 | 111 | 铹 | Rg | (280) |
| 79 | 金 | Au | 196.966 569 | 112 | 鿔 | Cn | (285) |
| 80 | 汞 | Hg | 200.59 | 113 | 鿭 | Nh | (284) |
| 81 | 铊 | Tl | 204.383 3 | 114 | 𫓧 | Fl | (289) |
| 82 | 铅 | Pb | 207.2 | 115 | 镆 | Mc | (289) |
| 83 | 铋 | Bi | 208.980 40 | 116 | 𫟷 | Lv | (293) |
| 84 | 钋 | Po | (209) | 117 | 鿬 | Ts | (294) |
| 85 | 砹 | At | (210) | 118 | 鿫 | Og | (294) |

(按原子序数排列)摘自 *CRC Handbook of Chemistry and Physics*. 99th ed. 2018.

# 附录 II
# 不同温度下水的饱和蒸汽压

| $t/℃$ | $p/\text{kPa}$ | $t/℃$ | $p/\text{kPa}$ | $t/℃$ | $p/\text{kPa}$ |
| --- | --- | --- | --- | --- | --- |
| 0.01 | 0.611 65 | 34 | 5.325 1 | 68 | 28.599 |
| 2 | 0.705 99 | 36 | 5.947 9 | 70 | 31.201 |
| 4 | 0.813 55 | 38 | 6.632 8 | 72 | 34.000 |
| 6 | 0.935 36 | 40 | 7.384 9 | 74 | 37.009 |
| 8 | 1.073 0 | 42 | 8.209 6 | 76 | 40.239 |
| 10 | 1.228 2 | 44 | 9.112 4 | 78 | 43.703 |
| 12 | 1.402 8 | 46 | 10.099 | 80 | 47.414 |
| 14 | 1.599 0 | 48 | 11.177 | 82 | 51.387 |
| 16 | 1.818 8 | 50 | 12.352 | 84 | 55.635 |
| 18 | 2.064 7 | 52 | 13.631 | 86 | 60.173 |
| 20 | 2.339 3 | 54 | 15.022 | 88 | 65.017 |
| 22 | 2.645 3 | 56 | 16.533 | 90 | 70.182 |
| 24 | 2.985 8 | 58 | 18.171 | 92 | 75.684 |
| 26 | 3.363 9 | 60 | 19.946 | 94 | 81.541 |
| 28 | 3.783 1 | 62 | 21.867 | 96 | 87.771 |
| 30 | 4.247 0 | 64 | 23.943 | 98 | 94.390 |
| 32 | 4.759 6 | 66 | 26.183 | 100 | 101.42 |

摘自 CRC Handbook of Chemistry and Physics. 99th ed. 2018.

# 附录Ⅲ
# 危险药品的分类、性质和管理

## 一、危险药品

危险药品是指受光、热、空气、水或撞击等外界因素的影响,可能引起燃烧、爆炸的药品,或具有强腐蚀性、剧毒性的药品。常用危险药品按危害性可分为以下几类来管理,见表1。

**表1 危险药品的分类、性质和注意事项**

| 类别 | 举例 | 性质 | 注意事项 |
| --- | --- | --- | --- |
| 易燃液体 | 汽油、丙酮、乙醚、甲醇、乙醇、苯 | 沸点低、易挥发,遇水则燃烧,甚至引起爆炸 | 存放阴凉处,远离热源。使用时注意通风,不得有明火 |
| 易燃固体 | 赤磷、硫、萘、硝化纤维 | 燃点低,受热、磨擦、撞击或遇氧化剂,可引起剧烈连续燃烧、爆炸 | 同上 |
| 易燃气体 | 氢气、乙炔、甲烷 | 因撞击、受热引起燃烧。与空气按一定比例混合,则会爆炸 | 使用时注意通风。如为钢瓶气,不得在实验室存放 |
| 遇水易燃品 | 钠、钾 | 遇水剧烈反应,产生可燃气体并放出热量,此反应热会引起燃烧 | 保存于煤油中切勿与水接触 |
| 自燃物品 | 白磷 | 在适当温度下被空气氧化、放热,达到燃点而引起自燃 | 保存于水中 |
| 易爆炸品 | 硝酸铵、苦味酸、三硝基甲苯 | 遇高热磨擦、撞击时会产生猛烈爆炸放出大量气体和热量 | 存放于阴凉、低下处。轻拿、轻放 |
| 氧化剂 | 硝酸钾、氯酸钾、过氧化氢、过氧化钠、高锰酸钾 | 具强氧化性,遇酸、受热、与有机物、易燃品、还原剂等混合时,因反应引起燃烧或爆炸 | 不得与易燃品、爆炸品、还原剂等一起存放 |
| 剧毒药品 | 氰化钾、三氧化二砷、升汞、氯化钡、六六六 | 剧毒,少量侵入人体(误食或接触伤口)引起中毒,甚至死亡 | 专人、专柜保管,并应设有使用登记制度。现用现领,用剩的剩余物都应交回保管人 |
| 腐蚀性药品 | 强酸、氟化氢、强碱、溴、酚 | 强腐蚀性,触及物品造成腐蚀、破坏,触及人体皮肤,引起化学烧伤 | 不要与氧化剂、易燃品、爆炸品放在一起 |

## 二、剧毒物品

中华人民共和国公安部 1993 年发布并实施中华人民共和国公共安全行业标准 GA58—1993。将剧毒药品分为 A、B 两级,见表2 和表3。

表2  A级无机剧毒药品名表

| 品名 | 别名 | 品名 |
|---|---|---|
| 白磷 | 黄磷 | 氟 |
| 液化二氧化硫 | 亚硫酸酐 | 一氧化氮 |
| 氟化氢(无水) | 无水氢氟酸 | 二氟化氧 |
| 磷化镁 | 二磷化三镁 | 二氧化氮 |
| 磷化氢 | 磷化三氢,膦 | 磷化钾 |
| 液化氯 | 液氯 | 磷化铝 |
| 氯化汞 | 氧化高汞,二氯化汞 | 磷化铝农药 |
| 氯化氰 | 氰化氯,氯甲腈 | 磷化钠 |
| 氰化汞 | 氰化高汞 | 六氟化钨 |
| 氰化汞钾 | 氰化钾汞,汞氰化钾 | 六氟化硒 |
| 氰化钴钾 | 钴氰化钾 | 氯化溴 |
| 氰化钠 | 山奈 | 氢氰酸 |
| 氰化镍 | 氰化亚镍 | 氰 |
| 氰化镍钾 | 氰化钾镍 | 氰化钡 |
| 液化氰化氢 | 无水氢氰酸 | 氰化钙 |
| 氰化铜 | 氰化高铜 | 氰化镉 |
| 氰化溴 | 溴化氰 | 氰化钴 |
| 氰化银钾 | 银氰化钾 | 氰化钾 |
| 三氯化砷 | 氯化亚砷 | 氰化金钾 |
| 三氧化二砷 | 白砒、砒霜、亚砷酸酐 | 氰化铅 |
| 砷化氢 | 砷化三氢,胂 | 氰化铈 |
| 四氟化硅 | 氟化硅 | 氰化锌 |
| 液化四氧化二氮 | 二氧化氮 | 氰化亚钴 |
| 羰基镍 | 四羰基镍,四碳酰镍 | 氰化亚铜 |
| 锑化氢 | 锑化三氢 | 氰化银 |
| 五羰基铁 | 羰基铁 | 三氟化氯 |
| 五氧化二砷 | 砷酸酐 | 三氟化氯 |
| 溴化羰 | 溴光气 | 四氟化硫 |
| 亚砷酸钠 | 偏亚砷酸钠 | 五氟化磷 |
| 氧氯化硒 | 氯化亚硒酰,二氯氧化硒 | 硒化氢 |
| 氧氰化汞 | 氰氧化汞 | 硒酸钾 |
| 迭氯化钠 |  | 硒酸钠 |
| 叠氮化钡 |  | 亚砷酸钾 |
| 叠氮酸 |  | 亚硒酸钾 |
| 氧化镉 |  | 亚硒酸钠 |
| 五铌化氯 |  |  |

表3　剧毒物品急性毒性分级标准和半致死量

| 中毒途径 | A级 | B级 |
| --- | --- | --- |
| 口服 | $\leq 5\ mg\cdot kg^{-1}$ | $5\sim 50\ mg\cdot kg^{-1}$ |
| 皮肤接触 | $\leq 40\ mg\cdot kg^{-1}$ | $40\sim 200\ mg\cdot kg^{-1}$ |
| 吸入粉尘或烟雾 | $\leq 0.5\ mg\cdot L^{-1}$ | $0.5\sim 2\ mg\cdot L^{-1}$ |
| 吸入蒸汽或气体 | $\leq 1\,000\ mg\cdot L^{-1}$ | $\leq 3\,000\ mg\cdot L^{-1}$ |

## 三、化学实验室毒品管理规定

1. 实验室使用毒品和剧毒品（无论A或B类毒品）应预先计算使用量，按用量到毒品库领取，尽量做到用多少领多少。使用后剩余毒品应送回毒品库统一管理。毒品库对领出和退回毒品要详细登记。

2. 实验室在领用毒品和剧毒品后，由两位教师（教辅人员）共同负责保证领用毒品的安全管理，实验室建立毒品使用账目。账目包括：

药品名称，领用日期，领用量，使用日期，使用量，剩余量，使用人签名，两位管理人签名。

3. 实验室使用毒品时，如剩余量较少且近期仍需使用须存放实验室内，此药品必须放于实验室毒品保险柜内，钥匙由两位管理人员掌管，保险柜上锁和开启均须两人同时在场。实验室配制有毒药品溶液时也应按用量配制，该溶液的使用，归还和存放也必须履行使用账目登记制度。

# 附录 Ⅳ
# 标准缓冲溶液

| 标准缓冲溶液名称 | 标准缓冲溶液配制 | pH(25℃) |
|---|---|---|
| $0.05\ mol \cdot L^{-1}$ 四草酸氢钾溶液 | 称取(54±3)℃下烘干4~5h的四草酸氢钾12.61g,溶于蒸馏水,稀释至1L | 1.679 |
| $0.05\ mol \cdot L^{-1}$ 邻苯二甲酸氢钾溶液 | 称取已在(115±5)℃下烘干2~3h的邻苯二甲酸氢钾10.12g,溶于蒸馏水,稀释至1L | 4.008 |
| 约$0.034\ mol \cdot L^{-1}$ 25℃饱和酒石酸氢钾溶液 | 在玻璃磨口瓶中装入蒸馏水和过量的酒石酸氢钾粉末(约20g/L),温度控制在(25±5)℃下,剧烈摇动20~30min,溶液澄清后,用倾泻法取其清液备用(如用0.02级的仪器,饱和温度应控制在(25±3)℃) | 3.557 |
| $0.025\ mol \cdot L^{-1}$ 磷酸二氢钾 $-0.025\ mol \cdot L^{-1}$ 磷酸氢二钠混合溶液 | 分别称取已在(115±5)℃下烘干2~3h的磷酸氢二钠3.53g和磷酸二氢钾3.39g溶于蒸馏水,稀释至1L(如用0.02级的仪器,蒸馏水应预先煮沸15~30min) | 6.865 |
| $0.008\ 665\ mol \cdot L^{-1}$ 磷酸二氢钾 $-0.030\ 32\ mol \cdot L^{-1}$ 磷酸氢二钠混合溶液 | 分别称取已在(115±5)℃下烘干2~3h的磷酸二氢钾1.179g和磷酸氢二钠4.30g溶于蒸馏水,稀释至1L(如用0.02级的仪器,蒸馏水应预先煮沸15~30min) | 7.413 |
| $0.01\ mol \cdot L^{-1}$ 硼砂溶液 | 称取硼砂3.80g(注意不能烘),溶于蒸馏水,稀释至1L(如用0.02级的仪器,蒸馏水应预先煮沸15~30min) | 9.180 |
| $0.025\ mol \cdot L^{-1}$ 碳酸氢钠 $-0.025\ mol \cdot L^{-1}$ 碳酸钠混合溶液 | 分别称取已在270~300℃干燥至恒重的碳酸钠2.65g和在硫酸干燥器中干燥约4h的碳酸氢钠2.10g,溶于蒸馏水,稀释至1L(如用0.02级的仪器,蒸馏水应预先煮沸15~30min) | 10.012 |
| 约$0.020\ mol \cdot L^{-1}$ 25℃饱和氢氧化钙溶液 | 在玻璃磨口瓶或聚乙烯塑料瓶中装入蒸馏水和过量的氢氧化钙粉末(约5~10g·$L^{-1}$),温度控制在(25±5)℃下,剧烈摇动20~30min,迅速用抽滤法滤取清液备用(如用0.02级的仪器,饱和温度应控制在(25±1)℃) | 12.454 |

摘自顾庆超,新编化学用表,5-5 缓冲溶液,p901-902,江苏教育出版社

# 附录 Ⅴ
# 参 考 书 目

[1] 北京大学化学与分子工程学院分析化学教学组. 基础分析化学实验. 北京:北京大学出版社,2010.
[2] 北京大学化学系物理化学教研室. 物理化学实验. 4版. 北京:北京大学出版社,2002.
[3] 北京师范大学无机化学教研室. 无机化学实验. 4版. 北京:高等教育出版社,2014.
[4] 曹凤歧,刘静. 无机化学实验与指导. 南京:东南大学出版社,2013.
[5] 陈虹锦. 实验化学(上册). 北京:科学出版社,2007.
[6] 陈焕光. 分析化学实验. 广州:中山大学出版社,2003.
[7] 复旦大学. 物理化学实验. 3版. 北京:高等教育出版社,2004.
[8] 郭伟强. 大学化学基础实验. 2版. 北京:科学出版社,2010.
[9] 国家药典委员会. 中华人民共和国药典(四部). 北京:中国医药科技出版社,2015.
[10] 李雪华. 基础化学实验. 北京:人民卫生出版社,2014.
[11] 吕苏琴,张春荣,揭念芹,等. 基础化学实验(Ⅰ). 北京:科学出版社,2001.
[12] 孟凡德. 医用基础化学实验. 北京:科学出版社,2001.
[13] 钱可萍. 无机及分析化学实验. 北京:高等教育出版社,1989.
[14] 清华大学化学系物理化学实验组. 物理化学实验. 北京:清华大学出版社,1991.
[15] 南京大学无机及分析化学实验组. 无机及分析化学实验. 4版. 北京:高等教育出版社,2006.
[16] 刘春丽. 物理化学实验. 北京:化学工业出版社,2017.
[17] 中南民族大学分析化学实验组. 分析化学实验. 北京:化学工业出版社,2017.
[18] 王少云. 分析化学与药物分析实验. 济南:山东大学出版社,2004.
[19] 魏祖期,李雪华. 基础化学实验. 北京:人民卫生出版社,2014.
[20] 无机及分析化学实验编写组. 无机及分析化学实验. 武汉:武汉大学出版社,2001.
[21] 武汉大学化学分子科学中心编写组. 无机及分析化学实验. 武汉:武汉大学出版社,2005.
[22] 曾慧慧. 现代实验化学(上册). 北京:北京大学医学出版社,2004.
[23] 周锦兰,张开诚. 实验化学. 武汉:华中科技大学出版社,2005.
[24] 冯清. 医学基础化学实验(双语版). 武汉:华中科技大学出版社,2007.
[25] 徐伟亮. 基础化学实验. 北京:科学出版社,2010.
[26] 中山大学. 无机化学实验. 3版. 北京:高等教育出版社,2015.
[27] Gershon J. Shugar. Chemical technician's ready reference handbook. 5th ed. New York: McGraw-Hill Book Company, 2011.
[28] Stanley Marcus. Experimental General Chemistry. New York: McGraw-Hill Education, 1999.